THE
EARTH
BOOK

FROM THE BEGINNING TO THE END OF OUR PLANET,
250 MILESTONES IN THE HISTORY OF EARTH SCIENCE

Jim Bell

STERLING
New York

STERLING
New York

An Imprint of Sterling Publishing Co., Inc.
1166 Avenue of the Americas
New York, NY 10036

ISBN 978-1-4549-2910-9

Distributed in Canada by Sterling Publishing Co., Inc.
c/o Canadian Manda Group, 664 Annette Street
Toronto, Ontario M6S 2C8, Canada
Distributed in the United Kingdom by GMC Distribution Services
Castle Place, 166 High Street, Lewes, East Sussex BN7 1XU, England
Distributed in Australia by NewSouth Books
University of New South Wales, Sydney, NSW 2052, Australia

For information about custom editions, special sales, and premium and corporate purchases,
please contact Sterling Special Sales at 800-805-5489 or specialsales@sterlingpublishing.com.

Manufactured in the United States of America

2 4 6 8 10 9 7 5 3 1

sterlingpublishing.com

Photo credits—see page 528

This book is dedicated to those who, from the deep past to the far future,
try to figure out how it all works, and then who try to make it work better . . .

"The top of Mount Everest is marine limestone."
—Author JOHN MCPHEE, when asked to sum up
all of geology in one compelling sentence.

"Look again at that dot. That's here. That's home. That's us. On it everyone you love, everyone you know, everyone you ever heard of, every human being who ever was, lived out their lives . . . on a mote of dust suspended in a sunbeam."
—CARL SAGAN, *Pale Blue Dot*, 1994

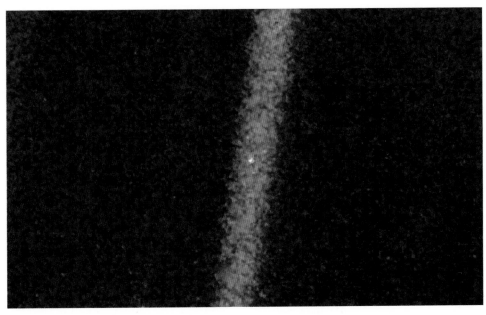

The famous "Pale Blue Dot" photo of the Earth, floating in a beam of scattered sunlight, taken by the Voyager 1 space probe on February 14, 1990 from a vantage point beyond the orbit of Neptune, more than 40 times the distance from the Earth to the Sun.

Contents

Introduction

It's a daunting task to set out to chronicle the history of the world in a single book. And I don't mean just the history of human beings and our various achievements and downfalls, I mean the entire history of the planet, from its formation in the spinning clouds of gas and dust from which our Sun and solar system came to be some 4.5 billion years ago, to the inevitable destruction of our planet in the eventual death throes of that same benevolent star some 5 billion years from now. Of all the things that have happened on, in, and around our world, which of them warrant consideration as being among the 250 most important milestones in the history of Earth?

I've taken a stab at answering that question, based on my own training as a geologist and planetary scientist; my own background and experience in field work, remote sensing, and computational analysis of data; and my own biases. For example, much of my professional teaching and research work focuses on planetary and space science—studies of other planets and solar system bodies, such as Mars and the Moon, using earth science and the study of our own planet as a basis for studying others (and vice versa!). So, I tend to think of the Earth not just as our home and the home of millions of other species, but as a member of a family of planets, moons, asteroids, and comets that orbit the Sun in our cosmic neighborhood. Indeed, studying the Earth from space, and studying other planets to learn about our own, is a major way that we've learned much of what we know about our home planet.

When I teach earth science, I make sure to point out that studying the Earth is like studying a series of nested, intertwining spheres. There is the *lithosphere*, the rocky surface and interior of our planet; the *atmosphere*, the thin layer of gases that warms the surface and sustains life; the *magnetosphere*, the magnetic bubble that protects our world from harmful solar radiation; the *hydrosphere*, a thin surface shell of water mostly in the oceans but also in seas, lakes, rivers, glaciers, and polar caps; and finally the *biosphere*, the collection of all living things on our planet. Each of these spheres has been critically important to the history of our planet, and they are all interrelated in complex ways that cannot easily be untangled. To truly understand the Earth as a system requires one to understand *all* these spheres of influence.

Thus, the history of the Earth spans topics in physics, chemistry, biology, astronomy, astrobiology, geology, mineralogy, planetary science, life science, public policy, atmospheric science, climate science, engineering, and many other scientific and social disciplines and subfields. I've tried to capture milestone events and discoveries that span all these fields, and to thus hopefully give a sense of the breadth of experience and expertise that goes into figuring out how our world came to be the way it is, and what will happen to it in the future.

Along the way, I've singled out about 120 individuals—out of the many thousands of scientists, explorers, inventors, and others who have made it a career to learn about our planet—who have made or contributed significantly to events and discoveries that I have flagged as notable milestones. Some of these people are well known (like Plato, Leonardo da Vinci, Magellan, Newton, Pasteur, Lewis & Clark, Darwin, Cousteau, and Goodall), others are famous within their academic or exploration circles but not so well known to the general public (like Steno, Hutton, Bowen, Wegener, Carrington, Agassiz, Humboldt, Dobson, Amundsen, Peary, and Van Allen). And still others have made critical contributions to the understanding of our world but for some reason have been left relatively obscure in the sometimes-fickle annals of history (like Chladni, Brock, Anning, Nadar, Dokuchaev, Bascom, Griggs, Angel, Norgay, and Lehman, all of whom you will learn more about soon).

As part of my research into these milestones, I was especially impressed by the significant role that many women have played in advancing our understanding of the planet. For most of the history of science, it has been a male-dominated career, with active traditions and other barriers established to keep women out of the club. That began to change—slowly—in the nineteenth and twentieth centuries, although the obstacles were huge early on. Pioneering research and discoveries about the Earth as well as its inhabitants made by women such as Florence Bascom, Dorothy Hill, Inge Lehman, Mary Leakey, Rachel Carson, Dian Fossey, Kathleen Sullivan, Sylvia Earle, and many others attest to the fact that women are just as capable as men in the pursuit of, and success in, science. Still, the fact that women do not yet make up 50 percent of the world's population of professional scientists means that many gender-based barriers and biases—conscious and unconscious—still exist in earth science and other fields. There is still much work to do.

Another aspect of diversity that I have attempted to capture among these milestones is geographic diversity, both on and within the planet. For example, I point out examples of major mountain belts on all the continents, sampling and celebrating the many different styles of mountain-building (*orogenesis*) that have occurred throughout our planet's history. I also point out examples of the major kinds of rocks and minerals that make up our planet, including how they are formed and what role(s) they may also have played in human history. The onion-skin-like layers of the interior of our planet—*core*, *mantle*, and *crust*—also each deserve special attention, for they each play a specific role in the way our planet gives off its internal heat, the way Earth generates a strong magnetic field, and the way the continents and oceans change over time. And you'll see a major thread run through the book related to the theory of *plate tectonics*—the way the Earth's crust is divided into a few dozen major pieces and the ways that they interact with each other to form new continents, ocean basins, and islands, as well as potentially catastrophic earthquakes and volcanic eruptions. Plate tectonic theory provides the foundation of our modern understanding of the way Earth's surface changes over time.

I also thought it important to mark, as major milestones, the boundaries of Earth's major periods of geologic time, as reconstructed by modern geologists. These boundaries are part of the internationally accepted *Geologic Time Scale*, a copy of which is included in an Appendix here for handy reference, courtesy of the Geological Society of America. For example, by far most (almost 90 percent) of the history of our planet was in a single geologic time span called the Precambrian, about which we know relatively little because there are so few rocks and fossils preserved on the surface of our dynamic planet from more than about 550 million years ago. Starting around then, however, in a significant milestone known as the Cambrian explosion, marine organisms began to create hard exoskeletons that remained preserved as fossils after the organisms died and fell to the seafloor. Fossils have provided key milestone markers in Earth's geologic history ever since, including evidence for at least five episodes of mass extinctions, where huge fractions of all the species on Earth died off, rather quickly. One of these milestones, the disappearance of the ancient dinosaurs and many other species around 65 million years ago, has been linked to the climatic and food chain catastrophe created by the impact of a large asteroid. The other mass extinction milestones remain mysteries, however, with multiple hypotheses like impacts, extensive volcanism, and rapid climate change being explored for their origin in ongoing research and debate.

Yet another daunting task for this research was trying to assign specific chronological dates to many of the discoveries and events chosen here as milestones. When did the Atlantic Ocean form, exactly? When did flowers first appear on the Earth? When will the next ice age begin? In the case of many events in Earth's history, especially in the deep past (or far future), there is considerable uncertainty or debate about the timing. Thus, in cases where the chronological timing of key events is uncertain or broad or both, I have indicated the best-known approximate date or range of dates with a "*c.*" (the Latin abbreviation for *circa*, meaning *about*) in front.

I've also chosen, in some cases, relatively modern milestone dates for past events or features about the Earth that might not be amenable to precise pinning down. For example, the discussion of many of Earth's *biomes* (specific ecological zones) as important aspects of the Earth system comes up throughout the book, but when did tundra first appear on the planet? Or the first tropical rainforest? Or for that matter, when was the first hurricane, tornado, wildfire, or landslide? For those kinds of temporally nebulous milestones, I've picked important dates that we humans have since related to those specific events or ecological zones, such as the hurricane that devastated Galveston, Texas in 1900, the 1973 creation of the Monteverde Cloud Forest Reserve in Costa Rica, or the United Nations designation as 2011 as the "International Year of Forests."

Finally, you might notice that many of the milestones that I've chosen have to do not just with the Earth, but more specifically with the development of life on our planet. When did life appear? How and when did photosynthesis originate? When did the first mammals appear? How about the first *Homo sapiens*? All of these and many other highlights in life sciences are milestones worthy of mention, not just for our species, but for our planet. While we now know that there is a short list of other places in our solar system which might be or might have been habitable at one time (places like Mars, Jupiter's large moon Europa, or Saturn's small moon Enceladus, for example), Earth is the only place that we know of so far that is not only habitable, but also *inhabited*.

For all we know, the origin and evolution of life on Earth could be a completely unique occurrence in the entire universe. Or, because we now know that there are likely countless Earth-like worlds in our galaxy and beyond, and that the components and conditions that make a planet like Earth habitable and inhabited are relatively common in the cosmos, perhaps the universe is teeming with life on similar habitable worlds, with each occurrence supremely tuned and adapted to its own unique environment following universal principles like evolution and natural selection. Either way, life on Earth is still special, and understanding the milestones related to life here will help us in the search for life elsewhere. As my mentor and hero Carl Sagan was fond of saying, "We are a way for the universe to know itself."

It is perhaps more important now than it has ever been in human history for us to understand our home world as a complex, interdependent set of systems, and especially for us to understand the special role that our species plays in that set of systems. We are not the first species on Earth to change the overall climate of our planet (the first were the cyanobacteria—blue-green algae—that developed the remarkable ability to "poison" the early Earth's atmosphere with massive amounts of oxygen via a brand-new innovation called photosynthesis, starting around 3.4 billion years ago). However, we are the first species with the ability to recognize that we are doing so, and with the power to do something about it. What will we collectively do with that knowledge? How will we collectively wield that power, if at all? These are profound questions that are at the heart of our relationship with our home planet.

Yes, it is indeed daunting to try to sum up everything there is to know about our planet in such a limited space. But just think of what we'll learn. Read about our planet, follow the Notes and Further Reading links and pointers to learn more, and ponder your role in it all. . . . Enjoy!

Acknowledgments

I am grateful to so many people for their work, wisdom, advice, and support before and during the research and writing of this book. Perhaps most relevant is to acknowledge the countless contributions made by my colleagues—known and unknown to me—in the expansive mega-field of Earth science. Figuring out how our planet works is a never-ending process, but it is built upon the foundation of field observations, laboratory studies, and computer modeling that has been forged by many hundreds of years of scientific curiosity and experimentation, as well as by the incessant desire of our species to explore the unknown. I had a lot of fun assembling the photographs for this book, and want to extend an enormous thanks to the many nameless, unsung heroes who create public-domain graphics for NASA and other governmental agencies, or who post their own incredible photos and artwork on public domain share spaces like Wikimedia Commons. Indeed, I want to give a shout out to Wikipedia in general (which I contribute to, financially), as it has been an outstanding starting resource from which to launch deeper research in a huge number of topics relevant to this book. I thank Michael Bourret at Dystel, Goderich, & Bourret for never-ending confidence in my writing abilities, Meredith Hale at Sterling for her editorial guidance and cheery patience and support, and Linda Liang, Clare Maxwell, and Christopher Bain at Sterling for enormous amounts of help and spot-on advice regarding the research for and choice of the photos here. Finally, my deepest thanks and love go to my darling companion Jordana Blacksberg, who has put up with my devotion to this project for far too long. It's time for me to redirect my devotion back where it belongs. . . .

Jim Bell
Mesa, Arizona
June 2018

Earth Is Born

Evidence from both meteorites and stellar astrophysics tells us that the Sun and all the planets of our solar system were born around the same time, some 4.5 billion years ago, from the collapse of a huge spinning cloud of hot interstellar gas and dust. Earth, our home world, is the largest of the rocky, terrestrial planets that orbit relatively close to the Sun, and is the only terrestrial planet with a large natural satellite. To a geologist, it's a rocky volcanic world that has separated its interior into a thin low-density crust, a thicker silicate mantle, and a high-density partially molten iron core. To an atmospheric scientist, it's a planet with a thin nitrogen-oxygen water-vapor atmosphere buffered by an extensive liquid-water ocean and a polar ice-cap system, all of which participate in large climate changes on seasonal to geologic time scales. To a biologist, it's heaven.

Earth is the only place in the Universe where we know life exists. Indeed, evidence from the fossil and geochemical record says that life on Earth began almost as soon as it could, when the early solar system's violent rain of asteroid and comet impacts quieted down. Earth's surface conditions appear to have remained relatively stable over the past four billion years; this stability, combined with our planet's favorable location in the so-called habitable zone, where temperatures remain moderate and water remains liquid, has enabled life to thrive and evolve into countless unique forms. Earth's crust is divided into a few dozen moving tectonic plates that essentially float on the upper mantle. Exciting geology—earthquakes, volcanoes, mountains, and trenches—occurs at the plate boundaries. Most of the oceanic crust (70 percent of Earth's surface area) is very young, having erupted from mid–ocean-ridge volcanoes spanning the time from a few hundred million years ago to today.

The high amounts of oxygen, ozone, and methane in Earth's atmosphere are signs of life that could be detected by alien astronomers studying our planet from afar. Indeed, these gases are exactly what Earth's astronomers are looking for today among the panoply of newly discovered earth-like extrasolar planets orbiting other Sun-like stars. Are there more Earths out there, waiting to be found and explored?

SEE ALSO Birth of the Moon (c. 4.5 Billion BCE), Late Heavy Bombardment (c. 4.1 Billion BCE), Plate Tectonics (c. 4–3 Billion BCE?), Life on Earth (c. 3.8 Billion BCE?)

In the violent early inner solar system of 4.54 billion years ago, small rocky bodies grew relatively quickly by accretion—crashing into each other and sometimes sticking together—into protoplanets, and eventually full-fledged planets. Mercury, Venus, Earth, and Mars all formed this way early on.

Earth's Core Forms

Frequent collisions in the early history of the solar system no doubt resulted in the catastrophic disruption and even vaporization of many asteroids and planetesimals (small protoplanetary objects that could grow into full-size planets). Sometimes these collisions must also have led to the accretion (growth) of some of these bodies. And as some of these lucky survivors grew, their gravity became stronger, helping them to attract even more incoming materials to help them grow even larger. Above a certain size (usually thought to be around 250–375 miles or 400–600 km across), a planetesimal's self-gravity pulls it into a relatively spherical shape. The resulting overlying pressures cause the temperatures inside those bodies to start to increase with depth, while at the same time considerable energy is continuing to be added from additional impact events at the surface.

The result of this accretion and internal heating is that growing bodies can *differentiate*, or segregate, with denser elements and minerals (like those dominated by iron) sinking to the interior and less-dense elements and minerals (like those dominated by silicon) floating to the top. Geophysical models show that this process—which ultimately led to the formation of the typical core/mantle/crust structure that we see in rocky and icy planetary bodies today—happened relatively quickly to many growing planets early in the history of the solar system.

In growing rocky planets like our own, additional internal heat was provided by the radioactive decay of certain elements (such as certain isotopes of aluminum or uranium, for example) that release heat as they decay. The resulting internal buildup of heat eventually entirely melted the iron-rich cores of some of these planets. Spinning, molten, electrically conducting cores led to the formation of strong magnetic fields on some of these worlds; on Earth, those fields would ultimately help to make the surface habitable by shielding the surface from much of the most harmful radiation from the Sun.

The Earth's core has continued to evolve with time. While the outer core is still molten, the inner region of the core is thought to have cooled and solidified about 1 to 1.5 billion years ago.

SEE ALSO Earth's Mantle and Magma Ocean (c. 4.5 Billion BCE), Continental Crust (c. 4 Billion BCE), Magnetite (c. 2000 BCE), Solar Flares and Space Weather (1859), Earth's Core Solidifies (~2–3 Billion)

Artist's concept cutaway view of the very young Earth early in the history of the solar system, continually bombarded by impacts but still segregating into a core, mantle, and crust.

Birth of the Moon

Earth is unique among the terrestrial planets in having a very large natural satellite. But where did our Moon come from? One idea is that the Moon formed in orbit around our planet at the same time and in the same way the Earth formed: by the slow growth from collisions ("accretion") of rocky and metallic planetesimals condensed out of the warm inner regions of the spinning cloud of gas and dust known as the solar nebula. Another idea is that the early (molten) Earth was spinning so fast that a blob of it shed off (fissioned) and went into orbit, forming the Moon. Yet another hypothesis proposes that the Moon was formed somewhere else in the inner solar system and was later captured by Earth's gravity.

These ideas competed for supremacy until the Apollo missions of the late 1960s and early 1970s brought Moon rocks and other information back to Earth and revealed that none of those hypotheses fit the Moon's actual physical and compositional data. The accretion model predicted that the Moon would have the same basic age and composition as Earth, but it does not: the Moon has a much lower density, much less iron, and appears to have formed 30–50 million years after Earth and the other planets formed. The fission model required the early Earth to be spinning too fast, and the capture model suggested that there was no way to dissipate all the energy a free-flying Moon would have had to lose in order to get captured into Earth orbit.

In the 1990s, planetary scientists proposed another idea: the giant impact model. If the early Earth had been struck just right—at an oblique angle—in a giant impact by a Mars-size protoplanet, computer simulations showed that enough of Earth's low-density, iron-poor mantle could have been melted and ripped off into orbit to eventually cool, grow, and form the Moon. The proto-Earth's entire surface would have been melted in that giant impact as well, causing a major catastrophe for our young planet. Even though it seems rather *ad hoc*, the giant-impact model is still the best explanation for the origin of our Moon, because the composition, density, and even age of the Moon match the model's predictions.

SEE ALSO Earth Is Born (c. 4.54 Billion BCE), Late Heavy Bombardment (c. 4.1 Billion BCE), Leaving Earth's Gravity (1968), Geology on the Moon (1972), Last Total Solar Eclipse (~600 Million)

Artist's conception of the grazing impact of a Mars-size body with the proto-Earth around four and a half billion years ago. Debris from a giant impact like this is believed to have led to the formation of our Moon.

Earth's Mantle and Magma Ocean

The interior of the early Earth was heated to extreme temperatures by a variety of factors, such as the heat created from super-high pressures deep beneath the surface, the heat released by the decay of radioactive elements like uranium, and the heat brought in by frequent impacts from comets, asteroids, and young growing planetesimals. All this heat eventually melted at least parts of Earth's interior, leading to differentiation into the basic core/mantle/crust structure that Earth still exhibits today. Many geophysicists think that the melting of the early Earth's interior could have been much more extensive, however. Specifically, evidence exists that the early Earth's mantle might have been wholly or partially molten, forming a sort of underground "magma ocean" beneath our planet's thin crust ("magma" is the geologic term for molten subsurface rock, as opposed to "lava," which is the geologic term for molten rock on the surface).

The evidence for an early-Earth magma ocean comes from lab experiments on the kinds of dense iron and magnesium silicate minerals that occur deep in the Earth. The mineral bridgmanite was studied extensively, in particular, because it is the most abundant mineral in Earth's mantle. Scientists using special diamond anvils designed to reproduce the high pressures of Earth's interior found that when bridgmanite melts, it turns into a denser iron-magnesium silicate that sinks below the less-dense crystalline bridgmanite. The melted material doesn't sink below Earth's even-denser molten iron-nickel core, however, helping to maintain the basic core/mantle/crust interior structure.

Earth's early magma ocean would not have been a calm and static place. Melting of the upper mantle would have created denser blobs that sank down to the core, and intense heating of the lower parts of the ocean from core-supplied heat would have created less-dense, buoyant blobs that worked their way up through the ocean. These motions could have set up convection cells that would have kept the mantle in constant motion for hundreds of millions of years as the mantle and magma ocean slowly cooled and solidified. The surface expression of all that violence in the mantle was likely to have been a bone-dry world with frequent volcanic eruptions. Certainly not a pleasant environment for life to take hold in!

SEE ALSO Earth Is Born (c. 4.54 Billion BCE), Earth's Core Forms (c. 4.54 Billion BCE), The Hadean (c. 4.5–4 Billion BCE), Continental Crust (c. 4 Billion BCE), Olivine (1789)

Top: Heat from the continuing barrage of asteroids and comets like in this artist's rendering helped a "magma ocean" form during our planet's early history. **Bottom:** *A tiny chip of volcanic rock heated inside a diamond anvil reveals evidence that the young Earth's interior was mostly molten.*

The Hadean

The young planet Earth was a violent, hellish place—partially or perhaps even wholly melted in the interior, constantly bombarded by high-speed impactors that imparted even more heat, and with a hot, dry surface crust continually disrupted by volcanic eruptions. Geologists have a name for this first 500 million years of Earth's history: the Hadean eon (from Hades, the Greek god of the hellish underworld).

Because of the constant impact bombardment and the eruption of new materials from the hot interior, the crust of the Hadean Earth was constantly being renewed and recycled. Once the impact rate abated and the volcanism slowed down, the cooled crust was relatively quickly modified through weathering and erosion, much of it related to the subsequent formation of Earth's oceans. Thus, very little evidence of the Hadean still exists. Nonetheless, geologists have found some evidence for heavily metamorphosed Hadean rocks preserved in some of the oldest remnants of Earth's continental crust. Minerals preserved in these rocks point to hellish conditions in the early Hadean, but also to a transition to a more habitable world over time.

Volatile materials such as hydrogen, water vapor, and carbon dioxide that were incorporated into the growing early Earth were released from the interior by melting and convection, likely forming a thick, hot, steamy early atmosphere on our planet. Some areas of hot liquid water may have been stable on the surface because of the high atmospheric pressures, but it would take significant cooling of the surface and condensation of the atmospheric water vapor to eventually lead to the kind of global liquid-water ocean that characterizes the Earth to this day.

The fact that very little evidence remains on the surface from the first 500 million years of Earth's history is one motivator for studying the ancient surfaces of other bodies in our solar system. The Moon, for example, preserves evidence of the violent impact history of Earth's Hadean eon (including evidence for the Late Heavy Bombardment), as do the ancient highland regions of Mars and the ancient crust of Mercury. By examining information gleaned from those worlds, we may yet be able to piece together the details of Earth's Hadean puzzle.

SEE ALSO Earth Is Born (c. 4.54 Billion BCE), Earth's Core Forms (c. 4.54 Billion BCE), Earth's Mantle and Magma Ocean (c. 4.5 Billion BCE), Late Heavy Bombardment (c. 4.1 Billion BCE), Continental Crust (c. 4 Billion BCE), Earth's Oceans (c. 4 Billion BCE)

Artist's impression of the surface of the Earth during the Hadean. The Moon was much closer to the Earth back then, and thus it appeared much larger in the early Earth's sky.

Late Heavy Bombardment

All the planets and other bodies in our solar system, including Earth, have been hammered by a veritable rain of asteroids and comets throughout geologic history. The rate of such catastrophic-impact events back in the early days of the solar system was many orders of magnitude higher than it is now. The record of that early cosmic impact history is not preserved on Earth, however, because most of our planet's surface is covered by younger volcanic deposits or has been eroded away by the action of wind, water, ice, or plate tectonics. The surface of the Moon, on the other hand, is much more revealing, and the huge number of lunar impact craters and large impact-created basins provides a stark reminder of just how battered Earth's surface must have once been.

One of the major legacies of the Apollo missions is the ability to determine the absolute ages of specific impact-cratering events using radioactive dating of lunar samples. The results indicate ages for large lunar impact events of around 4.1 to 3.8 billion years BCE—a surprisingly "young" discovery, considering that all the major planets formed significantly earlier than that, around 4.5 billion years ago. Many planetary scientists believe that the simplest explanation is that the Moon—and, by inference, the Earth—went through a period of intense impact-cratering about 400 to 700 million years after their initial formation. But why?

Some have speculated that Jupiter is to blame. As the largest planet in the solar system, Jupiter exerts the most gravitational influence on the other planets, asteroids, and comets. Planetary scientists have recently hypothesized that slow changes in the orbits of Jupiter and the other giant planets early in the solar system's history caused occasional "resonances" among the planets, especially when Jupiter and Saturn were aligned just right in their orbits. These resonances "pumped" gravitational energy throughout the early solar system, disrupting the orbits of other planets and especially of smaller asteroids and comets. Many of those small bodies could have been diverted into the inner solar system. If this model is right, the resulting cataclysm certainly wreaked havoc on the terrestrial planets and no doubt had a profound influence on the development and stability of life on our home world.

SEE ALSO Plate Tectonics (c. 4–3 Billion BCE?), Life on Earth (c. 3.8 Billion BCE?), Radioactivity (1896)

Enormous impact basins, like the 578-mile-wide (930 km) Orientale Basin on the Moon shown here, provide evidence for a renewed period of intense planetary bombardment some 400 to 700 million years after the Earth formed. The colors here show lower (blue) and higher (red) gravity on the inner and outer parts of the basin.

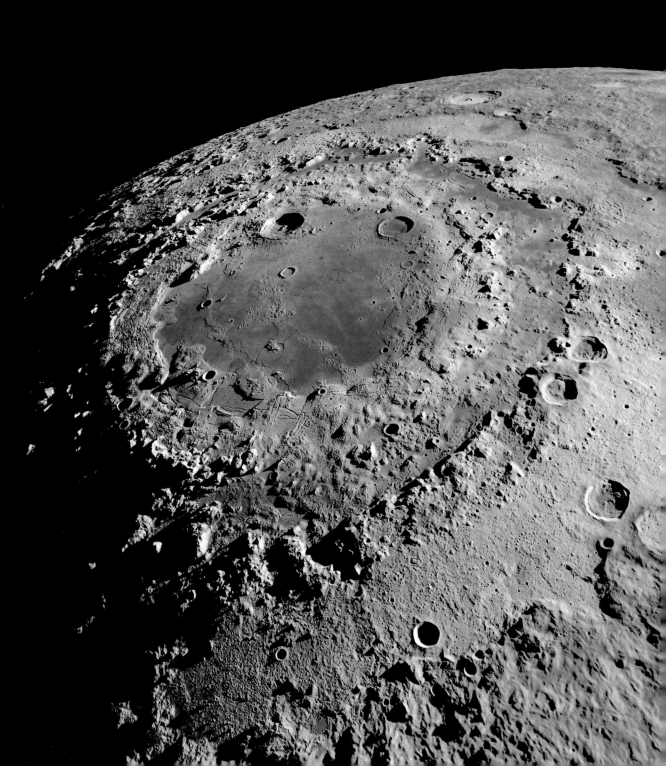

Continental Crust

Norman L. Bowen (1887–1956)

The rocks that erupted onto Earth's surface in the Hadean came from the dense, molten magnesium- and iron-rich rocks that made up the early mantle. Geologists call these kinds of rocks *mafic basalts*, where *mafic* is a hybrid of "magnesium and iron (Fe) rich," and *basalt* is a kind of fine-grained volcanic rock that is relatively low in silicon, sodium, and potassium relative to other kinds of volcanic rocks. The rocks were constantly re-melted, re-processed, and re-cycled through the Hadean crust.

When basaltic rocks are re-melted, especially if they are in a confined underground volume such as within a magma chamber under a volcano, or within a large subsurface volume oozing or "intruding" into other surrounding rocks (what geologists call a *batholith*), they can start to undergo a mini-version of the differentiation process that happened within the Earth's interior as a whole: that is, the heavier elements sink to the bottom and the lighter elements rise to the top. This is because as the melt begins to cool and solidify, minerals crystallize out in a specific way, with more mafic minerals like olivine coming out of the melt first and settling to the bottom, and higher silicon minerals such as feldspar and eventually even quartz crystallizing out of the melt at the end, and thus concentrating near the top of the chamber or batholith. Geologists call these high-silica components *felsic* minerals (from "feldspar rich"). The order that these minerals undergo *fractional crystallization* out of subsurface melts was first figured out by Canadian petrologist Norman L. Bowen in 1928; this is now known as Bowen's Reaction Series.

The result of the recycling of mafic rocks is the creation of younger felsic rocks that are less dense and thus "float" on the denser mafic basalt crust. Over time, felsic rock "islands" floating on the mafic crust built up to form the first pieces of the Earth's protocontinents. Geologists call these ancient, earliest-formed central cores of the continents *cratons* or, where widely exposed, *shields*. Only a dozen or so major shield regions have survived from the early history of our world; but, via plate tectonics, these regions helped seed the growth and accretion of much more continental crust, which now covers about 40 percent of the surface of our planet.

SEE ALSO Earth's Core Forms (c. 4.54 Billion BCE), Earth's Mantle and Magma Ocean (c. 4.5 Billion BCE), The Hadean (c. 4.5–4 Billion BCE), Late Heavy Bombardment (c. 4.1 BCE), The Archean (c. 4–2.5 Billion BCE), Plate Tectonics (c. 4–3 Billion BCE?), Feldspar (1747), Olivine (1789)

Fractured, metamorphosed remnants of ancient continental crust found along the eastern shores of Hudson Bay, in the Canadian craton. These rocks formed by the re-melting of basaltic rocks originally erupted in the Hadean.

Earth's Oceans

During the Hadean (hellish) eon between about 4.5 and 4 billion years ago, huge quantities of molten volcanic lava erupted onto Earth's surface. Along with that liquid rock, a lot of gases were also "erupted," including hydrogen, ammonia, methane, carbon dioxide (CO_2), sulfur dioxide (SO_2), and water vapor, forming Earth's steamy earliest atmosphere. As the Earth cooled and transitioned into the Archean eon around 4 billion years ago, the pressure and temperature conditions began to allow large quantities of water to routinely be stable as a liquid on the surface (rather than as steam in the atmosphere). Earth's oceans were born.

Where did all these volatile gases come from? Perhaps the comets and especially asteroids/planetesimals that collided and eventually grew to become the Earth contained water, which became trapped in the subsurface as the Earth grew, but slowly escaped via Hadean volcanism. Or perhaps the steady rain of comets and asteroids that have been crashing into our planet during and since the Hadean delivered a so-called "late veneer" of water to the Earth, which condensed and became the oceans. Both ideas have merit. Even today, for example, large quantities of water (and CO_2, SO_2, and other gases) can be measured coming out of active volcanoes. Comets and asteroids also continue to strike the Earth (though rarely); studies of surviving meteorites show that some of them are indeed water-rich. Perhaps Earth's oceans came from both internal and external sources of water.

CO_2 and ammonia dissolve in water, and Earth's early oceans quickly became a "sink" for dissolving enormous quantities of these compounds, removing most of them from the atmosphere. Many scientists believe that the result was an Archean atmosphere that could have had abundant hydrogen and methane and much less free oxygen (much like the atmosphere of Saturn's large moon Titan today). Scientists refer to this as the *reducing* (as opposed to *oxidizing*) model of early Earth's atmosphere. Experiments dating back to the 1950s have shown that when liquid water in contact with such an atmosphere is exposed to energy sources like lightning or solar UV radiation, the result can be the formation of abundant organic molecules, including simple amino acids and other essential building blocks of life.

SEE ALSO The Hadean (c. 4.5–4 Billion BCE), The Archean (c. 4–2.5 Billion BCE), Plate Tectonics (c. 4–3 Billion BCE), Earth's Oceans Evaporate (~1 Billion)

Artist's concept of the Earth near the end of the Hadean/beginning of the Archean, when liquid-water seas and oceans began to be stable on the surface. Large impact basins, just like those still preserved on the Moon today, provide evidence of continuing bombardment by asteroids and comets.

The Archean

The second-oldest of Earth's four major eons of geologic time is known as the Archean (sometimes spelled Archaean, from the Greek for "beginning" or "origin"). The beginning of the Archean around 4 billion years ago, or about 500 million years after the Earth formed, marks the approximate age of the oldest radioactively datable rocks still preserved on the surface of our planet.

The early Archean Earth would be almost unrecognizable to us. Our planet was probably almost completely covered by a hot, mildly acidic ocean, broken up only occasionally by small regions of early continental crust (protocontinents) formed via the re-melting and re-working of older, denser oceanic crust. The atmosphere probably had little free oxygen, and likely kept the surface quite hot because of the presence of abundant greenhouse gases such as water vapor, CO_2, and others. The surface was also heated from below, as internal radioactive heating—as well as the residual heat from the accretion of the planet—drove a much higher rate of volcanism than today. In the early Archean, life on Earth was rare or perhaps even nonexistent.

In contrast, the changes to the Earth over the span of 1,500 million years, by the time of the late Archean around 2.5 billion years ago, were among the most dramatic ever experienced by our planet. The rates of impact cratering and volcanic eruptions slowed significantly. Plate tectonics began, and, partly as a result, continental landmasses began to grow from the roots of the older protocontinental cratons. Erosion from the new continents into the oceans helped to increase the salinity (salt content) and neutralize the acidity of the ocean. The atmosphere cooled, became more oxidizing, and the essential components of the modern ocean-land hydrologic cycle—evaporation, condensation, precipitation—began to develop. And by the end of the Archean, life was thriving on planet Earth.

Among the most profound (for us, at least) events in the Archean was the rise of a class of single-celled microorganisms called cyanobacteria (formerly blue-green algae) that developed the remarkable ability to generate oxygen via a brand-new innovation: photosynthesis. Over time, the buildup of free oxygen that started in the Archean would provide a potent energy source for more complex forms of life.

SEE ALSO The Hadean (c. 4.5–4 Billion BCE), Continental Crust (c. 4 Billion BCE), Earth's Oceans (c. 4 Billion BCE), Plate Tectonics (c. 4–3 Billion BCE?), Life on Earth (c. 3.8 Billion BCE?), Stromatolites (c. 3.7 Billion BCE), Photosynthesis (c. 3.4 Billion BCE), The Greenhouse Effect (1896)

With stable liquid water on the surface, abundant sources of heat and energy, and a rich supply of organic molecules building up over time, the Archean Earth (depicted here in an artist's conception) appears to have been a fertile environment for the formation of life.

Plate Tectonics

Earth's crust and upper mantle in the Hadean eon (4.5–4 billion years ago) were hot, mostly molten, violent, and unstable places. As the planet cooled and the seemingly endless rain of asteroid and comet impacts slowed considerably in the Archean (4–2.5 billion years ago), the outer layers of our planet began to take on their more familiar appearance. This included the formation of the oceans, the formation of the first pieces of lower-density continental crust that could "float" on the higher-density volcanic lavas that make up the seafloor, and the division of the upper mantle into a rigid, cooler outermost section called the *lithosphere*, and a hotter region just beneath that called the *asthenosphere*.

The asthenosphere (Greek for "weak" and "sphere") starts somewhere on average about 30 to 60 miles (c. 50 to 100 kilometers) beneath the surface, and varies in thickness from around 10 to more than 300 miles (a few tens to more than 500 kilometers), depending on temperature. The rocks there are *ductile*, meaning that they can easily deform or even slowly flow, unlike the colder, stiffer lithosphere above. Rocks in the warm asthenosphere are compelled to move by enormous convection plumes that carry molten rock and heat from the deep interior of the Earth up toward the surface. "Blobs" of hotter (or even molten) mantle rocks cause the asthenosphere to bulge, bend, and move laterally as the plumes ascend. This places enormous stress on the rigid rocks of the lithosphere.

During the Archean, sometime between 4 and 3 billion years ago (the timing is controversial and the subject of much active research), the rigid lithosphere fractured under the stress and broke into numerous (possibly hundreds or thousands) of individual plates, each of which remained semi-anchored to the moving asthenosphere below. These puzzle pieces were then free to move about, crashing into one another to create early mountain belts, or one diving underneath the other to create enormous trenches.

As the continents grew into larger plates, they became more formidable obstacles to the denser seafloor volcanic plates, which are also continuously growing at mid-ocean ridges. Earth now has about two dozen of these large lithospheric plates, and many of their boundaries mark zones of strong earthquakes and extensive volcanic eruptions.

SEE ALSO Earth's Mantle and Magma Ocean (c. 4.5 Billion BCE), The Hadean (c. 4.5–4.0 Billion BCE), Continental Crust (c. 4 Billion BCE), Earth's Oceans (c. 4 Billion BCE), The Archean (c. 4–2.5 Billion BCE), Island Arcs (1949), Mapping the Seafloor (1957), Reversing Magnetic Polarity (1963), Seafloor Spreading (1973)

The famous San Andreas fault, seen here running through southern California, is one of the most famous boundaries between Earth's lithospheric plates (the Pacific and North American).

Life on Earth

No one knows exactly how, when, or why life first appeared on planet Earth, but we know that almost as soon as it could, it did. The oldest signs of life on Earth are chemical, not fossil, and are inferred as evidence because all known life on this planet is based on a common chemical architecture. Specifically, certain biogeochemical processes and reactions that are common to all life on Earth create recognizable patterns in certain chemical elements. For example, changes in the relative abundance of isotopes (different forms of the same element that contain equal numbers of protons but different numbers of neutrons) of carbon, hydrogen, nitrogen, oxygen, phosphorus, and other trace elements can provide unique fingerprints implicating the presence of past life in ancient rock and mineral deposits, even if no actual fossils are preserved there.

Life prefers to use (and create) certain building blocks. Anomalous kinds of chemistry, such as the occurrence of extra amounts of the isotope carbon-12 (^{12}C) compared to the isotope carbon-13 (^{13}C), as found in some 3.8-billion-year-old rocks from Greenland or other extremely old preserved parts of Earth's crust, provide circumstantial but controversial so-called "chemofossil" evidence for life very early in our planet's history.

Recent studies of the very earliest period of Earth's history, the Hadean (4.5–4 billion years ago), provide evidence that oceans and at least protocontinents formed very early in Earth's history, and that conditions may have been suitable for life just a few hundred million years after our planet formed. However, the cataclysmic rain of asteroid and comet impacts during the Late Heavy Bombardment of 4.1 to 3.8 billion years ago could have killed off earlier life forms, or perhaps just frustrated their attempts to flourish.

Whatever the case may be, very soon after Earth's crust cooled, the oceans formed, the Late Heavy Bombardment ended, and Earth's surface environment became stable enough to consistently support life. The fact that it thrived and began to evolve into so many niches is remarkable. Now that we understand many of the starting conditions, as well as many of the requirements for habitability on a terrestrial planet, astronomers, planetary scientists, and astrobiologists are searching for evidence of life on other earthlike worlds.

SEE ALSO Earth Is Born (c. 4.54 Billion BCE), The Hadean (c. 4.5–4 Billion BCE), Late Heavy Bombardment (c. 4.1 Billion BCE), Continental Crust (c. 4 Billion BCE), Earth's Oceans (c. 4 Billion BCE), The Archean (c. 4–2.5 Billion BCE)

Research into the origin and evolution of life on Earth encompasses astronomy, astrophysics, biology, chemistry, geology, and many other scientific domains.

Stromatolites

The oldest known fossil evidence of microbial life on our planet is dated at around 3.7 billion years old and is preserved in the layers of ancient Archean *stromatolites*, which are rock and mineral structures built up by coordinated groups of simple single-celled organisms, especially cyanobacteria (formerly called blue-green algae).

While the details of stromatolite formation are the subject of active research, the basic outline seems to be that coordinated groups of microorganisms form threadlike biofilm structures called microbial mats that trap and ultimately cement together sedimentary grains, usually in shallow-water environments. Stromatolites built by organisms that rely on photosynthesis for their energy actively grow toward shallower water and more intense sunlight. The mobility and growth patterns of the microorganisms and the cemented grains that they build up over time vary with temperature and other environmental factors, tidal cycles, and/or sea level rise and fall, resulting in a variety of stromatolite shapes and sizes, including layers, domes, cones, branches, and columns.

The fossil record of stromatolites reveals that coordinated groups of microbes were among Earth's most prolific and successful life forms from about 3.7 billion years ago right up through the so-called Cambrian Explosion around 550 million years ago, when predatory shallow-water grazers appear to have significantly culled the stromatolite herd. During their heyday in the Archean and early Proterozoic (the second-youngest of Earth's four major geologic eons, from 2.5 billion years ago to the Cambrian Explosion), stromatolites formed by photosynthetic cyanobacteria were responsible for a massive increase in oxygen in the Earth's atmosphere.

The presence of actual fossilized microbes within ancient stromatolites is extremely rare, which leads to some controversy about the origin of many of these structures in the fossil record because there are a variety of non-biologic ways to create layered, domical, or other similar cemented sedimentary grain structures. Ultimately, strong similarities between the detailed shapes of ancient fossilized stromatolite structures and modern-day living stromatolites provide the strongest evidence for a biologic origin of specific ancient deposits. Indeed, stromatolites, also referred to more generally as *microbialites* in modern geobiology, still form in places such as Shark Bay in Western Australia or on the shores of the Great Salt Lake in Utah, making them among the oldest extant life forms on our planet.

SEE ALSO Earth's Oceans (c. 4 Billion BCE), The Archean (c. 4–2.5 Billion BCE), Photosynthesis (c. 3.4 Billion BCE), The Great Oxidation (c. 2.5 Billion BCE), Cambrian Explosion (c. 550 Million BCE)

Main image: *Modern-day stromatolite domes in shallow water at Shark Bay, Western Australia.*
Inset: *Cross-sectional view of a 2.4-inch-tall (6 cm) stromatolite fossil from the Old Range of Western Australia.*

Greenstone Belts

The Archean geologic eon spans an enormous range of the early history of our planet, from about 4 to 2.5 billion years ago. Because of this great age, very few Archean rocks are preserved on the surface of the Earth today, with most of them concentrated within about a dozen major shields (ancient continental crustal regions with low topographic relief) that represent about half of Earth's continental crustal surface area.

How exactly these ancient shields grew from their original late Hadean protocontinental cores is the subject of active geologic research, but the basic outline is that an early form of still poorly understood plate tectonics appears to have caused chunks of ancient oceanic crust to collide into and accrete (grow) onto the original cratons, over time forming larger and larger regions of lower-density continental crust.

The accreted terrains are called *greenstone belts* by geologists: "greenstone" because they contain green-hued metamorphic minerals such as chlorite, and "belts" because they often occur in multiple linear bands that appear to have been pasted on over time to pre-existing cratonic terrains. The mafic oceanic crustal rocks were heated or even completely melted during the collision with the craton, sometimes forming distinctive structures such as pillow lavas—blobs of lava erupted under water and formed into pillow-like shapes by being so rapidly cooled. While these rocks were being crushed, squeezed, and/or melted during the collision, they were also being mixed with more felsic, lower-density sedimentary rocks that were eroding off the continental crust and onto the adjacent seafloor.

Greenstone belts are a sort of geologic mess of rocks and minerals. Since, on average, these belts still have a lower density than the oceanic crust surrounding them, they have helped to grow the overall mass of continental crust over time. Putting together the pieces of the geologic puzzle is often daunting because greenstone belts are ancient Archean rocks that have been heavily modified during their formation as well as subsequently. Nonetheless, since so little of the Archean Earth is still around for us to study and explore, the fifty or so greenstone belt regions identified around the world so far have become an important focus for geologists trying to understand the earliest part of our planet's history.

SEE ALSO Continental Crust (c. 4 Billion BCE), The Archean (c. 4–2.5 Billion BCE), Plate Tectonics (c. 4–3 Billion BCE?)

Glacially smoothed Archean greenstone (metamorphosed basaltic pillow lava) from the Upper Peninsula of Michigan. The scene here is about 10 feet (3 meters) across.

Photosynthesis

Life on Earth requires liquid water, organic molecules, and a reliable source of energy. It is perhaps not surprising, then, that one of the most important innovations made by early life forms evolving on our world was the ability to turn readily available and reliable sunlight into energy to drive their internal biologic processes. *Photosynthesis—* the ability to turn sunlight into energy—has literally changed the world.

Geobiologists aren't sure exactly when photosynthesis began, because evidence for several kinds of similar precursor chemical reactions is preserved either in the fossil record or in the functioning cells of a number of existing organisms. However, fossilized remains of 3.4-billion-year-old microorganisms known as Filamentous Anoxygenic Phototrophs (FAPs) provide some of the earliest evidence for what we consider photosynthesis today. This early form of photosynthesis was "anoxygenic" because the byproduct of the harvesting of sunlight did not include free oxygen.

Via anoxygenic photosynthesis, Archean organisms such as FAPs and green sulfur bacteria took advantage of the hydrogen-rich, highly reducing atmosphere of the early Earth. Specifically, sunlight initiates the release of electrons in clusters of proteins, pigments, and other molecules called reaction centers, focusing especially on the breakdown of carbon dioxide and hydrogen sulfide to form more complex organic molecules that are ultimately processed into usable "food" in the form of glucose. Byproducts of anoxygenic photosynthesis are primarily water and elemental sulfur.

Later, toward the end of the Archean, other organisms such as cyanobacteria developed oxygenic photosynthesis—a similar process involving sunlight-induced electron transfers in molecular (including chlorophyll) reaction centers, but using carbon dioxide and water as inputs and yielding glucose and oxygen as byproducts. Over time, photosynthesis was incorporated into the cells of *eukaryotes—*organisms with complex internal cell structures—including, eventually, plants, via a process known as *endosymbiosis*.

The rise and rapid proliferation of oxygenic photosynthesizing cyanobacteria around the end of the Archean led to what geologists call "The Great Oxidation," the first (but not the last) example of life on Earth profoundly changing the atmosphere of our planet.

SEE ALSO Life on Earth (c. 3.8 Billion BCE?), Banded Iron Formations (c. 3–1.8 Billion BCE), The Great Oxidation (c. 2.5 Billion BCE), Eukaryotes (c. 2 Billion BCE), Advanced C4 Photosynthesis (c. 30–20 Million BCE), Endosymbiosis (1966), Rising CO_2 (2013)

In places like this shallow-water kelp forest off the coast of southern California, plants use oxygenic photosynthesis to convert sunlight into internal energy, such as glucose.

Banded Iron Formations

The evolutionary development of photosynthesis (extraction of energy from sunlight), and in particular of *oxygenic* photosynthesis, resulting in free oxygen as a "waste" product, began toward the end of the Archean eon. Until that time, oxygen was relatively rare in the Earth's atmosphere.

Free oxygen is not a particularly stable molecule in Earth's atmosphere. As the term implies, *oxidation* of many common rocks and minerals occurs rapidly in the presence of oxygen and relatively quickly depletes the supply. Indeed, evidence for such oxidation and subsequent depletion can be seen around the world in Banded Iron Formations, or BIFs, which are outcrops or beds of rocks exhibiting striped, semi-regular layers of reddish and then non-reddish rocks, stacked together like layers of meat and cheese in a sliced sandwich.

Reddish BIF layers are thought to form when enhanced local oxygen levels oxidize the dissolved or shallow-water sediment known as "pristine" iron—formed from volcanically erupted rocks—to form red or orange iron oxides. After the oxygen supply runs out, new beds that form retain the white, gray, or black tones characteristic of unoxidized precursor minerals.

Geologists have identified small numbers of BIF outcrops dating back to the early- to mid-Archean, potentially reflecting rare and local early periods of oxygen saturation. However, BIFs appear most abundantly in the geologic record in the period from about 2.5 to 1.8 billion years ago. This time span corresponds to the appearance and rise of photosynthetic cyanobacteria, which produce free oxygen. Oxygen would have built up in oceans and seas populated by abundant cyanobacteria until it reached a tipping point and oxidized the abundant and relatively *anoxic* (oxygenless) muds and other sediments on the seafloor, turning the exposed parts of them red. The process may have continued cyclically for millions of years of sedimentation, resulting in alternating oxidized and anoxic layers.

Many details of the formation and timing of BIFs are still poorly understood. Still, the relatively rapid appearance of extensive and almost rhythmic oxidative/anoxic episodes peaking from about 2.5 to 1.8 billion years ago, followed by the ubiquitous occurrence of thick red beds of rocks since then, strongly implicates the cyanobacteria as major players in this important geologic and atmospheric enigma.

SEE ALSO The Hadean (c. 4.5–4 Billion BCE), Earth's Oceans (c. 4 Billion BCE), The Archean (c. 4–2.5 Billion BCE), Photosynthesis (c. 3.4 Billion BCE), The Great Oxidation (c. 2.5 Billion BCE)

Photo showing stripes of iron-rich and silica-rich rocks known as Banded Iron Formations at Fortescue Falls, Karijini National Park, Western Australia.

The Great Oxidation

The rise and proliferation of photosynthetic organisms in the Archean, and especially of the oxygen-producing cyanobacteria (formerly known as blue-green algae) in the late Archean, had a profound impact on the composition of the Earth's atmosphere and oceans. Before the appearance of oxygenic photosynthesizing organisms, free oxygen (O_2) was a minor trace gas in the atmosphere, and a toxic poison to most forms of life on Earth. Since the rise of the cyanobacteria, however, oxygen has become a significant atmospheric component (now representing about 20 percent of Earth's atmosphere), and entirely new species of *aerobic* organisms (those that survive and thrive in an oxygenated environment) have appeared.

Evidence from the geologic record suggests that this change from Earth's *reducing* atmosphere (with abundant hydrogen-bearing molecules such as methane and little oxygen) to its current *oxidizing* atmosphere began rather abruptly around 2.5 billion years ago. "The Great Oxidation" fundamentally and permanently changed the chemistry of the oceans as well as, eventually, the land. Some geobiologists also refer to this change as "the Oxygen Catastrophe," because the ever-increasing levels of the oxygen that was dissolved in the oceans led to the mass extinction of enormous numbers of species of anaerobic organisms.

The global increase in Earth's atmospheric oxygen was not a sudden event, however. As oxygen levels slowly began to increase, oxygen was periodically removed from the ocean and atmosphere by the oxidative chemical weathering of seafloor and land sediments. Iron-bearing minerals in fine-grained sediments were particularly susceptible to weathering, and these episodes of rusting led to a peak in the creation of Banded Iron Formations also around 2.5 billion years ago. Only over another 2 billion years, after the rate of oxygen production by microbes (and eventually plants) outstripped the rate of its depletion by oxidation, could atmospheric oxygen abundance rise more rapidly to the present level. The detailed time history of atmospheric oxygen increases (and decreases) is the subject of much research and many competing hypotheses.

The presence of so much oxygen in Earth's atmosphere is a signal that could easily be interpreted as evidence of life on Earth by alien astronomers studying our planet. Indeed, oxygen is a key *biosignature* that astronomers on Earth are searching for on other worlds.

SEE ALSO The Hadean (c. 4.5–4 Billion BCE), Earth's Oceans (c. 4 Billion BCE), The Archean (c. 4–2.5 Billion BCE), Photosynthesis (c. 3.4 Billion BCE), Banded Iron Formations (c. 3–1.8 Billion BCE), Snowball Earth? (c. 720–635 Million BCE)

High-resolution microscope view of filamentary strands of cyanobacteria, viewed through a green filter.

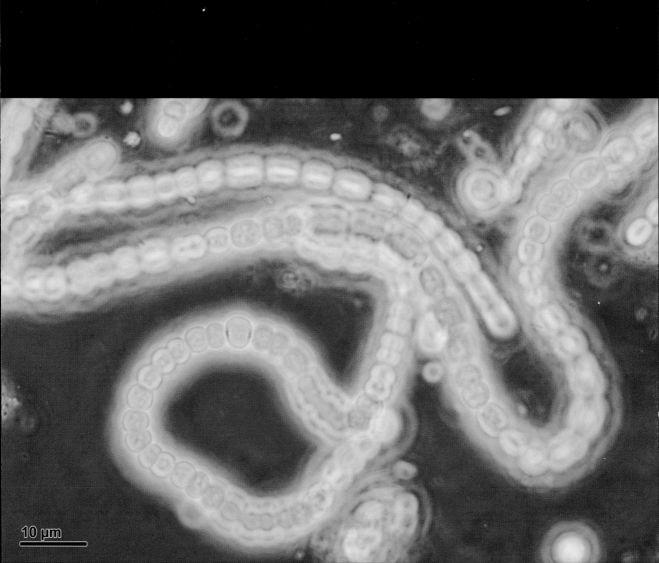
10 μm

Eukaryotes

From its origin perhaps 3.8 billion years ago and throughout the entirety of the Archean eon, life on Earth consisted of a modest variety of simple, single-celled, *prokaryotic* organisms. Prokaryotes are single-celled organisms that lack a cell nucleus or any of the other enclosed organelles (specialized structures) found in more complex organisms. Despite their relative simplicity, prokaryotes have still been quite capable of using relatively complex chemical reactions (like photosynthesis) to survive and evolve.

Around 2 billion years ago, however, in an innovative evolutionary step that would become a milestone in the history of life on Earth, some prokaryotes developed a membrane-enclosed nucleus and other membrane-enclosed organelles, becoming *eukaryotes*. For example, the nucleus of a eukaryotic cell became a special place to store the genetic material (such as DNA and RNA) needed for reproduction. Compartmentalized mitochondria inside eukaryotic cells became special places to generate chemical energy for the cell.

Eukaryotes represent a third unique branch on the tree of life on Earth. The other two kinds of life, *bacteria* and *archaea*, are prokaryotes with enough distinctive differences in their DNA, RNA, and structure to warrant separate branches on the tree. While existing bacteria and archaea have not evolved as much since their origin early in the history of our planet, eukaryotes would evolve into multicellular organisms of stunning form and complexity, including algae, plants, fungi, and animals. You, for example, are a eukaryote.

No one knows exactly how eukaryotic cells came into existence, but there are many competing hypotheses. In one model, indentations in large precursor prokaryotic cells closed in on themselves to house specialized structures within the cell. A related idea is that some kinds of predator prokaryotic cells that surrounded others as prey could incorporate the prey into their structures to become specialized organelles—a process called *endosymbiosis*. For example, chloroplasts, the organelles within eukaryotic cells where photosynthesis occurs, could have evolved from the endosymbiotic synthesis of precursor cyanobacteria. Another hypothesis involving endosymbiosis is that precursor bacterial and archaean cells fused or merged somehow (chimerically) to become eukaryotic cells. Evolutionary biologists continue to search for the answer.

SEE ALSO The Archean (c. 4–2.5 Billion BCE), Life on Earth (c. 3.8 Billion BCE?), Photosynthesis (c. 3.4 Billion BCE), Complex Multicellular Organisms (c. 1 Billion BCE), Endosymbiosis (1966)

Tiny fossilized remains of microfossils called Eosphaera, found within the more than 3.4-billion-year-old Gunflint chert formation in Pilbara, Western Australia. These enigmatic microorganisms might represent transitional forms between prokaryotes and eukaryotes.

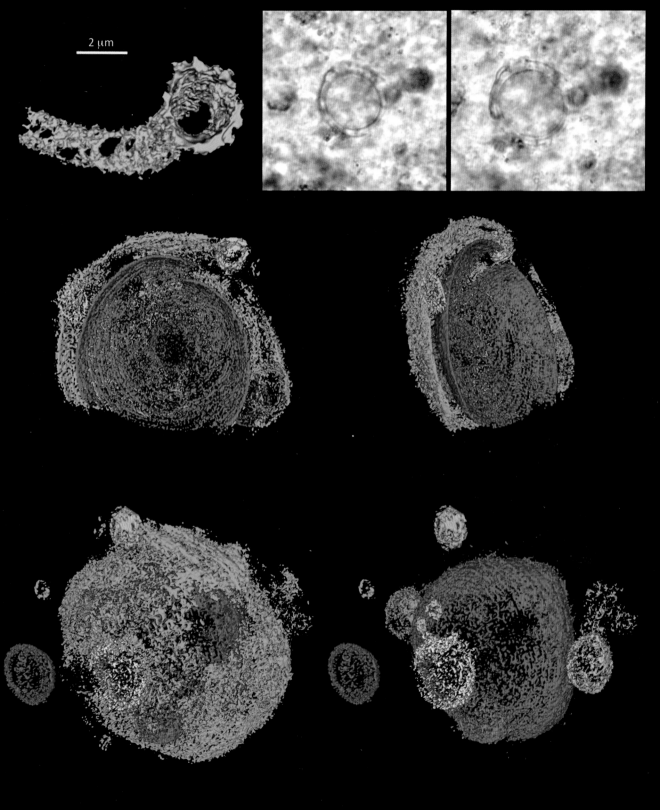

2 μm

The Origin of Sex

Eukaryotes were the first class of single-celled and multi-cellular organisms to appear on Earth to have complex internal structures such as a nucleus, mitochondria, and other similar organelles. For hundreds of millions of years after they appeared, eukaryotes reproduced *asexually*, using the biologic process of *mitosis*. In mitosis, chromosomes replicate inside a cell's nucleus, and the cell then divides into two daughter cells, each with one genetically identical nucleus from the mother cell. Asexual reproduction via mitosis is an efficient survival strategy because it ensures that every organism can bear its own young. It has its downsides, too, however, such as the inability to prevent the accumulation of genetic mutations in offspring.

Around 1.2 billion years ago, eukaryotes evolved a different kind of scheme for procreation. Using a new biologic process called *meiosis*, some organisms began engaging in *sexual* reproduction. In meiosis, chromosomes replicate inside a cell's nucleus, and the cell then goes through two rounds of cell division—first so that the replicated chromosome pairs can exchange genetic information, and then second to split up the replicated pairs into four new cells called *gametes* that each contains half the number of chromosomes as the original parent cell does. The key next step, then, is for gametes from one parent to fuse with the gametes from another parent—*fertilization*. And thus, sex was born.

It almost seems counterintuitive that sexual reproduction (involving the need to find a mate) would be evolutionarily more advantageous than asexual reproduction (which can be done all alone). In addition to being less efficient, sexual reproduction allows a parent to transmit only half its genetic material to its offspring. However, that might be part of the key to the success of sex. That is, splitting the chromosomes from the parents provides a mechanism not only to prevent the accumulation of deleterious genetic mutations, but also to enable the accumulation of advantageous mutations.

Sexual reproduction also turns out to be a powerful agent of *natural selection*—long-term changes in inherited traits of a population in order to increase (or decrease) their chances of survival. Propagation of advantageous mutations or competitive advantages, for example, can substantially increase a species' rate of adaptation to changing environments.

SEE ALSO Life on Earth (c. 3.8 Billion BCE?), Eukaryotes (c. 2 Billion BCE), Natural Selection (1858–1859), Endosymbiosis (1966), Genetic Engineering of Crops (1982)

The novel idea of merging the genetic information from two parents (like the male sperm and female egg shown here) to create unique offspring—sexual reproduction—was pioneered by early eukaryotic organisms around 1.2 billion years ago. It's turned out to be a pretty useful evolutionary innovation!

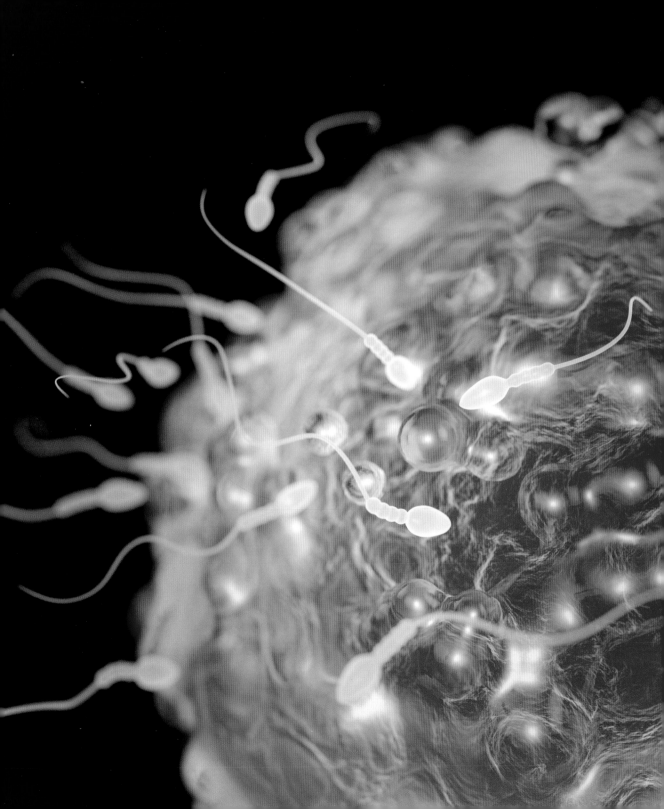

Complex Multicellular Organisms

For about 3 billion years after it first formed, almost all life on Earth was composed of single-celled organisms. First were the *prokaryotes* (simple single-celled organisms with no internal cellular structures), and then after a few billion years came the *eukaryotes* (more complex single-celled organisms with a cell nucleus and other specialized internal structures). According to the geochemical and fossil record of life, *multicellular* prokaryotes and eukaryotes appear to have evolved only rarely during life's first 3 billion years.

The first appearance of more complex and truly multicellular organisms, where an outer membrane encloses a diverse variety of different kinds of cells, appears to have occurred only about a billion years ago, but widely across the eukaryotic branch of the tree of life. Some of the best preserved fossils come from the 650-million-year-old Doushantuo Formation in southern China, including embryo-like clusters of fossilized cells that exhibit evidence for many different kinds of cells within the same overall structure. Other evidence indicates that only six distinct classes or *clades* of eukaryotes have separately evolved to exhibit complex multicellular structures: animals, land plants, two kinds of algae, and two kinds of fungi.

Just as evolutionary biologists are puzzled by the origin of the transition from prokaryotic cells to eukaryotic cells in the evolution of life, they are also challenged to develop hypotheses for the origin of complex multicellularity in eukaryotes. Nonetheless, details of the physical structures of preserved microfossils, as well as the evolutionary connections of the genes of ancient organisms and their more modern descendants, have allowed for a number of hypotheses. For example, one idea is that complex multicellular organisms formed from the fusion of single-celled colonies of the same organism. A similar hypothesis is that complex multicellular organisms formed from the fusion of single-celled organisms of different species that exhibited symbiotic functions. Another idea is that some single-celled organisms developed multiple nuclei, which then evolved to serve different specific functions.

New microfossil finds by paleontologists occasionally help to test these (and other) hypotheses, but the fact that these ancient organisms were small, simple, and lacking shells or other hard body parts makes their occurrence in the fossil record extremely rare, and thus the search for clues to their origins is extremely daunting.

SEE ALSO Life on Earth (c. 3.8 Billion BCE?), Eukaryotes (c. 2 Billion BCE), Endosymbiosis (1966)

Artist's concept of aggregated collections of multicellular eukaryotes. Enclosed within the outer membrane are many different kinds of cells designed to carry out different specific functions.

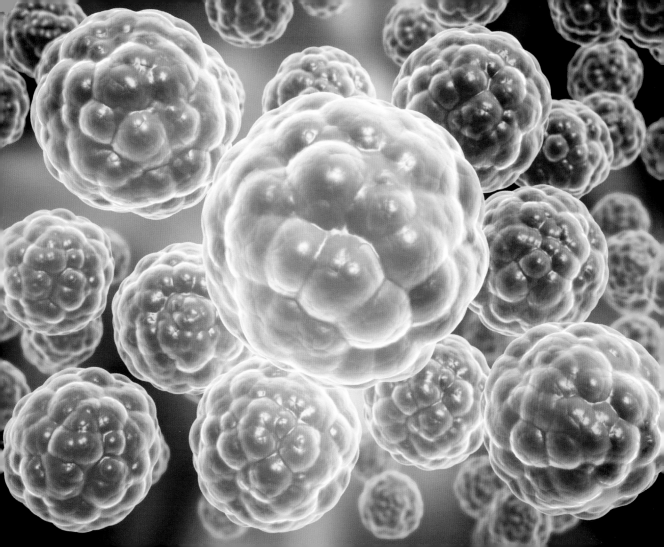

Snowball Earth?

Large swings in the planet's average surface temperature, measured in ice cores and driven by changes in the Earth's orbital parameters, tell us of the occurrence of ice ages roughly every 100,000 years during the last million years. These turn out to be relatively small events in time, however, compared to what appear to be at least five potentially much longer periods of extensive cold planetary conditions that are preserved in the more ancient rock record.

Perhaps the most widely studied and well-known of these super-long cold-climate episodes occurred during a span of about 85 million years, starting about 720 million years ago, in a period known as the Cryogenian (from the Greek "cold birth") on the geologic timescale. Geologists believe that the Earth may have experienced up to three very long periods of the most severe ice ages in the planet's history during this time.

How cold did it get? By mapping characteristic glacial deposits globally, and by reconstructing the positions of the continents during the Cryogenian, geologists have found that glaciers appear to have reached from the poles all the way to the equator, in both hemispheres. The vision that emerges of our planet during such times of extreme glaciation—of frozen oceans and completely snow-covered continents—has been dubbed "Snowball Earth." A similar deep cold Snowball Earth episode is hypothesized to have occurred even farther back in time, in the late Archean around 2.4 to 2.1 billion years ago.

What could have caused these large and long-term decreases in global temperature, and how did our planet recover from them? The Archean glaciation occurred just after the Great Oxidation event, and so many geologists hypothesize that increasing oxygen levels led to the breakdown of strong greenhouse-effect gases such as methane, resulting in a dramatic cooling of the atmosphere and surface. Consensus does not exist for a similar potentially direct cause for Cryogenian Snowball-Earth episodes, however. In terms of global climate recovery, most hypotheses focus on the re-accumulation of enough of those same atmospheric greenhouse gases to re-warm the planet. Paleoclimatologists remain puzzled, however, about the details of how the Earth thawed back out from these deepest of freezes.

SEE ALSO The Great Oxidation (c. 2.5 Billion BCE), Cambrian Explosion (c. 550 Million BCE), End of the Last "Ice Age" (c. 10,000 BCE), The Little Ice Age (c. 1500), Discovering Ice Ages (1837), The Greenhouse Effect (1896)

Artist's impression of "Snowball Earth," a time (or times) in the distant past when our planet was hypothesized to be mostly or entirely covered by snow and ice.

Cambrian Explosion

The first life forms to emerge on Earth were *prokaryotes*—simple, single-celled organisms lacking either a distinct nucleus with a membrane or specialized internal cellular structures that are common in more complex *eukaryotes*. For the first 3 billion years or so of Earth's history, life was dominated by such single-celled organisms. Even with the evolution of the first complex multicellular organisms about a billion years ago, it's still very hard to find evidence for these early forms of "soft" life in the fossil record.

Around 550 million years ago, however, in what is often called the *Cambrian Explosion* because it so dramatically marks the beginning of the Cambrian geologic period, the diversity of life on Earth began to drastically increase. Specifically, many eukaryotic organisms began to develop hard-shelled exoskeletons and other body parts, structures that could be preserved in sediments after the organisms died and fell to the seafloor. As such, many of the ancestors of today's modern plants and animals appear rather early in the fossil record, which really begins in earnest at the Cambrian Explosion. Biologists hypothesize that exoskeletons might have been an evolutionary response to adaptations by other competing organisms, such as the development of eyes and other advances in predation.

In addition to trying to understand the origin of large increases in species diversity, biologists are also trying to understand the reasons for at least five rather sudden, drastic mass extinctions of species found in the fossil record. The most dramatic of these occurred at the boundary between the Permian and Triassic geologic periods, around 250 million years ago. Across a span of perhaps only a million years, about 70 percent of all land species and 96 percent of all oceanic species died off, a period informally called the "great dying" and the "mother of all mass extinctions."

What caused such massive and relatively sudden loss of life on Earth? Geologists have implicated climate change, massive impact events, and enormous volcanic outpourings. Whatever the cause, it took more than 100 million years for the diversity of life on Earth that had begun to flourish at the Cambrian Explosion to once again reach pre-Permian levels.

SEE ALSO Life on Earth (c. 3.8 Billion BCE?), Stromatolites (c. 3.7 Billion BCE), Eukaryotes (c. 2 Billion BCE), Complex Multicellular Organisms (c. 1 Billion BCE), Dinosaur-Killing Impact (c. 65 Million BCE)

Fossilized remains (packed in lithified seafloor sediments) of the exoskeletons and other hard-shelled body parts from new organisms that began to appear around 550 million years ago. This "Cambrian explosion" marked an enormous increase in the diversity of life on Earth.

Roots of the Pyrénées

Geologists use the word *orogeny* to describe a period of mountain-building within the Earth's crust. Around 100 separate and distinct epochs of mountain-building have been identified around the world by geologists, ranging from some of the most ancient preserved continental terrains to some of the tallest peaks on Earth that are still being built today.

The geologic process of mountain building—*orogenesis*—is driven by the movement of the Earth's lithospheric plates. Specifically, when a continental plate is thrust or crumpled upward by colliding with another plate, one or more mountain ranges form along the collisional plate boundary. The tectonic forces involved are enormous, and the resulting deformation of the crust and lithosphere can extend far below the visible mountains on the surface.

Many of the tallest, most pristine-looking mountain ranges in the world are relatively young, as wind, rain, glaciers, and other erosive processes haven't had time to wear them down. However, many of these ranges have ancient roots upon which younger mountains are being built. A prime example is the Pyrénées mountain range that makes up much of the border between France, Spain, Portugal, and Andorra. The Pyrénées are young, with peaks rising more than 11,150 feet (3,400 meters) above sea level from the collision of the Iberian "microcontinent" with the European plate from about 55 to 25 million years ago. However, the roots of the Pyrénées, and thus many of the uplifted rocks that make up the mountains themselves, are ancient.

Specifically, the rocks from which the Pyrénées are built were originally formed starting around 500 million years ago, in an even earlier mountain-building episode called the Variscan orogeny. Back then, the large continental plates called Gondwana (which would eventually split into Africa, South America, Antarctica, Australia, and the Indian microcontinent) and Laurussia (which would eventually split into North America, Greenland, and Europe) were colliding to eventually form the supercontinent of Pangea. Extreme pressures and temperatures from the tectonic forces along the collisional boundary deformed, metamorphosed, and melted continental, ocean crust, and even some upper-mantle rocks along that zone. Many of these extensively fractured and folded ancient "basement" rocks can now be seen even high among the present-day peaks of the Pyrénées.

SEE ALSO Continental Crust (c. 4 Billion BCE), Plate Tectonics (c. 4–3 Billion BCE?), The Appalachians (c. 480 Million BCE), Pangea (c. 300 Million BCE), The Sierra Nevada (c. 155 Million BCE), The Rockies (c. 80 Million BCE), The Himalayas (c. 70 Million BCE), The Alps (c. 65 Million BCE)

View of the "young but ancient" Pyrénées Mountains within Ordesa National Park, Spain.

The Appalachians

Airline travelers passing over Pennsylvania on a clear day, or drivers snaking their way through switchback roads in the Smoky Mountains of Tennessee, can't help but notice the distinctive patterns of seemingly tortured and twisted rocks that they encounter along the way. These heavily forested, subdued rolling hills and mountains are all that remain of the Appalachian mountain range. Formed via multiple long episodes of continental-plate collisions that began some 480 million years ago, Appalachian peaks that once rose to the heights of the modern Alps and Rockies have since been worn down to much less dramatic topography that now rarely reaches above 6,700 feet (2,000 meters).

Prior to the formation of the Appalachians, that part of the North American craton was what geologists call a *passive margin*—a peaceful boundary between two plates where there is no relative motion. However, around 480 million years ago, the plate motions changed, and that boundary was activated into a *convergent margin*—a place where adjacent plates crunch into each other. The resulting mountain-building event, where a now long-gone dense oceanic plate collided with and began subducting, or descending, under the eastern edge of the less-dense North American continental plate, is known as the Taconic orogeny.

Rocks and sediments that had been deposited along the shoreline and on the seafloor of the formerly passive margin of eastern North America were uplifted, faulted, folded, and metamorphosed by the plate collision. Volcanoes erupted along the margin, too, as the subducting plate melted and blobs of less dense magma made their way to the surface. The action continued for some 40 million years, until the oceanic plate was completely consumed under the North American plate. The region then turned back into a passive plate margin, and the young, tall mountains began to erode.

This wasn't the end of Appalachian mountain-building. Similar kinds of orogenies occurred along that eastern plate margin for the next 250 million years. Each new mountain-building episode thrust and welded new continental and former marine sedimentary terrains onto the continent and renewed the grandeur of the mountains. Today, the plate margin is again a passive one, and we see only the heavily eroded remains of those ancient plate collisions.

SEE ALSO Continental Crust (c. 4 Billion BCE), Plate Tectonics (c. 4–3 Billion BCE?), Roots of the Pyrénées (c. 500 Million BCE), Pangea (c. 300 Million BCE), The Sierra Nevada (c. 155 Million BCE), The Rockies (c. 80 Million BCE), The Himalayas (c. 70 Million BCE), The Alps (c. 65 Million BCE)

Satellite image of the ancient, folded, eroded remnants of the Appalachian Mountains (brown bands), in the northeastern United States. New York City is visible just above center right.

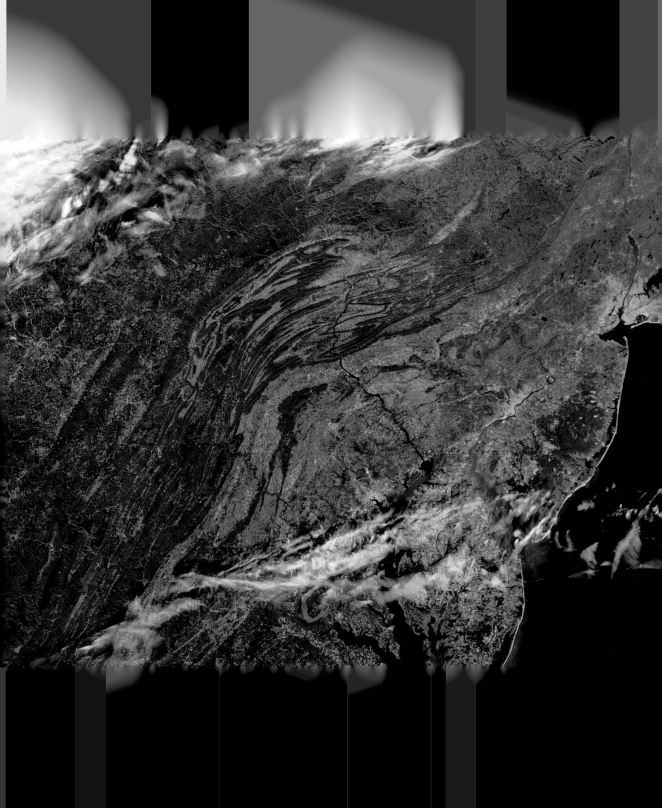

First Land Plants

For more than 85 percent of the history of life on Earth, almost everything alive was in the water. Perhaps the most notable exception in the fossil record were the *green algae*, a group of ancient photosynthesizing eukaryotic organisms that appear to have evolved the ability to survive along shorelines where they would occasionally get dried out. Evolutionary biologists believe that additional adaptations eventually helped these organisms evolve to become the first plants to permanently dwell on land.

Some of the oldest evidence for the first land plants comes from tiny, roughly 470-million-year-old fossilized spores (plant reproductive cells) with structures that resemble spores of the modern moss-like liverwort plant (*Marchantiophyta*), which are most commonly found today in moist, tropical areas. These earliest forms of land-dwelling plants were *non-vascular*—that is, they did not have the deep roots or tissues that later *vascular* (ducted) land plants would evolve to move water and minerals throughout the plant.

Plants developed a variety of evolutionary strategies and forms in order to battle the constant threat of drying out (desiccation). Some only grow close to water and thus generally try to avoid desiccation entirely. Others developed the ability to slow their metabolism greatly when they dry out, waiting for the next exposure to water to reactivate. And still others developed specialized structures such as vascular tissues and stomata (pores) to help store water and guard against evaporation while still enabling the atmospheric gas transfer needed for photosynthesis.

The establishment and rapid proliferation of permanent land-dwelling plants had a profound effect on the Earth's atmosphere and surface by significantly increasing the rate of accumulation of free oxygen (O_2), a waste product of photosynthesis. Among the most dramatic implications was a rapid increase in the rate of destruction of atmospheric methane by oxygen, significantly decreasing the greenhouse-gas warming of the atmosphere and potentially leading to multiple occurrences of "Snowball Earth" glaciation. Another important effect of rapidly increasing oxygen was the enabling, for the first time, of fire. Wildfires have since become an important part of the ecology and life cycles of many ancient—as well as modern—plants.

SEE ALSO Photosynthesis (c. 3.4 Billion BCE), The Great Oxidation (c. 2.5 Billion BCE), Eukaryotes (c. 2 Billion BCE), Snowball Earth? (c. 720–635 Million BCE), Big Burn Wildfire (1910)

The first plants to establish a foothold on land were probably very similar to mosses like these, growing on rocks close to the water.

Mass Extinctions

The fossil record reveals that at least five times, life on Earth has experienced widespread and relatively rapid decreases in diversity—catastrophic events known as mass extinctions. The first one of these recognized events occurred around 450 million years ago. During the Ordovician–Silurian or "End Ordovician" mass extinction, somewhere between 70 and 85 percent of all species on Earth went extinct, over a period of perhaps only a few million years or less. This has turned out to be the second-largest mass extinction event in Earth's history, eclipsed only by the so-called "Great Dying" extinction event at the end of the Permian period, around 250 million years ago.

Enormous numbers of marine organisms were decimated during the End Ordovician mass extinction. For example, fossils from many species of creatures such as brachiopods, bryozoans, conodonts, and trilobites rather abruptly disappeared around 450 million years ago. The magnitude of the carnage varied considerably with the lifestyles of the organisms. In particular, species that primarily dwelled in shallow water, or in the planktonic zone of the ocean relatively close to the surface, were much more likely to become extinct than species that primarily dwelled in deep water.

While the exact cause of the End Ordovician mass extinction is not known, the correlation between species extinction and lifestyle in shallow or sunlight-rich waters strongly suggests that the cause of the extinction was climatic in nature, perhaps related to a dramatic change in temperature and/or sunlight reaching the surface. Such a dramatic change in climate could have been caused by many kinds of events. For example, the dust and wildfire-created soot and ash from a large-scale impact event (such as the one strongly linked to the extinction of the dinosaurs around 65 million years ago) could have cut off sunlight and substantially modified the climate. Large volcanic eruptions could have had similar effects. Or perhaps Snowball Earth–like cooling and glaciation decimated plankton and other creatures at the base of the food chain. Yet another hypothesis is that a burst of high-energy gamma rays from a relatively nearby stellar supernova explosion killed large numbers of organisms and initiated a mass extinction. Whatever the cause, it would not be the last time that life on Earth faced such a major crisis.

SEE ALSO Complex Multicellular Organisms (c. 1 Billion BCE), Snowball Earth? (c. 720–635 Million BCE), The Great Dying (c. 252 Million BCE), Triassic Extinction (c. 200 Million BCE), Deccan Traps (c. 66 Million BCE), Dinosaur-Killing Impact (c. 65 Million BCE), Arizona Impact (c. 50,000 BCE)

Massive volcanic eruptions, intense continental glaciations, asteroid impacts, and/or changes in ocean chemistry are all possible explanations for the End Ordovician and late Devonian mass extinctions of 450 million and 360 million years ago, each of which wiped out 70 to 85 percent of all species then on Earth.

First Animals on Land

The diversity of complex multicellular life increased dramatically around the time of the so-called Cambrian Explosion 550 million years ago. In addition to the first organisms to develop shells and other hard body parts, the first *vertebrates* (organisms with either a cartilaginous or bony spine and a well-defined head and tail) also evolved around this time. The first vertebrates were filter feeders that lacked a jawbone. That evolutionary innovation would soon appear in the Devonian period of the geologic timescale (about 420 to 360 million years ago), when jawed fish first appeared.

Yet another dramatic evolutionary step seen in the fossil record is the appearance of the first *tetrapods*, or four-limbed amphibious vertebrates, which evolved from particular kinds of lobe-finned fishes in the Devonian. Tetrapods became the first vertebrates to use primitive legs to swim in the ocean as well as to move on the land. One of the most interesting tetrapod fossils identified is from a vertebrate known as *Ichthyostega*, which is thought to be one of the first fish-like creatures to walk on land, starting about 375 million years ago. *Ichthyostega* likely lived in swamps or other shallow-water environments, and, based on its heft and rather weak and stubby legs, most likely spent most of its time in water rather than on land. Still, an important innovation that *Ichthyostega* exhibited was the ability to breathe air directly, using lungs rather than gills.

The details of the transition from fish to tetrapod amphibians is the topic of intense debate among evolutionary biologists, who await new Devonian fossil finds in the hope of discovering direct evidence of the nature of this transition. Regardless of whatever evolutionary and/or environmental pressures drove a branch of the lobe-finned fish to evolve legs and become able to survive out of the water, the result has been profound for life on Earth: over the next tens to hundreds of millions of years, the success of the tetrapods would eventually lead to the evolutionary development of specialized tetrapod subclasses called amphibians, reptiles, and mammals.

SEE ALSO Cambrian Explosion (c. 550 Million BCE), First Land Plants (c. 470 Million BCE), Reptiles (c. 320 Million BCE), Mammals (c. 220 Million BCE), Age of the Dinosaurs (c. 200–65 Million BCE)

Artist's concept of Ichthyostega, a 5-foot-long (1.5 m) "transitional" vertebrate somewhere between fish and tetrapods, combining a fishlike tail and gills with an amphibian skull and limbs.

The Ural Mountains

Like all the Earth's other continental land masses, Asia is a "supercontinent" that was built, piece by piece, from pasted-on portions of oceanic and continental crust that collided with an original ancient core or *craton* formed back in the Archean. High mountain ranges are often formed at the convergent zones where plates collide. Over time, however, extensive erosion and/or the subsequent rifting and splitting of the plates have helped to wear down the oldest mountain ranges until they are reduced to just a shadow of their former selves.

This is not so with the ancient Ural Mountains, however—a tall, jagged range of peaks that run along a nearly north-south line from modern central Kazakhstan through western Russia, near the Europe-Asia border. The Urals are one of the oldest surviving mountain ranges on Earth, having formed in a mountain-building event called the *Uralian orogeny* over a period of about 90 million years, starting back in the late Paleozoic era of the geologic time scale, about 320 million years ago. Back then, the eastern edge of one continental plate collided with the western edge of another, causing enormous compression, crumpling, and uplift in the resulting north-south collisional zone. The sedimentary and volcanic rocks of the plate margins were fractured, folded, metamorphosed, and melted in the collision, and the two formerly independent continental plates were welded together—a continental growth process that geologists call *accretion*.

The evolution of the Ural Mountains since the Paleozoic has been quite different from the evolution of many of the world's other mountain ranges formed back then, however, because of the long-term stability of the Asian supercontinent. Over time, most supercontinents on Earth have rifted apart after they initially formed, because of the continuing dynamic nature of lithospheric plate motions across the globe. The region of Asia where the Urals were originally formed has not rifted apart, however, sparing the Urals from the tectonic and erosive effects that have worn down other mountain ranges. More "normal" erosive forces (wind, water, and glaciers) have acted on the mountains over time, of course, but have only been able to wear them down to heights of 5,000 to 6,200 feet (1,500 to 1,900 meters)—impressive reaches for mountains so old.

SEE ALSO Continental Crust (c. 4 Billion BCE), Plate Tectonics (c. 4–3 Billion BCE?), Roots of the Pyrénées (c. 500 Million BCE), The Appalachians (c. 480 Million BCE), Pangea (c. 300 Million BCE), The Rockies (c. 80 Million BCE), The Himalayas (c. 70 Million BCE), The Alps (c. 65 Million BCE)

Tall, craggy peaks in the Ural Mountains, near the village of Saranpaul in central Russia, make one of the oldest mountain ranges on the Earth seem young.

Reptiles

The evolution about 375 million years ago of animals capable of living on dry land—at least temporarily at first—quickly led to the development of specialized sub-species of amphibious *tetrapod* vertebrates that could exploit newly available niches for food and safety. Among the first of those new classes of animals to appear were the reptiles. Unlike their amphibian ancestors, reptiles developed the ability to lay soft-shelled amniotic (fluid-filled) eggs on dry land, thus avoiding the need to return to the water for reproduction like many other animals had.

Some of the earliest known unambiguous fossils of reptiles are of a now-extinct animal known as *Hylonomus* (from the Latin for "forest dweller"), a vertebrate tetrapod about 8–10 inches (20–25 centimeters) long that lived in the Pennsylvanian geologic period (about 320 to 300 million years ago) and that would have looked a lot like a small modern-day lizard. Based on fossil finds, *Hylonomus* appears to have dined on insects and lived among the rotten trees and leaves of forested coastal woodlands.

Permanent life on land presented both opportunities and challenges for early (and modern) reptiles. Early reptiles didn't have to worry about other predators on land, for example, although once the reptile population grew to fill the niche of available food, reptile species would eventually evolve into predators of one another. Reptiles also had to adapt to dramatically changing temperatures or other environmental conditions on land compared to in the water—a significant concern for cold-blooded animals that could become easy lethargic prey in cold temperatures, or while exposed to the sunshine to warm themselves up.

Despite the challenges, reptiles would continue to successfully adapt to life on land in terms of evolutionary developments, and would eventually become the largest animals ever to roam the land. Among the descendants of the first reptiles such as *Hylonomus* were the ornithischian ("bird-hipped") and saurischian ("lizard-hipped") dinosaurs, turtles, tortoises, crocodiles, alligators, snakes, lizards, and—yes—birds. Tens of thousands of modern reptilian species have been cataloged, but tens of thousands more have gone extinct in the past—many from slow natural-selection forces, but many (such as the ancient dinosaurs) in dramatic and rather sudden extinction events caused by meteorite impacts, volcanic eruptions, or other large-scale climate-changing events.

SEE ALSO Snowball Earth? (c. 720–635 Million BCE), Cambrian Explosion (c. 550 Million BCE), First Land Plants (c. 470 Million BCE), First Animals on Land (c. 375 Million BCE), The Great Dying (c. 252 Million BCE), Triassic Extinction (c. 200 Million BCE), Deccan Traps (c. 66 Million BCE), Dinosaur-Killing Impact (c. 65 Million BCE), Arizona Impact (c. 50,000 BCE)

According to the fossil record, early reptiles such as Hylonomus *probably didn't look very different from modern lizards like this one.*

The Atlas Mountains

Perhaps one of the most important geologic events that Africa has gone through occurred around 300 million years ago, when relative plate motions caused a collision between Africa's northern margin and the eastern margin of North America. The resulting continent-on-continent collision caused enormous compression, fracturing, deformation, and uplift of the crust, building enormous mountains (comparable to the Alps and Rockies today) in an event that geologists call the *Alleghenian orogeny*. When Africa and North America eventually rifted apart later with the opening of the Atlantic Ocean, the parts of those mountains that stayed with North America eventually became the Appalachian Mountains, and the parts that stayed with Africa eventually became the Atlas Mountains, a chain of peaks that extend mostly east–west through Morocco, Algeria, and Tunisia.

During and after the opening of the Atlantic, these ancient parts of the Atlas Mountains were worn down considerably by erosion. Today, those parts of the range are called the Anti-Atlas mountains. More recently (roughly over the past 60–70 million years), relative plate motions changed again, and mountain-building in the Atlas range was again re-activated when the northern part of the African continent collided with the southwestern part of the European plate (the Iberian Peninsula, where Spain and Portugal are today). The collision once again uplifted the mountains (and their associated sedimentary and seafloor deposits) to great heights, creating the modern High Atlas and other tall sub-ranges, with many jagged young peaks rising above 13,000 feet (4000 meters).

While many puzzles and much uncertainty remain about the specific timing and styles of plate motions involved, the same geologically recent African–European continental collision that created the High Atlas and other young mountain ranges in north Africa also appears to have created the modern Pyrénées and Alps in Europe, and to have closed the Strait of Gibraltar to enable the modern Mediterranean Sea to eventually form.

SEE ALSO Continental Crust (c. 4 Billion BCE), Plate Tectonics (c. 4–3 Billion BCE?), Roots of the Pyrénées (c. 500 Million BCE), The Appalachians (c. 480 Million BCE), Pangea (c. 300 Million BCE), The Atlantic Ocean (c. 140 million BCE), The Alps (c. 65 Million BCE), The Mediterranean Sea (c. 6–5 Million BCE)

Ancient peaks and canyons in the Atlas Mountains, which extend about 2,500 kilometers (1,600 miles) through Morocco, Algeria, and Tunisia.

Pangea

Alfred Wegener (1880–1930)

Less dense, higher-silicon continental crust has been "floating" on top of denser, more iron-rich oceanic crust since the first cores of the continents (*cratons*) began to form back in the Archean. Earth's conveyor belt of plate tectonics moves these continental "iceberg" lithospheric plates around the globe over time, and occasionally they collide with oceanic crustal plates or, more dramatically, with other continental icebergs.

When continental plates collide, the resulting battle is rather epic, because neither plate is willing to yield. When a continental plate runs into an oceanic plate, the latter usually dives (*subducts*) under the former because it has a much higher density. But in a continent–continent collision, each plate has comparable strength and density. So they tend to crumple up, buckle, warp, fold, and uplift huge mountains along the collisional boundary, all in a slow-motion drama that can take millions of years to unfold. Most of the time, when continents collide they end up sticking together, building a bigger supercontinent.

Such supercontinental construction appears to have occurred starting more than 300 million years ago, when a series of continent–continent collisions over many tens of millions of years caused essentially all the world's continental land masses to stick together and become a single supercontinent named Pangea (from "all" and "Earth"), named by German geophysicist and meteorologist Alfred Wegener, an early proponent of the theory of continental drift. Pangea is hypothesized to have remained a single continent for more than 125 million years, when intracontinental rifting started tearing the supercontinent apart into the smaller pieces that would soon go on to become the seven major continents that we recognize around the world today.

Even though it is long gone, evidence that Pangea existed as a single, continuous landmass comes from a variety of sources. For example, fossils of the same species and of similar age have been identified in regions that are now widely separated across the globe. Many aspects of matching geology (composition, mountain ranges) occur in areas that must once have been adjacent but are now widely separated. And we now know that the continents can indeed move relative to one another over geologic time scales because we can measure those tiny motions today using satellites and other modern technology.

SEE ALSO Continental Crust (c. 4 Billion BCE), The Archean (c. 4–2.5 Billion BCE), Plate Tectonics (c. 4–3 Billion BCE?), The Atlantic Ocean (c. 140 million BCE), East African Rift Zone (c. 30 million BCE), Continental Drift (1912)

A global artistic map showing the supercontinent of Pangea as it existed about 250 to 300 million years ago. The eastern edge of what would become the continent of North America is just starting to break away from what would become the northern edge of the continent of Africa.

The Great Dying

Life on Earth has experienced at least five known large-scale *mass extinction* events, when the majority of species rather suddenly disappeared from the geologic record. The first of these occurred around 450 million years ago, at the end of the Ordovician geologic time period, when somewhere between 70 and 85 percent of all species on Earth went extinct over a period of perhaps only a few million years or less. While this was a profound and catastrophic milestone in the history of life on Earth, it pales in comparison to the largest known mass extinction, which occurred at the end of the Permian geologic period, about 252 million years ago. The End Permian mass extinction was so severe that it nearly led to the end of all life on Earth. Thus, it is often called "The Great Dying."

The Great Dying was responsible for about 96 percent of all ocean-dwelling species going extinct, and about 70 percent of all land-dwelling vertebrate species. Some species appear to have died out over a short timespan of just a few thousand to tens of thousands of years; others appear to have dwindled over a few million years.

Hypotheses for the origin of The Great Dying fall into two categories: abrupt events such as meteor impact or massive volcanic eruptions leading to dramatic changes in sunlight and/or temperature, versus more gradual events such as changes in seawater acidity, sea level, or atmospheric oxygen abundance, which could also lead to dramatic (slower) climate changes. These and other hypotheses are being actively explored for all five major mass-extinction events, with the impact hypothesis being preferred only for one of them so far, the dinosaur-killing impact at the end of the Cretaceous some 65 million years ago.

Life on Earth was *almost* entirely wiped out 252 million years ago. Luckily for us, some species survived and went on to thrive in newly available evolutionary niches. It took tens of millions of years, however, for life to recover its prior biodiversity.

SEE ALSO Complex Multicellular Organisms (c. 1 Billion BCE), Snowball Earth? (c. 720–635 Million BCE), Cambrian Explosion (c. 550 Million BCE), Mass Extinctions (c. 450 Million BCE), Triassic Extinction (c. 200 Million BCE), Deccan Traps (c. 66 Million BCE), Dinosaur-Killing Impact (c. 65 Million BCE)

A tabulate (table-like) coral fossil from just before the end of the Permian geologic time period, when these kinds of organisms and 96 percent of all other marine species went extinct.

Mammals

The first land-dwelling *amniotes* (tetrapod, or four-legged, vertebrates that lay their eggs on land) appeared in the Pennsylvanian geologic period (about 320 to 300 million years ago), and quickly dominated many available niches for survival and reproduction. Natural selection also relatively quickly diversified early amniotic species, however, and very shortly after amniotes appeared, an important division of them also appears in the fossil record, between the *sauropsids* (precursors of the reptiles and birds) and the *synapsids* (all other kinds of terrestrial vertebrates). Over the next 50 million years or so, one particular group of synapsids, called the *therapsids* (large four-legged carnivorous animals), appears to have become the dominant land vertebrate based on fossil evidence. A specific sub-class of therapsids called *cynodonts* ("dog-teethed") proved to be an especially successful species by the late Permian geologic time period.

Everything changed 252 million years ago, however, as the End Permian mass extinction ("The Great Dying") wiped out about 70 percent of all land-dwelling vertebrate species, ending the dominance of the large, carnivorous therapsids. Cynodonts were one of the few classes of therapsids to survive the mass extinction, and relatively soon thereafter (about 220 million years ago) the first fossil evidence of a new, nocturnal, insectivore class of cynodonts appears: mammals.

Mammals are vertebrate amniotes that have hair, complex middle-ear bones, a neocortex in the brain, and other evolutionary innovations; female mammals nurse their young with milk secreted from mammary glands. A larger brain, superior sense of smell, generally smaller size, and nocturnal lifestyle all likely helped early mammals find specific and successful survival and evolutionary niches in a Triassic world increasingly dominated by large carnivore and herbivore predators whose lineage had also survived the Great Dying, including the dinosaurs and other predatory reptiles.

Unbeknownst to the dwellers of the Triassic-to-Cretaceous time period (c. 250 to 65 million years ago), several more large-scale mass-extinction events were coming, the most recent one likely initiated by a large meteor impact that defines the end of the Cretaceous period. These events would end the dominance of the ancient dinosaurs (although their lineage still survives in modern birds) and would open up new evolutionary niches that would, eventually, enable mammals to become the dominant predators on Earth.

SEE ALSO Cambrian Explosion (c. 550 Million BCE), First Land Plants (c. 470 Million BCE), First Animals on Land (c. 375 Million BCE), Reptiles (c. 320 Million BCE), The Great Dying (c. 252 Million BCE), Dinosaur-Killing Impact (c. 65 Million BCE).

Artist's rendering of a small, furry, shrew-like early mammal called Megazostrodon, *startled by a scorpion.* Megazostrodon *first appears in the fossil record about 200 million years ago.*

Triassic Extinction

The second-most recent of Earth's five known large mass extinction events occurred at the end of the Triassic geologic time period, just over 200 million years ago. Across a relatively short span of geologic time (tens of thousands of years), at least half the species known to be living on Earth became extinct. Particularly hard-hit on land were the *archosaurs*, a group of sauropsid vertebrate amniotes that had evolved to become the dominant land vertebrates in the Triassic period.

In the seas, over a third of the *genera* (the taxonomic class above *species*) went extinct, including many of the large amphibians and the entire class of eel-like vertebrates known as *conodonts*. The rapid wiping out of the conodonts, precursors of which had been around for hundreds of millions of years and which had survived the even more catastrophic End Permian mass extinction, is puzzling. Most all that is left of those sea animals are abundant and sharp fossilized teeth (among the first biostructures built from hydroxyapatite, a calcium-rich mineral that remains a key component of our own bones and teeth today).

The event(s) or circumstances responsible for the End Triassic mass extinction remain as controversial as the explanation for The Great Dying and several of the other large extinction events in the history of life on Earth. A large asteroid impact event is an obvious candidate to explain the rather sudden loss of species globally, but no "smoking gun" impact crater or other obvious geologic signal has been found for this event. Massive volcanic eruptions and ejection of large amounts of greenhouse gases into the atmosphere has also been suggested, perhaps associated with large-scale mantle upwelling that also led to the hypothesized start of the breakup of the supercontinent of Pangea around the same time. Other climate-change ideas have been floated, but many act too slowly to fit this scenario.

Surviving classes of archosaurs included precursors of the dinosaurs that became the dominant land vertebrates for more than 130 million years, from the Jurassic until the next mass extinction event at the end of the Cretaceous. Another species that survived the End Triassic extinction, the mammals, continued to evolve traits and behaviors that would ultimately enable them, unlike the non-avian dinosaurs, to survive that next great catastrophe.

SEE ALSO Cambrian Explosion (c. 550 Million BCE), First Animals on Land (c. 375 Million BCE), Pangea (c. 300 Million BCE), The Great Dying (c. 252 Million BCE), Mammals (c. 220 Million BCE), The Atlantic Ocean (c. 140 Million BCE), Dinosaur-Killing Impact (c. 65 Million BCE)

High-resolution microscope view of a millimeter-sized fossilized tooth from a jawless, eel-like Triassic vertebrate called a conodont.

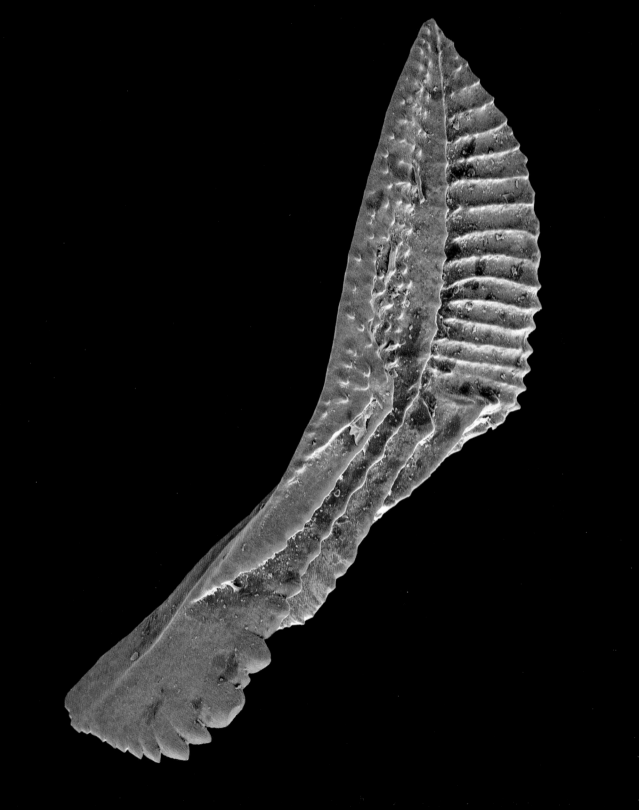

Age of the Dinosaurs

As mass extinctions at the end of the Permian and Triassic periods wreaked havoc on what were before then the dominant classes of vertebrate land animals on Earth, new ecologic and evolutionary niches for survival were opened up to what had previously been relatively minor, or perhaps not-yet-existing, kinds of creatures. Such was the case with the ancient dinosaurs, a class of animals unlike anything the world had seen.

Dinosaurs are amniotic vertebrate reptiles that first appear in the fossil record in the middle Triassic geologic period (around 240 million years ago), as a class of animals that diverged from their contemporaneous archosaur predecessors. Archosaurs were the dominant land predators during the Triassic, but only a few of their classes, including the dinosaurs, remained after the End Triassic mass extinction. Fossils of the oldest-known Triassic dinosaurs indicate that they probably started as relatively small (dog-sized) bipedal predators—certainly not the giant, dominant terrestrial animals they would soon become.

Why did early Jurassic dinosaurs evolve to larger sizes and the dizzying array of body styles preserved in the fossil record? A leading hypothesis is that since many of their early lineages survived the End Triassic mass extinction, they found themselves in a world where the major land predators had been wiped out, and amid a flora and fauna that was undergoing a rapid and prolific recovery. Rather suddenly, dinosaurs became the top competitors for abundant food and prey resources in the Jurassic world. In their heyday, dinosaurs evolved into *at least* 700 different species. They ranged in size from 2.5-foot-long (75-centimeter) *Eoraptor* to the 130-foot-long (40-meter) *Argentinosaurus*, and included what must have been among the most terrifyingly powerful and capable predators to ever live, such as *Tyrannosaurus Rex*.

The Age of the Dinosaurs ended rather abruptly, however, some 65 million years ago, when a large asteroid is hypothesized to have impacted the Earth and created a climate and food-chain catastrophe that would wipe out most of the dinosaurs, as well as perhaps 75 percent of all other species of life on Earth. One specific class of dinosaurs did survive the End Cretaceous mass extinction, however, and they are still with us today: more than 10,000 species of birds now descend from ancestors that were feathered dinosaurs.

SEE ALSO First Animals on Land (c. 375 Million BCE), The Great Dying (c. 252 Million BCE), Mammals (c. 220 Million BCE), Triassic Extinction (c. 200 Million BCE), The First Birds (c. 160 Million BCE), Dinosaur-Killing Impact (c. 65 Million BCE)

The size and diversity of the dinosaurs, the world's dominant land-dwelling vertebrate class for more than 135 million years, was stunning. In this artist's rendering, 13-foot-tall (4 m) plant-eating Diplodocus wade in the shallow waters below flying Pterosaurs.

The First Birds

During the Age of the Dinosaurs (about 200 to 65 million years ago), large numbers of carnivorous and herbivorous animal species evolved to fill available ecological niches created by several large prior mass-extinction events. But dinosaurs in general had first appeared much earlier, in the mid-Triassic period (around 240 million years ago). Back then, a subset of dinosaurs that had feathers appeared relatively soon in the fossil record. Feathered dinosaurs were more generally part of a subclass of dinosaurs called *therapods*, characterized by hollow bones and three-toed limbs.

Many therapod species would survive the End Triassic extinction and then go on to evolve into dominant predators in the Jurassic. Much controversy exists about whether famous therapods such as *Velociraptor* or *Tyrannosaurus rex* were also feathered, and indeed much uncertainty exists in general among paleontologists about why some dinosaur classes evolved feathers at all. Perhaps they provided excellent insulation to help regulate body temperature. Perhaps they provided advantageous camouflage to hide from predators, or display plumage to attract mates. Maybe feathers just tasted bad to non-feathered predators. Perhaps importantly, most possible advantages of feathers as opposed to bare scale-like skin could have been advantageous to *all* classes of dinosaurs.

Indeed, when analyzing the skeletal and presumed muscular structure of the oldest fossils of feathered bird-like dinosaurs—such as the 160-million-year-old *Archaeopteryx*, from a class of dinosaurs known as *avialans* ("bird wings")—many paleontologists are not sure that those animals could actually fly. Feathers could serve many functions.

Nonetheless, feathered dinosaurs that clearly flew have been found in the younger fossil record, which reveals that the subgroup of avialans known as *aves* (birds) diversified dramatically in the Cretaceous period, developing structures and traits that would help them to more successfully fly, hunt, and take advantage of their much higher degree of mobility compared to their land-dwelling ancestors. Perhaps extreme mobility helped them to become the only remaining lineage of the ancient dinosaurs to survive the catastrophic mass extinction at the end of the Cretaceous. Since then, birds have continued to evolve into a diverse multitude of species, and are among the most intelligent animals on Earth.

SEE ALSO First Animals on Land (c. 375 Million BCE), The Great Dying (c. 252 Million BCE), Mammals (c. 220 Million BCE), Triassic Extinction (c. 200 Million BCE), Age of the Dinosaurs (c. 200–65 Million BCE), Dinosaur-Killing Impact (c. 65 Million BCE)

Imprint of an early Archaeopteryx *fossil, from around 160 million years ago. Such feathered, bird-like dinosaurs are the precursors to modern bird species. This typical specimen was about the size of a common raven.*

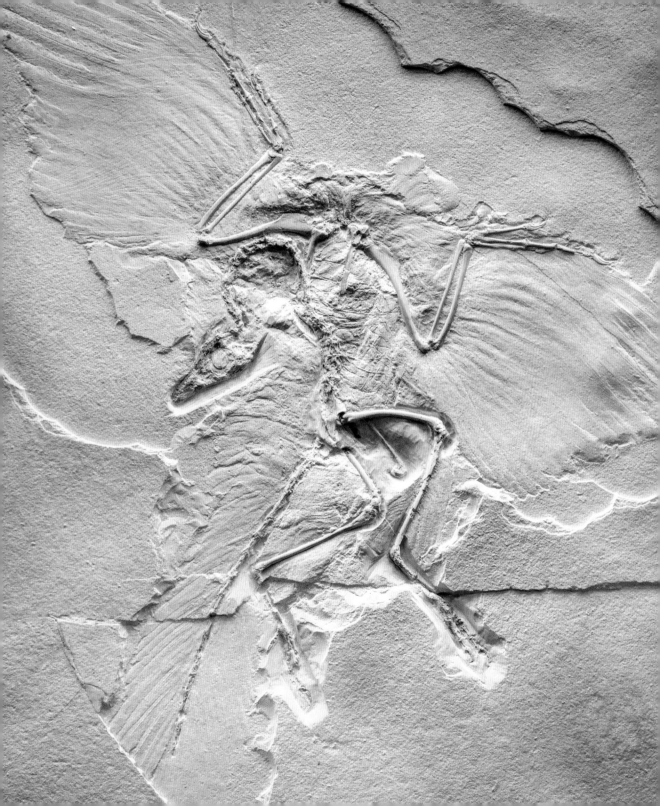

The Sierra Nevada

Even though it is not at all obvious, much of the Earth's most intense volcanic activity occurs deep underground. In the elevated temperatures of the mantle and upper lithosphere, hot "plastic" (deformable) rock can flow in enormous convection currents. And in many places, that rock can also be melted. Molten rock that flows underground is called *magma*. (Molten rock that makes it to the surface is called *lava*.) Enormous volumes of magma can pool and collect in huge underground pockets called magma chambers. From there, magma can flow along fractures or faults and *intrude* other rocky layers, sometimes making linear zones or swarms of light-colored rocks cross-cutting darker ones (or vice versa). These kinds of igneous intrusive landforms are called *plutons*.

While the eastern edge of the North American plate was transitioning to a passive continental margin in the Cretaceous period, the western edge was becoming more dynamic. Specifically, a dense oceanic plate called the Farallon plate began subducting under the continent around 155 million years ago. The eventual deep sinkage and melting of the leading edge of the plate yielded huge volumes of subsurface magma that collected in enormous roughly north–south-aligned magma chambers underneath the western edge of North America, along what is now eastern California. Because of fractional crystallization of the melted basaltic Farallon plate (mixed with melted surrounding continental crust), much of the igneous rock that formed in these magma chambers is high in silicon, and thus high in granite. The collected set of these now-solidified magma chambers is called the Sierra Nevada *batholith*, and these once deeply buried rocks now find themselves at the surface of the Earth, high above sea level, and eroding dramatically as part of the Sierra Nevada mountain range.

How did magma chambers become mountains? About 20 million years ago, the tectonic environment changed again along western North America, and the far-western part of the continent began to be stretched into the characteristic basin-and-range geology still seen today. Mantle volcanic heating helped to uplift the Sierra Nevada batholith region. But glaciers and rivers wore down the mountains and sediments above the batholith, eventually exposing the huge volumes of granitic plutons to the surface. The resulting geology, especially in spectacular places like Yosemite National Park, is stunning.

SEE ALSO Continental Crust (c. 4 Billion BCE), Plate Tectonics (c. 4–3 Billion BCE?), Roots of the Pyrénées (c. 500 Million BCE), The Appalachians (c. 480 Million BCE), Pangea (c. 300 Million BCE), The Atlantic Ocean (c. 140 million BCE), The Rockies (c. 80 Million BCE), The Himalayas (c. 70 Million BCE), National Parks (1872), Island Arcs (1949), Basin and Range (1982)

Spectacular view of Yosemite Valley from Glacier Point, including the iconic Half Dome mountain (tallest peak).

The Atlantic Ocean

The reconstructed history of the relative motions of the Earth's several dozen major lithospheric plates is based on evidence from the fossil record of plants and animals, as well as the geologic record of mountain-building. That history tells us that the Earth's continental plates have probably come together into single supercontinents at least several times since the Archean. The last and best known such supercontinent was called Pangea, which started forming around 300 million years ago.

The Earth's lithospheric plates are dynamic, however, and so it wasn't too long (only about 100 million years) until the same kinds of internal forces that brought it together began rifting Pangea apart. The fact that the start of the breakup of Pangea occurred right around the same time and in the same region as an episode of massive volcanic eruptions associated with large-scale mantle upwelling is not a coincidence, nor is it a coincidence that the catastrophic End Triassic mass-extinction event also occurred around the same time. The breakup of Pangea was a global, game-changing event.

The result of the creation of a rift zone in the Pangean crust and the pulling apart of the North American and African plates, the African and South American plates, and the North American and Eurasian plates, was the creation of a deep basin into which ocean waters from other parts of the globe could flow. The Atlantic Ocean was born. The mantle plume that first initiated the breakup of Pangea began to spew forth enormous volumes of volcanic rock to create a new mountain belt within the widening basin between the American plates and the African/Eurasian plates. This mid-Atlantic ridge became a divergent plate boundary running roughly north–south, pushing out new oceanic plate volcanic rocks to the east and west. The volcanism and plate motion continues to this day.

The mid-Atlantic ridge is part of the largest chain of mountains on Earth, extending from northern polar latitudes near Iceland to southern polar latitudes near Antarctica. For most of human history, these volcanically active mountains were unknown. It took modern seafaring technology such as sonar to eventually map the seafloor in the 1950s through 1970s in order to reveal these mountains, and in order to finally understand plate tectonics.

SEE ALSO Continental Crust (c. 4 Billion BCE), The Archean (c. 4–2.5 Billion BCE), Plate Tectonics (c. 4–3 Billion BCE?), The Atlas Mountains (c. 300 Million BCE), Pangea (c. 300 Million BCE), Triassic Extinction (c. 200 Million BCE), The Himalayas (c. 70 Million BCE), Mapping the Seafloor (1957), Reversing Magnetic Polarity (1963), Seafloor Spreading (1973)

Sunrise seagull over the Atlantic Ocean, which formed some 140 million years ago as a rift in the supercontinent of Pangea that separated what are now the North American and African tectonic plates.

Flowers

Imagine a world without flowers. We take them for granted, from spectacular bouquets in florist shops to lowly daffodils growing in a pavement crack. But, remarkable as it seems, most of our planet's history has been flower-free. Flowers are new! At least, relatively new. Land plants began to proliferate about 470 million years ago, dominated by simple non-vascular moss-like plants such as liverwort, and followed soon thereafter by vascular plants that could stray farther from the shore by wicking up water with their roots and storing it in their stems and other structures.

Early sea plants reproduced by sending out *spores* containing their genetic material that would be carried through the water and hopefully encounter other similar spores, fertilize, and grow elsewhere. Spore reproduction was less efficient on land (spores could dry out), and so some land plants evolved *pollen* and *seeds* to protect their genetic materials during harsher weather conditions or over potentially long periods of drought. But why would plants develop flowers? Animals might be the answer.

As far back as 350 million years ago, many early plants had developed symbiotic relationships with insects and other animals. Pollen grains in the group of seed-producing plants called *gymnosperms*, such as conifers, for example, must be physically transferred from special pollen cones to the ovule cones of other plants for fertilization. The wind can do this randomly, but insects can help dramatically increase fertilization rates.

That kind of symbiosis was exploited dramatically (and beautifully) when the first flowering plants (*angiosperms*) appeared in the Cretaceous period. Flowers are the reproductive organs of angiosperm plants, and their bright colors and delicate structures appear to be specifically designed to attract insects, birds, and other *pollinators* who can help get the needed genetic materials in contact with each other. Even more remarkably, specific colors and patterns of flowers appear to attract specific kinds of pollinators, thus helping to ensure that pollen will travel from one plant to another of the same species. Such species–specific pollination appears to have helped flowering plants to adapt and evolve to changing environmental circumstances more rapidly, and thus to become the most diverse and widespread group of land plants in the world today.

SEE ALSO The Origin of Sex (c. 1.2 Billion BCE), First Land Plants (c. 470 Million BCE), First Animals on Land (c. 375 Million BCE), The First Birds (c. 160 Million BCE), Genetic Engineering of Crops (1982)

Top: *An approximately 125-million-year-old fossil of* Archaefructus liaoningensis, *among the first known flowering plants to exist on Earth.* **Bottom:** *A bee collects pollen from a Zinnia flower.*

The Rockies

Areas that make up the central cores of stable continental crust—cratons—date back to the earliest era of crustal formation in the early Archean. Once they've grown to true continental size by billions of years of accretion along their edges, these central regions are generally considered to be geologically "quiet" areas. Thus, it is perhaps unusual to see evidence of relatively young large-scale mountain-building so far from the edges of stable continental interiors. And yet there they are, far from the western boundary of the North American plate: the Rocky Mountains. How did they get there?

The original rocks that make up the ancestral roots of the Rocky Mountains were made from pre-existing layers of ancient continental crust and shallow seafloor sediments that were fractured, deformed, metamorphosed, melted, and uplifted by the continental-plate collisions that formed the supercontinent of Pangea about 300 million years ago. After then, however, those ancestral Rockies were mostly worn down by erosion for a few hundred million years, while all of the exciting new mountain-building took place far away, on the western margin of the plate.

But the events farther west eventually did have a local effect. Specifically, the Farallon oceanic plate was sinking (subducting) under the western edge of the North American plate at a shallow angle, causing its tectonic and volcanic effects to be felt far inland. Around 80 million years ago, the friction and compressive stresses from the oceanic plate sliding underneath and pushing up on the continent began causing enormous faulting and uplift in the region of the ancestral Rockies. Once again, huge peaks were raised up, in 20 million years of mountain-building that stretched from modern-day northern British Columbia down to New Mexico. When the bulging of the continent ceased, the highest resulting mountain peaks of the newborn Rockies were likely to have been over 20,000 feet (6,000 meters) tall, as high as the Himalayas are today.

Over the last 60 million years, the part of the North American craton around the Rocky Mountains has gone relatively quiet again, with wind, water, and glaciers eroding the Rockies down by 30 percent or so to their present (still impressive) heights.

SEE ALSO Continental Crust (c. 4 Billion BCE), Plate Tectonics (c. 4–3 Billion BCE?), The Appalachians (c. 480 Million BCE), Pangea (c. 300 Million BCE), The Sierra Nevada (c. 155 Million BCE), The Atlantic Ocean (c. 140 million BCE), The Himalayas (c. 70 Million BCE), Basin and Range (1982), Yellowstone Supervolcano (~100,000)

A Storm in the Rocky Mountains, Mt. Rosalie, *painted by Albert Bierstadt in 1866, captures the dramatic geologic and meteorologic nature of this relatively young mountain belt.*

The Himalayas

Throughout Earth's history, collisions among the lithosphere's several dozen major tectonic plates have been the primary agent of volcanism, mountain-building (*orogeny*), and continental growth. In particular, continent-to-continent collisions have been responsible for most of the world's major orogenies, with compressed, folded, and crumpled-up crust producing spectacular chains of peaks like those in the ancient Appalachian, Ural, and Atlas ranges and in the modern Rockies, Alps, and Pyrénées mountains.

When oceanic crust runs into continental crust, the former almost always dives under (subducts) the latter, because oceanic crust is made of dense high-iron basaltic volcanic rocks, while continental crust is made of less-dense, high-silicon volcanic and sedimentary rocks. Subduction causes melting of the oceanic plate, leading to the creation of deep ocean tranches, tall volcanic peaks, and some uplift of the continent in general. But by plate-collision standards, such interactions are relatively gentle. Continental–continental collisions are more of a titanic clash of equals, with neither crustal plate compelled to subduct under the other. The result, especially in head-on plate collisions, is enormous compression and uplift of the crust in the collision zone. Voilá, the world's tallest mountains are formed.

Indeed, that is precisely what is happening in Nepal and surroundings. Starting around 70 million years ago, the relatively small Indian continental plate, which had broken off the supercontinent of Pangea around 140 million years ago, began to head toward a head-on collision with the Eurasian continental plate. The continents began fully plowing into each other about 10–20 million years ago, and they continue to compress and crumple each other up to this day.

The result is a spectacularly tall, young, and pristine mountain range, the Himalayas, which includes a whopping 50 peaks above 23,600 feet (7,200 meters), including the world's tallest peak, Mt. Everest (29,000 feet, or 8,850 meters). The specific mechanism for building the mountains is known as thrust-faulting—converging blocks of crust are continually thrust upward and stacked on top of each other. The forces are so extreme that they have lifted what were once seafloor deposits from far below sea level to the top of the world.

SEE ALSO Continental Crust (c. 4 Billion BCE), Plate Tectonics (c. 4–3 Billion BCE?), Roots of the Pyrénées (c. 500 Million BCE), The Appalachians (c. 480 Million BCE), The Rockies (c. 80 Million BCE), The Himalayas (c. 70 Millio

A view of the Himalayas, looking south from over the Tibetan Plateau, taken by astronauts onboard the International Space Station in 2004. Mt. Everest is in the middle of this scene.

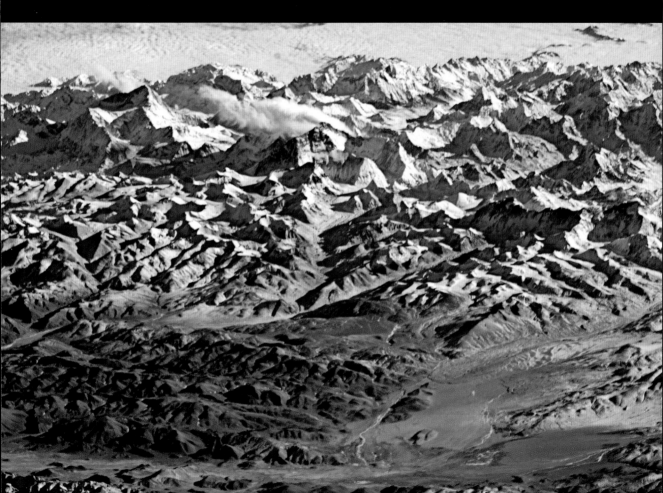

Deccan Traps

Many eighteenth- and nineteenth-century geologists were divided into two schools of thought. *Gradualists* thought changes in the Earth's geology, biology, and climate came about slowly, over millennia to eons, for example via the action of slow plate motions or generations of natural selection. *Catastrophists*, however, thought change came in short, catastrophic bursts of activity associated with storms, volcanic eruptions, earthquakes, genetic mutations, or impact events. Arguments between the camps were intense. Today, though, modern geologists realize that major changes can actually come both ways.

Volcanic eruptions are examples of catastrophic events that can effect relatively sudden and dramatic changes to landscapes and climate. An impressive example comes from one of the largest episodes of volcanic activity in Earth's history, an enormous outpouring of lava that erupted on the small continental plate of India as it was traveling (across what is now the Indian Ocean) toward a Himalaya-creating collision with the Eurasian plate. Over less than 50,000 years, lava flows spewed forth to eventually cover half the microcontinent to a thickness of more than 6,600 feet (2,000 meters).

The area covered by these voluminous lavas is called the Deccan Traps because of the stair-step nature of the thick layered lava flows ("deccan" from the Sanskrit "southern," and "traps" from the Scandinavian word for stairs—"trappa"). A leading hypothesis for the origin of the Traps is that the Indian plate passed over and tapped into a hot mantle plume under the Indian Ocean, enabling huge volumes of magma to ascend to the surface.

In addition to molten rock, the volcanic outpouring also released enormous amounts of greenhouse gases. Another hypothesis thus links the Deccan Traps eruption to the mass extinction that occurred soon thereafter, at the end of the Cretaceous period (c. 65 million years ago). While this idea competes with the impact (followed by short-term climate change) hypothesis to explain the extinction of the ancient dinosaurs and many other species, in the spirit of the way the Gradualism–Catastrophism debate was resolved, many geologists and paleontologists are now more accepting of the idea that perhaps *both* near-simultaneous catastrophic events tipped the scales toward the mass extinction.

SEE ALSO Late Heavy Bombardment (c. 4.1 Billion BCE), Snowball Earth? (c. 720–635 Million BCE), The Great Dying (c. 252 Million BCE), Triassic Extinction (c. 200 Million BCE), Dinosaur-Killing Impact (c. 65 Million BCE), Arizona Impact (c. 50,000 BCE)

A pastoral landscape surrounded by layered volcanic hills in the Western Ghats Mountains in southern India— part of the large igneous province known as the Deccan Traps.

The Alps

One effect of the breakup of the Pangean supercontinent around 200 to 150 million years ago was the creation of the mid-Atlantic ridge's east–west spreading plate boundary and the eventual collision (starting around 65 million years ago) of the African and Eurasian continental plates. As in all major continent–continent collisions since the origin of plate tectonics, the result was the creation of some impressive mountains, including the Alps.

Before Africa and Eurasia collided, a large and deep basin existed between them that harbored the now-gone Tethys Sea. Like any good marine basin, the Tethys Sea collected enormous volumes of limestone and mudstone sediments on its floor, along with salty mineral deposits (evaporites) along its shores. All these rocks and minerals were compressed, folded, metamorphosed, and uplifted as the African and Eurasian plates moved closer over time. Eventually, the Tethys Sea was completely pinched off, and former marine sedimentary deposits were uplifted and stacked together to form the Alps, like a rumpled tablecloth being pushed across a table. Indeed, the thrust blocks making up the Alps are called *nappes*—French for "tablecloths"—by geologists.

Mountain-building in Europe along the roughly 750-mile (1,200-kilometer) length of the Alps involved a variety of processes, occurring in starts and stops over tens of millions of years. Nappes of metamorphosed sedimentary and volcanic rocks have all been created and mixed by the incessant compressional forces from the colliding continents. Much of the resulting terrain of the Alps is a geologic mess that can be difficult to untangle. The base of the famous Matterhorn peak in Switzerland, for example, consists of ancient metamorphosed Eurasian continental crust, but the tip of the Matterhorn is made of ancient rocks from the African plate. Sandwiched in between the base and the peak are compressed and folded sediments from both Africa and Eurasia that were once laid down in the Tethys Sea.

The mountain-building event or *orogeny* that created the Alps continues to uplift them today, at a rate estimated to be between 1 millimeter and 1 centimeter per year. Snow, rain, and glaciers erode the mountains down, however, at roughly the same rate.

SEE ALSO Continental Crust (c. 4 Billion BCE), Plate Tectonics (c. 4–3 Billion BCE?), Roots of the Pyrénées (c. 500 Million BCE), The Ural Mountains (c. 320 Million BCE), The Atlas Mountains (c. 300 Million BCE), Pangea (c. 300 Million BCE), The Himalayas (c. 70 Million BCE), Discovering Ice Ages (1837), Birth of Environmentalism (1845)

Summertime view of the rugged and relatively geologically young Italian Alps soaring high above spectacular high mountain lakes.

Dinosaur-Killing Impact

The role of large impacts in catastrophically changing Earth's climate and biosphere was not fully appreciated until smoking-gun evidence was found that the impact of a large asteroid with Earth was probably responsible for causing the extinction of the dinosaurs and many other species about 65 million years ago, at the boundary between the Cretaceous and Paleogene geologic periods. The key was the discovery of a thin global layer of sediment enriched in the rare element iridium. Iridium is a heavy metal in the platinum family that often bonds with iron in rocks and minerals. Most of Earth's heavy metals sank into the deep mantle and core when Earth was forming, so a globally distributed iridium-rich deposit in the crust is quite an anomaly. Geologists hypothesized that the iridium came from a large metal-bearing asteroid that impacted Earth and vaporized, dramatically changing the climate and wreaking havoc on most plant and animal species.

The impact would have lifted vaporized rock and dust into the atmosphere and set off large-scale fires that filled the sky with soot and smoke, blotting out the Sun and lowering global surface temperatures for years. While the effect on Earth's life was not as large as during the Permian-Triassic extinction of 252 million years ago, species such as the ancient dinosaurs that ultimately depended on sunlight and photosynthesis to nurture the base of their food chain were still decimated. Some species, such as mammals and birds capable of burrowing or subsisting on insects, carrion, or other non-plant food-chain staples, weren't driven to extinction by the event, however.

The idea that the ancient dinosaurs were killed off by an asteroid impact, and that other large impact events could also have led to mass extinctions at other times in Earth's history, is a hypothesis that is constantly being tested. Other geologic and climatic effects, such as dramatic changes in atmospheric oxygen, large sea-level changes, or massive eruptions of volcanic rock and gases, have also occurred throughout Earth's history, and sometimes around the same times as hypothesized extinction-level impact events. Thus, multiple events that seem to have been conspiring may have contributed to the environmental conditions that led to Earth's major mass extinctions.

SEE ALSO Earth's Core Forms (c. 4.54 Billion BCE), Late Heavy Bombardment (c. 4.1 Billion BCE), Cambrian Explosion (c. 550 Million BCE), The Great Dying (c. 252 Million BCE), Arizona Impact (c. 50,000 BCE), The Tunguska Explosion (1908), Extinction Impact Hypothesis (1980)

Artist's concept of a large asteroid crashing into the Earth marks the precise moment of the end of the Cretaceous and the beginning of the Paleogene (formerly Tertiary) period of geologic history, about 65 million years ago.

Primates

Despite surviving at least two large-scale mass-extinction events, early mammals still had to develop behavioral strategies and advantageous evolutionary adaptations that would help them to survive in a world typically dominated by larger and stronger predators. Some mammals developed burrowing capabilities to help them live furtive lifestyles hidden from most predators. Others, such as the precursors of modern primates, learned to take to the trees.

Primates are mammals that have hands, hand-like feet, and forward-facing eyes. They range in size from tiny 1-ounce (30-gram) lemurs to hulking 440-pound (200-kilogram) gorillas, and include many species in between, such as marmosets, tarsiers, monkeys, apes, and of course humans. Most primates are skilled tree-dwellers, with the most notable exception being humans. An arboreal lifestyle led to primates evolving to rely more on their sense of vision than smell (the dominant sense for most other mammals), with sharper stereoscopic vision, color perception, and opposable thumbs also being highly advantageous adaptations.

The earliest evidence for primates dates back to about 60 to 50 million years ago, with animals such as *Darwinius masillae*. These now-extinct animals looked similar to modern-day lemurs, but had significant differences in the nature of their claws and teeth. Exactly how and why the primate lineage diverged from the rest of the mammals is the subject of intense study and debate. Some genetic evidence suggests, for example, that small numbers of early primates or primate precursors may have become geographically separated from their main population and forced to re-establish themselves in different ecological niches.

The earliest fossil evidence of simian primates (monkeys and apes) dates back to about 40 million years ago. They appear to have quickly spread from their hypothesized origin in Asia to settle in (then) tropical environments in Africa, Europe, and even North and South America, somehow crossing the narrower-than-today but still significant width of the Atlantic ocean.

Some primates live in solitude, others in mated pairs, still others in large social structures. Primates develop more slowly from infant to adult than other similarly sized mammals, and have longer lifespans. Studies of the behavior of many modern primates also reveal high intelligence and an apparently long history of inventing and using tools.

SEE ALSO First Animals on Land (c. 375 Million BCE), Mammals (c. 220 Million BCE), First Hominids (c. 10 Million BCE), *Homo sapiens* Emerges (c. 200,000 BCE), *Gorillas in the Mist* (1983), *Chimpanzees* (1988)

Visitors at the Natural History Museum in London view the 47-million-year-old skeleton of one of the most complete fossil primates ever found, known formally as Darwinius masillae, *and informally as "Ida." The animal was about the size of a small cat with a long tail.*

Antarctica

We tend to think of Antarctica as a cold and distant place, an inhospitable island at the bottom of the world. In reality, though, Antarctica is a full-fledged continent that has experienced some of the most dramatic changes of any of Earth's landmasses over time. A world of history lies under all that ice and snow.

The continental crust that would eventually become Antarctica started out as one of a half-dozen or so ancient low-density cratons that merged to form the supercontinent of Gondwana in the late Proterozoic (around 750 to 600 million years ago). Gondwana was the largest landmass on Earth for hundreds of millions more years, until it merged with the other large landmass on Earth, Laurasia (consisting of the North American and Eurasian plates), to form the new supercontinent of Pangea. Eventually, starting around 175 million years ago, Pangea started to break up into smaller constituent plates, one after another peeling off to become the continents that we recognize in the world today. Among the last parts of the breakup was the separation of Australia and Antarctica starting around 35 million years ago, with Australia moving to the north and Antarctica to the south.

Antarctica is the third smallest of the continents (just larger than Europe, and just smaller than South America), with about 9 percent of Earth's continental landmass. As Antarctica got closer to the South Pole, glaciers and ice sheets began to form and grow as the average annual temperatures dropped. The continent also eventually became encompassed by circumpolar ocean currents. The constant flow of colder waters completely around the continent helped to lock in the deep freeze, maintaining and even increasing the ice, snow, and glaciers until they completely covered the land starting around 15 million years ago.

When it was part of Gondwana, Antarctica had been a tropical to mid-latitude place. Today, however, most of the geologic and biologic secrets of the continent lie buried under miles of ice. While fieldwork and drilling have revealed some of its volcanic, tectonic, and sedimentary history, a full understanding of the continent's history may have to await its eventual migration to warmer climes.

SEE ALSO Continental Crust (c. 4 Billion BCE), Plate Tectonics (c. 4–3 Billion BCE?), Pangea (c. 300 Million BCE), Continental Drift (1912), Exploration by Aviation (1926), International Geophysical Year (IGY) (1958), Lake Vostok (2012)

NASA satellite composite of the continent of Antarctica, centered on the South Pole. The continent itself is covered with ice, snow, and glaciers, and sea ice covers major bays and inlets.

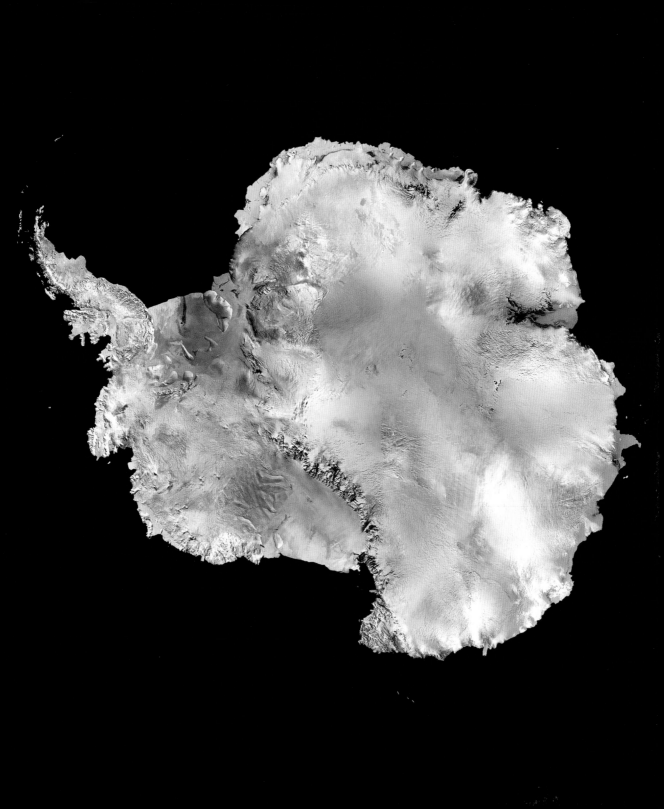

East African Rift Zone

The breakup of the supercontinent of Pangea starting around 200 to 175 million years ago was the result of *rifting*, or tearing apart, of its continental crust to form smaller pieces carried away from one another by large-scale lithospheric plate motions. Computer models support the hypothesis that *mantle upwelling*, the rising of hot, buoyant plumes of magma from the deep mantle to the shallow mantle, is likely the culprit behind rifting. According to this model, a rising mantle plume bulges, warps, and partially melts the overlying crust, causing it to weaken. And as the hot plume releases its heat close to the surface, its margins cool and begin to sink, dragging the overlying crust laterally along with the sinking plume.

Continental rifting caused the breakup of Pangea, continent-sized piece by piece, over a time span of about 150 million years. Because this was all happening on a spherical planet, however, the continents could only move apart for so long before starting to crash into one another again in new places. Some of the most recent collisions are responsible for the formation of the Himalayas and the Alps. Collisions like this, plus the continuing effects of deep, hot mantle plumes, can work to thin and weaken continental crust and form new rifting.

That's one hypothesis for what started happening under the eastern edge of the African continent about 30 million years ago. Regardless of the cause, however, as the crust thinned and weakened, it began to rift apart along a 3,700-mile-long (6,000-kilometer) valley stretching from Ethiopia down to Mozambique. The rifting has been accompanied by significant numbers of earthquakes and volcanic eruptions, attesting to the enormous heat and energy that is driving the rupture of the African plate.

The East African Rift Zone is the largest seismically active rift system on Earth today, and the spreading of the crust is occurring at a rate of about 0.24–0.28 inches (6–7 millimeters) per year. If things keep moving this way, in perhaps as little as 10 million years Africa will split into two new plates (already being called the Somali plate to the east and the Nubian plate to the west by geologists), with a new ocean basin formed between them.

SEE ALSO Continental Crust (c. 4 Billion BCE), Plate Tectonics (c. 4–3 Billion BCE?), Pangea (c. 300 Million BCE), The Atlantic Ocean (c. 140 Million BCE), The Himalayas (c. 70 Million BCE), The Alps (c. 65 Million BCE), East African Rift Zone (c. 30 Million BCE), Continental Drift (1912)

Illustrative depiction of the Earth's crust being pulled apart by plate tectonics in the East African Rift Zone. The Horn of Africa and Somalia are at right, and the Nile River is at left. In as little as 10 million years, a new ocean basin will form here, splitting Africa into two separate plates.

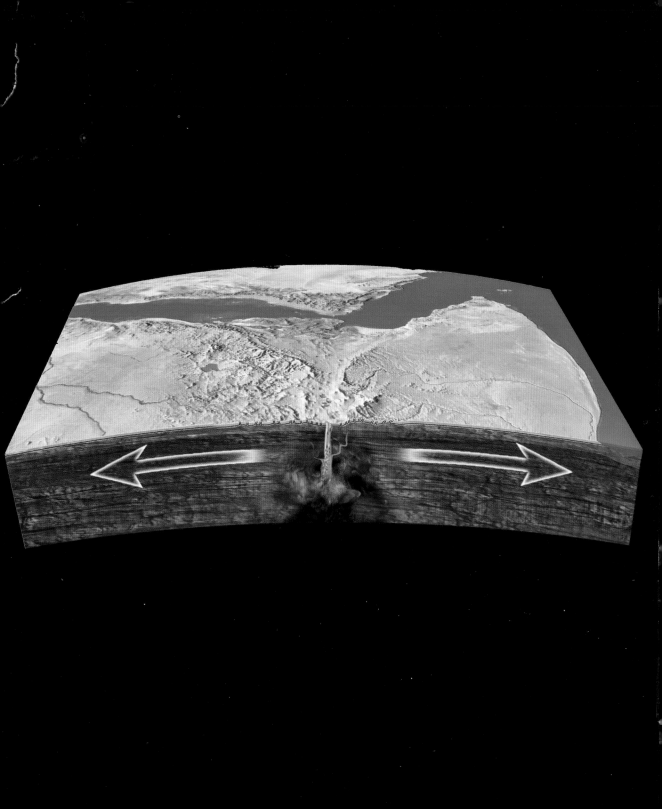

Advanced C$_4$ Photosynthesis

Photosynthesis, the conversion of sunlight into energy for cellular metabolism, was one of life's earliest evolutionary innovations, around 3.4 billion years ago. In essence, atmospheric carbon dioxide (CO_2) and water vapor (H_2O) react in the presence of sunlight to produce glucose ($C_6H_{12}O_6$) and free oxygen (O_2) as byproducts. Cells then process the glucose as fuel. This process is referred to as C$_3$ *carbon fixation*. A thick, O_2-poor early atmosphere with ample volcanically provided CO_2 made for optimal conditions for early life forms to exploit plentiful sunlight as a fuel source.

The appearance and rapid proliferation of the first land plants around 470 million years ago began to dramatically decrease the amount of CO_2 in the atmosphere, making C$_3$ carbon fixation less efficient over time. Simple, single-celled photosynthetic organisms didn't really care much (and still don't) because their energy needs are modest, and more than 95 percent of modern plants that rely on C$_3$ carbon fixation grow in regions where sunlight intensity and temperatures are moderate and groundwater is plentiful. However, for plants to grow in sunnier, hotter, and dryer environments requires higher photosynthetic efficiency, because their energy needs still must be met regardless of decreasing atmospheric CO_2 levels.

Steadily decreasing atmospheric CO_2 levels between about 470 and 300 million years ago thus provided significant pressures on some plants to become more efficient photosynthesizers. A variety of plants did just that, starting around 30 to 20 million years ago, by developing a new scheme to harness the Sun's energy called *advanced C$_4$ photosynthesis*. Plants using this process have evolved specialized photosynthetic cells within their leaves that more effectively trap and concentrate CO_2 and water for eventual delivery to chloroplast cells, where higher rates of "normal" C$_3$ photosynthesis then get carried out using more CO_2, even though the atmosphere outside the leaf has less CO_2 over time.

C$_4$ plants have advantages over C$_3$ plants during droughts, heat waves, or low CO_2 conditions. While they represent less than 5 percent of the plant species today, they account for almost 25 percent of the fixation of atmospheric CO_2, making them a critical sink for the *biosequestration* of CO_2, which would otherwise further warm our current climate.

SEE ALSO Photosynthesis (c. 3.4 Billion BCE), The Great Oxidation (c. 2.5 Billion BCE), First Land Plants (c. 470 Million BCE), Flowers (c. 130 Million BCE)

Close-up of a leaf, the primary site of photosynthesis in plants.

Cascade Volcanoes

A map of the world's active volcanoes and earthquake zones is essentially a map of the boundaries of the world's several dozen large tectonic plates. Those boundaries are thus where some of the most interesting geology is happening. When two continental plates collide, high mountain ranges are formed as the plates crumple into each other. When an oceanic and a continental plate collide, however, the interaction is quite different, because rather than just crumpling, the denser oceanic plate will usually slip underneath and sink (*subduct*) below the continental plate. When the front edge of the sinking plate begins to melt, buoyant magma and gases rise up and melt and mix with the continental crust above, bulging the crust upward like a dome and creating violent volcanic eruptions when the magma and gas burst through the surface.

This is precisely the kind of collisional plate interaction that has been occurring along the Pacific Northwest coast of the North American continental plate. Specifically, three remaining fragments of the former ancient Farallon oceanic plate have been getting pushed eastward relative to the North American plate, in the process getting subducted under the continent and causing extensive earthquakes and volcanic eruptions along a gently curving path called the Cascade volcanic arc. Very little remains of these three fragments, as most of their former extent has been destroyed by subduction.

The effects of these subducting oceanic plates on the surface geology of the Pacific Northwest have been profound. Perhaps most dramatic are the 20 major volcanic peaks and more than 4,000 volcanic vents that stretch over 700 miles (1,100 kilometers) along an arc roughly parallel to the coast, but about 100 miles (160 kilometers) inland. These are the Cascade Volcanoes, and they occur within a broader line of subduction-related uplift called the Cascade Mountains. The tall volcanoes in this region of the world—a dozen are over 10,000 feet (3,000 meters)—are called *stratovolcanoes* or *composite volcanoes*, conical mountains built up by layers of ash and lava erupted in often highly explosive events. Mt. Vesuvius and Mt. Fuji are among the most famous stratovolcanoes in the world, and the Cascade volcano Mt. St. Helens is well known in the US.

SEE ALSO Plate Tectonics (c. 4–3 Billion BCE?), The Sierra Nevada (c. 155 Million BCE), The Himalayas (c. 70 million BCE), The Andes (c. 10 Million BCE), Island Arcs (1949), Mount St. Helens Eruption (1980), Volcanic Explosivity Index (1982)

Aerial view of the Cascade volcanic peaks in Oregon, looking north. The Three Sisters are in the foreground, followed by Mts. Washington, Jefferson, Hood, and Adams to the north.

c. 28 Million BCE

Hawaiian Islands

John Tuzo Wilson (1908–1993)

The Earth's mantle releases internal heat in giant cycles of convection—rising of hot deformable or molten rock to the surface, where it cools, becomes less buoyant, and sinks back down. Geologists have a special name for places where the hot, rising plume of magma reaches close to—or breaches—the surface: *hotspots*. Hotspots are places where volcanoes erupt seemingly "in the middle of nowhere," compared to other more common volcanic eruptions along the highly active boundaries of converging or diverging plates. Examples of famous past and present hotspot volcanoes include the Deccan Traps, the Yellowstone Caldera, Iceland, and the Hawaiian Islands.

The Big Island of Hawaii and its neighboring islands and undersea mountains ("seamounts") are aligned in two long segments, first a 1,500-mile-long (2,400-kilometer) chain of large islands and seamounts running from the Big Island to the northwest, and second a 1,000-mile-long (1,600 kilometer) archipelago of seamounts running much more north–south all the way to the Aleutian Islands of Alaska. The linear nature of the alignment of the islands, including the sharp bend in the alignment toward the north, provided a clue to geologists: perhaps there was a hotspot in the middle of the Pacific Ocean that the Pacific plate has slowly been moving on top of over time.

The second important clue to this hotspot hypothesis for the formation of the Hawaiian chain comes from the ages of the volcanic rocks on these islands. On the Big Island, fresh volcanic rock from the Kilauea and Mauna Loa volcanoes is erupting onto the surface and growing that island today. But the rocks on the next island west, Maui, are about a million years old. Continuing west from there, the rocks on Molokai, Oahu, and Kauai are about 2, 3, and 5 million years old, respectively. The pattern was clear to Canadian geologist John Tuzo Wilson, who first figured it out in 1963: older island and seamount rocks heading northwest mean that the Pacific plate is slowly moving to the northwest over the relatively stationary Hawaiian hotspot. The oldest exposed island, tiny Kure atoll, is around 28 million years old, marking the approximate time the Hawaiian Islands first appeared above sea level.

SEE ALSO Deccan Traps (c. 66 Million BCE), East African Rift Zone (c. 30 Million BCE), Yellowstone Supervolcano (~100,000), Loihi (~100,000–200,000)

NASA satellite view of the Hawaiian Island chain, from the Big Island of Hawaii in the southeast to Kauai and tiny Ni'ihau in the northwest. The full chain of smaller islands and atolls above sea level extends more than another 620 miles (1000 km) to the northwest.

The Andes

One of the longest continuous tectonic plate-collision zones under continental crust occurs along the western edge of the South American plate. For more than 90 million years, new oceanic crust being created at submerged spreading centers has been riding on a conveyor belt and crashing into South America. The collision has gone through a number of distinct episodes and orientations. Starting around 10 million years ago, a piece of the former Farallon plate called the Nazca plate began to collide into western South America, forming the high peaks and volcanoes of the modern Andes Mountains.

Just like along the Cascade Volcanoes stretch of the subduction zone of former Farallon plate fragments in North America, subduction of the Nazca plate under South America has led to extensive mountain-building and explosive volcanism. The Andes volcanic zone extends along a belt running through Colombia, Ecuador, Peru, Bolivia, Chile, and Argentina, from north to south; the zone arguably even includes volcanoes in Antarctica, which was once attached to South America. Hundreds of active and extinct volcanoes occur along the belt.

The Andes also have a special place in the hearts of *petrologists*, geologists who study the composition, mineralogy, and formation conditions of rocks. Because of the long history of oceanic plate subduction along the western South American plate boundary, the continental crust has thickened considerably, and thus magma rising from oceanic slabs through the overlying continental crust melts and assimilates significant amounts of silica and alkaline elements that are more common in continental than in ocean crust. The resulting ash and lava that makes it to the surface has a distinctly different composition than mid-ocean ridge basalts. In honor of the mountains in which they occur, petrologists call this composition *andesite*.

The active volcanoes of the Andes form the southeastern section of the *Pacific Ring of Fire*, a hemispheric-scale zone of intense seismic and volcanic activity that includes the Pacific coasts of Central and North America (including the Cascade Volcanoes), up through the Aleutian Islands, and back around through Japan, Southeast Asia, Indonesia, and New Zealand. The Ring of Fire is one of the largest continuous geologic structures on Earth.

SEE ALSO Plate Tectonics (c. 4–3 Billion BCE?), The Sierra Nevada (c. 155 Million BCE), Cascade Volcanoes (c. 30–10 Million BCE), Island Arcs (1949), Seafloor Spreading (1973)

Aerial view of Carbajal Valley in the Southern Andes, Tierra del Fuego, Argentina.

First Hominids

Although originally referring only to humans and our closest extinct relatives, the term *hominids* is now widely understood to refer to the entire family of so-called "great apes," including gorillas, orangutans, chimpanzees, modern humans, and their extinct ancestors. The origins of the hominid family date back to around 10 million years ago, based on the fossil record. During most of the Miocene (which spans about 22 to 3 million years ago), primates adapted to a variety of mostly arboreal lifestyles and ecological niches, especially within the expansive tropical environments that were common in the equatorial and mid-latitudes during that time. Near the end of the Miocene, however (starting around 10 to 8 million years ago), the fossil and geologic records reveal that the tropical zones began to shrink considerably, replaced by more temperate grassland savanna ecological zones. As plants and other animals began to adapt to this and other new environments, primates were forced to come down out of the trees to find food.

Hominids are generally large, tail-less primates. Males are generally larger than females (*sexual dimorphism*), and most species are *quadrupeds* (walk on four legs, though some of course are *bipeds*). All have opposable thumbs and use their hands to hunt or gather food (most species primarily eat fruits, but a few are omnivores) and, in some cases, to make tools. Gestation periods for hominid females are around eight or nine months, and all typically give birth to a single child at a time. All hominid babies are born helpless and require significant care through infancy and adolescence, generally not reaching sexual maturity until they are between about 8 and 15 years old, depending on the species.

Exactly when the hominid genetic family diverged from the rest of the primates is not precisely known. Nonetheless, the available evidence, plus genetic differences mapped among extant hominid species, indicate that over the past c. 10 million years, first orangutans, then gorillas, then chimpanzees diverged from the line of common hominid ancestors that eventually led to humans. Chimpanzees and humans, for example, share about 98.4 percent of their DNA, proving the close and relatively recent linkage of these two primate hominid species.

SEE ALSO Mammals (c. 220 Million BCE), Primates (c. 60 Million BCE), *Homo sapiens* Emerges (c. 200,000 BCE), *Gorillas in the Mist* (1983), Chimpanzees (1988), Savanna (2013)

Bonobos (a hominid species similar to chimpanzees) in their natural habitat, in the Democratic Republic of Congo. Africa.

Sahara Desert

One way of characterizing the geographic, biologic, and climatic diversity of planet Earth has been to divide the surface into about ten major kinds of ecological community types or *biomes*. Many biomes are distinguished by specific ranges of average annual temperature and rainfall. Deserts are the driest of the biomes, but can occur in areas ranging from the coldest to the hottest places on Earth. The coldest deserts are found at high latitudes in the Arctic and Antarctic, where air and ocean circulation patterns around the poles block the arrival of moist air from the tropics. In contrast, the hottest deserts on Earth are close to or along the equator, where intense sunlight and mountain ranges prevent rainfall from occurring over wide swaths of land.

The largest hot desert on Earth is the Sahara, which covers most of the northern continent of Africa. The land area of the Sahara Desert is comparable to that of the United States or China, making it the largest non-polar desert in the world. Average summertime high temperatures exceed 104°F (40°C), and the Sahara witnessed the highest recorded temperature in history, 136°F (58°C). Average annual rainfall is extremely low in the enormous central region of the desert—less than 0.4 inches (10 millimeters) per year. Even the "wet" regions of the desert only see about 10 inches (250 millimeters) of rain per year.

Because of plate tectonics, the world's major continental landmasses like Africa have changed positions dramatically over time. Back in the late Miocene (about 7 million years ago), for example, the collision of the African and European plates that formed the Alps also caused the closing of what had been a great ocean basin (the Tethys Sea) between the continents. Without the ocean's influence, the once-lush and tropical plains and mountains of northern Africa began to dry out, and the Sahara was born.

Today, as in most desert environments on Earth, wind is the major agent of geologic change in the Sahara Desert, slowly sandblasting and breaking down the quartz-rich continental rocks and former ocean sediments into enormous "seas" of sand that now cover the land.

SEE ALSO Plate Tectonics (c. 4–3 Billion BCE?), Pangea (c. 300 Million BCE), The Atlas Mountains (c. 300 Million BCE), The Himalayas (c. 70 Million BCE), The Alps (c. 65 Million BCE), Tropical Rain/Cloud Forests (1973), Temperate Rainforests (1976), Tundra (1992), Boreal Forests (1992), Grasslands and Chaparral (2004), Temperate Deciduous Forests (2011), Savanna (2013)

Spectacular sand dunes in the Sahara Desert, the largest non-polar desert on Earth.

The Grand Canyon

Liquid water flowing over the surface of the Earth is a major force for geologic change. Over time, ocean waves can wear down the seashore, and rivers and streams can carve deep gullies and canyons into even the hardest continental bedrock. The formation of canyons, in particular, can be further accelerated if that rock is being uplifted and is thus more susceptible to fluvial (stream- and river-related) erosion. That is exactly the situation going on within one of the Earth's most spectacular geologic structures, the Grand Canyon.

The main part of the Grand Canyon of the Colorado River is just under 300 miles (480 kilometers) long and up to 20 miles (32 kilometers) wide, running mainly through Arizona but also in parts of Nevada, Utah, Colorado, and Wyoming. At its deepest, the Canyon extends nearly 6,100 feet (1,860 meters)—more than a mile—above the raging river below. The canyon is perhaps most famous for its numerous colorful layers of sedimentary, volcanic, and metamorphic rocks, as well as the fact that nearly 2 billion years of Earth's geologic history is exposed to view in the Canyon's walls.

There is considerable debate among geologists about when the Grand Canyon actually formed. Parts of the Canyon system appear to have formed as far back as the late Cretaceous (70–65 million years ago), probably as part of the overall regional uplift that formed the modern Rocky Mountains and the Colorado Plateau. The general consensus among geologists, based on rock and fossil ages and other clues, is that older Canyon networks were merged with newer, deeper ones created by more recent erosion only about 5 or 6 million years ago. The Grand Canyon appears to be a relatively young feature.

People have lived around and especially within the Grand Canyon for thousands of years, using the water, vegetation, and animal life along the more temperate regions near and along the river as resources for survival in what is otherwise a harsh desert environment outside the Canyon itself. Modern civilization also uses the Canyon for resources, now focusing on generation of hydroelectric power, tourism, and ecological/geological studies.

SEE ALSO Plate Tectonics (c. 4–3 Billion BCE?), The Sierra Nevada (c. 155 Million BCE), The Rockies (c. 80 Million BCE), Exploring the Grand Canyon (1869)

The view from Yavapai Point across the Grand Canyon exposes nearly 2 billion years of Earth's geologic record for detailed study.

The Mediterranean Sea

The collision of the African and Eurasian continental plates, which started some 70–60 million years ago, closed off a great ocean basin that had existed between the continents. That basin, filled by the now-gone Tethys Sea, was part of an oceanic plate that was subducted and melted underneath the colliding African and Eurasian plates. The collision led to significant regional volcanism, as well as a zone of mountain-building across northern Africa and southern Europe.

The collision of Africa and Eurasia appears to have cut the remains of the Tethys Sea off from the rest of the Atlantic Ocean around 6 million years ago, beginning an era of many hundreds of thousands of years of drying out. Evaporation of the landlocked seawater resulted in the formation of thick, extensive layers of salt over much of the original and newly formed parts of the Tethys basin. In some places, the salt deposits are more than 1.9 miles (3 kilometers) thick. This "salinity crisis" in the basin between Africa and Eurasia nearly dried up the former Mediterranean Sea over the course of more than 600,000 years.

The crisis ended rather suddenly, however, about 5.3 million years ago when the natural dam between the Atlantic Ocean and the Mediterranean basin was catastrophically breached at the current Strait of Gibraltar. Ocean waters rushed back into the basin at a rate estimated to have been about 1,000 times the discharge rate of the Amazon River today, refilling parts of the basin by up to 30 feet (10 meters) per day. After only a few months of this deluge, much of the modern Mediterranean Sea had been refilled.

What had been the humid, subtropical climate conditions of that region of the world changed starting a few million years ago, when fossil evidence records a shift to the drier "Mediterranean climate" conditions of today. The region became heavily wooded with coniferous trees (such as the cedar on the flag of Lebanon) and other plants that could handle the hot and dry summer conditions. Since then, the ecology of the Mediterranean region has been dramatically altered, however, by thousands of years of human influence.

SEE ALSO Continental Crust (c. 4 Billion BCE), Plate Tectonics (c. 4–3 Billion BCE?), Roots of the Pyrénées (c. 500 Million BCE), Pangea (c. 300 Million BCE), The Atlas Mountains (c. 300 Million BCE), The Atlantic Ocean (c. 140 Million BCE), The Alps (c. 65 Million BCE), Sahara Desert (c. 7 Million BCE), The Caspian and Black Seas (c. 5.5 Million BCE)

Artistic rendering, based on geophysical data, of the Mediterranean basin just before the collision between Africa and Southern Europe cut it off from the Atlantic Ocean.

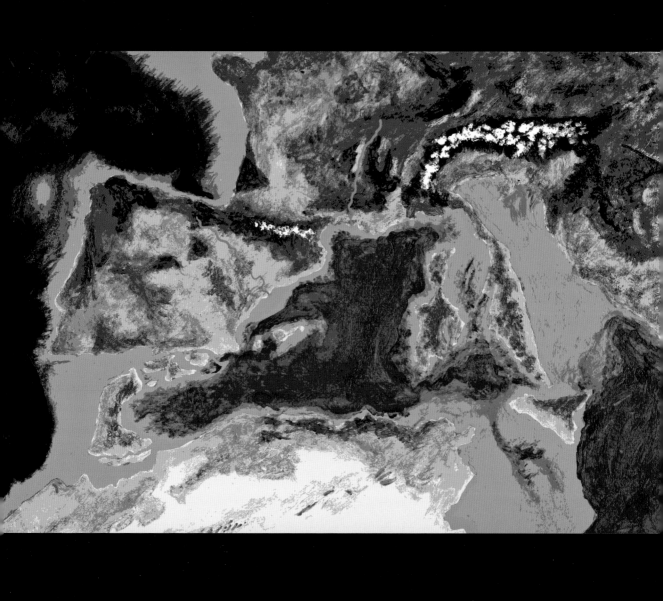

The Caspian and Black Seas

The breakup of the supercontinent of Pangea, including the opening of the Atlantic Ocean starting around 140 million years ago, set in motion a series of other eventual plate collisions that would shape Earth's modern geography to this day. Among those were the collision of the African and Eurasian continental plates, which closed off a large former ocean basin (the Tethys Sea), created mountain ranges, and ultimately led to the formation of the modern Mediterranean Sea.

The plate collision also helped to uplift the Caucasus and other mountain ranges from southeastern Europe well into Asia. The result was the pinching off of additional regions of the original Tethys Basin, and the creation of the largest inland lakes on the planet: the Caspian Sea and the Black Sea.

Both the Caspian and the Black seas formed around the same time as the modern Mediterranean, but because the Caspian is not directly connected to the rest of the global ocean system, and the Black Sea only occasionally is (when the water level is especially high), they are both dominated primarily by freshwater coming from rivers and streams flowing out of the surrounding mountains and plains. They are salty, however (about a third the salinity of ocean water), because they have re-hydrated some of the huge salty deposits left behind when the original Tethys Sea evaporated more than a half million years before these new seas began to form.

The Caspian and Black seas together capture an enormous fraction of the world's lacustrine, or lake-related, freshwater. The deepest parts of these enclosed basins stretch 3,300 to 6,600 feet (around 1,000 to 2,200 meters) below sea level, and the waters at such depths are *anoxic*—they do not mix with the upper layers of water that mix with oxygen from the atmosphere. As a result, shipwrecks, prehistoric settlements, and other ancient artifacts from many thousands of years of human habitation and commerce along the current and former shores of these inland seas are remarkably well preserved, making them exciting areas for deepwater archaeological exploration and research.

SEE ALSO Plate Tectonics (c. 4–3 Billion BCE?), Roots of the Pyrénées (c. 500 Million BCE), The Atlas Mountains (c. 300 Million BCE), Pangea (c. 300 Million BCE), The Atlantic Ocean (c. 140 Million BCE), The Alps (c. 65 Million BCE), Mediterranean Sea (c. 6–5 Million BCE), The Great Lakes (c. 8000 BCE)

A view from orbit of the Caspian Sea as imaged by the MODIS sensor on the NASA Terra satellite in 2003.

Galápagos Islands

Charles Darwin (1809–1882)

Most, but not all, of the seismic and volcanic activity on Earth is concentrated at the convergent (collisional) boundaries between the major tectonic plates. However, some special regions within the interiors of the plates, far from the boundaries, do experience extensive volcanic and seismic activity. These special regions occur where a large plume of molten magma material—a hotspot—gets close to the surface. Among the most famous of these regions is the Hawaiian hotspot. Another famous location is along the equator just west of Ecuador, where the Galápagos hotspot has been churning out volcanic lavas and new tropical islands for the past 5 million years or so.

One special aspect of the Galápagos hotspot is that it occurs so close to the divergent (spreading apart) boundaries of three plates just off the coast of the South American continental plate. The Cocos, Nazca, and Pacific plates are all moving apart from one another near the region where the Galápagos Islands are forming, producing mid-ocean ridge-spreading centers where the oceanic crust is particularly new and thin. The mantle hotspot beneath the islands has exploited the thinness of the plates there to punch some volcanic lava through to the surface. The result is a broad plateau built onto the seafloor, and about 20 islands that have been built up—some of them to over 5,000 feet (1,500 meters) above sea level—by volcanic eruptions that continue to this day.

The specific location and isolation of the Galápagos Islands have helped to create a unique ecological niche there. A variety of species of birds and reptiles, for example, has evolved from their earlier continental ancestors to adapt to the specific climatic and food-supply environment of the islands. Among the first to recognize the role of isolation and specific environmental circumstances in helping these species adapt to this specific environment was the naturalist Charles Darwin, whose study of finches and other animals among the islands of the Galápagos would prove crucial to the development of the theory of evolution driven by natural selection.

SEE ALSO Plate Tectonics (c. 4–3 Billion BCE?), Hawaiian Islands (c. 28 Million BCE), Natural Selection (1858–1859)

Main image: *A blue-footed boobie, native denizen of the Galápagos Islands.*
Inset: *True-color image of the Galápagos Islands from space, acquired in 2002 from NASA's Terra satellite.*

The Stone Age

Historians, anthropologists, and others who study the history of human civilization often refer to our time as the Space Age. More generally, cultural and historical anthropologists have historically divided the prehistoric era of modern humans and their recent predecessor species into three broad technological and societal periods, known from oldest to youngest as the Stone Age, Bronze Age, and Iron Age.

The Stone Age is the longest of these prehistoric periods, and is thought to have begun about 3.4 million years ago, based on the archaeological discovery of the oldest fragments of stone- and bone-based tools and implements in what is modern-day Ethiopia. While the term *Stone Age* may evoke images of families and clans of animal-skin–clad humans living in caves and hunting wooly mammoths, archaeologists have discovered that the vast majority of Stone Age peoples lived in the extensive grassland savanna environments that extended during this time from southern Africa to north through the Nile Valley and then to the east into Asia and modern China. Grasslands do not support sustainable arboreal lifestyles, which is one of the factors that helped guide the evolution of many species of primates out of the trees.

Life on the savanna also proved highly advantageous to species that could use readily available natural materials—rocks, wood, bone, and so forth—to fashion high-tech tools such as spears, knives, flints, levers, and mortars and pestles that significantly enhanced hunting and food preparation. Cultural anthropologists are quick to point out that referring to peoples of this time as "primitive" is another unfortunately inappropriate interpretation of "Stone Age." On the contrary, Stone Age tool-makers, hunters, and gatherers appear to have been extremely skilled with the use of their available technology, mastering not only specialized tools, but also fire, early watercraft, and even the creation of impressive religious or astronomical structures made from earthworks or boulders.

Starting around 6,000 years ago, some human societies developed the ability to melt ores of copper, tin, and other metals, ushering in the Bronze and then Iron Ages. Other societies continued to thrive on Stone-Age technologies, however, almost to the present day.

SEE ALSO Primates (c. 60 Million BCE), *Homo sapiens* Emerges (c. 200,000 BCE), The First Mines (c. 40,000 BCE), The Bronze Age (c. 3300–1200 BCE), Stonehenge (c. 3000 BCE), The Iron Age (c. 1200–500 BCE), Savanna (2013)

Typical Stone Age tools, from a late-nineteenth-century German engraving.

Paläolithische Feuersteingeräte aus französischen Fundstätten.

(Kjökkenmöddinger.)

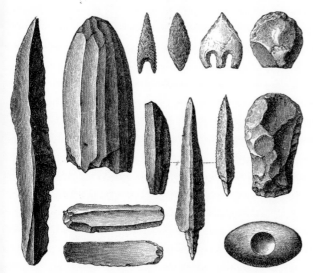

Feuersteinnucleus, Messer, Pfeilspitzen und Schaber.

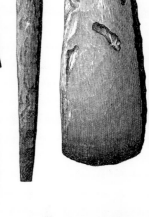

Feuersteinäxte und Schleifsteine.

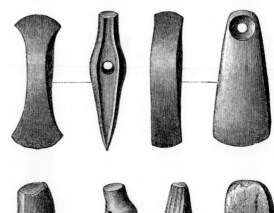

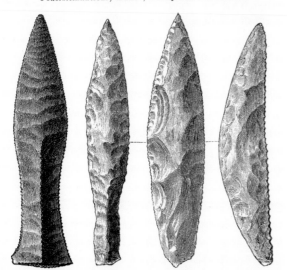

Feuersteindolche, Lanzenspitze und Säge.

Durchbohrte Steinhämmer, Steinäxte und Hammeraxt.

The Dead Sea

The breakup of continental landmasses creates enormous valleys or rifts in the crust. The effects of the East African rift extend much farther than just the African plate. A complex zone of interactions among the African, Arabian, Indian, and Eurasian plates extends to the north of the East African rift zone, along the Red Sea and into the Middle East and beyond.

In particular, continuing compression between the African and Eurasian plates to the west and extension of the African plate in the major rift zone to the south have produced a region of roughly north–south valleys within parts of the Arabian Peninsula, especially regions encompassing modern-day Israel, Jordan, and Saudi Arabia. The Dead Sea, the lowest elevation lake (not really a "sea") anywhere on Earth, lies within one of these extensional valleys, the full extent of which runs from the Sea of Aqaba to the south and up to the Sea of Galilee to the north.

The surface waters of the Dead Sea are currently at about 1,400 feet (430 meters) below sea level, and the deepest parts of the sea are about 1,000 feet (300 meters) below that. The volume of water in the Dead Sea is about the same as that in Lake Tahoe in the western United States, but the Dead Sea is *hypersaline*—extremely salty—compared to most other inland lakes or seas on Earth. With a salinity of about 34 percent (nearly 10 times saltier than ocean water), very few plants and animals can thrive in and around the Dead Sea.

How did the Dead Sea get so salty? A leading hypothesis is that the Arabian Peninsula has been undergoing multiple cycles of uplift and subsidence over the past tens of millions of years and has occasionally been inundated with seawater. The region apparently underwent a last major episode of uplift around 3 million years ago, cutting off the basin from the rest of the Mediterranean and creating a true large inland sea that was the precursor to the Dead Sea. As those waters evaporated, enormous salt deposits were formed and the salt concentration of the remaining waters increased dramatically—earning the shrinking lake its unfortunate but relatively accurate "dead" moniker.

SEE ALSO Pangea (c. 300 Million BCE), The Atlantic Ocean (c. 140 Million BCE), East African Rift Zone (c. 30 Million BCE), The Mediterranean Sea (c. 6–5 Million BCE), The Caspian and Black Seas (c. 5.5 Million BCE), Death Valley (c. 2 Million BCE)

Salt-encrusted rocks line the shore of the slowly evaporating Dead Sea.

Death Valley

The melted remains of the Farallon tectonic plate under western North America are buoyant, rising blobs of magma that push upward on and bulge the overlying continental plate. Especially in the Southwest—from northern Nevada through northwest Mexico—the extensional (stretching) forces on the bulging North American plate have caused deep, nearly vertical ("normal") fault cracks to form, aligned roughly north–south. Over the past few million years, the crustal blocks between adjacent normal faults have down-cropped, creating deep valleys that geologists call *graben,* with high parallel ridges on either side that geologists call *horst.* The undulating, semi-regular repeated patterns of horst and graben across much of the Southwest defines a region known as the *Basin and Range.*

The deepest graben among the Basin and Range is a 140-mile-long (225-kilometer), 5- to 15-mile-wide (8–24 kilometer) crustal block just west of Las Vegas known as Death Valley. The block has dropped so much that even though the valley floor has been filled by a large volume of sediments from the surrounding mountains, Death Valley is still the lowest elevation point in North America, at just under 300 feet (90 meters) below sea level. Ample fossil and geologic evidence (such as huge salt and borax deposits) reveal that there used to be abundant surface water (rivers, lakes, and small inland seas) in what is now the Mojave Desert, but continued regional uplift and changing climate conditions have driven that water away. The result is a spectacular landscape of rocky sediments and colorful mineral deposits eroding from rugged, barren mountains onto an expansive, relatively flat valley floor.

Multiple mountain ranges to the west of the Mojave Desert effectively siphon water vapor out of the air, and decreasing snowfall in mountains to the north and east contributes to dwindling river and groundwater flows, making Death Valley among the driest places in the world, with average annual rainfall of only around 2 inches (50 millimeters). It also has the dubious distinction of holding the record for highest reliably recorded air temperature at the surface of the Earth: 134°F (57°C).

SEE ALSO The Sierra Nevada (c. 155 Million BCE), The Rockies (c. 80 Million BCE), Cascade Volcanoes (c. 30–10 Million BCE), The Andes (c. 10 Million BCE), The Dead Sea (c. 3 Million BCE), Basin and Range (1982).

Sand dunes and parched mountains within Death Valley National Park.

Lake Victoria

By definition, rift valleys such as the East African rift zone are regions where continental crust is being torn apart. Rifting apart of continental plates occurs over a wide zone of faulting or cracking of the continental crust, in linear zones perpendicular to the direction of the opening rift. In East Africa, like in Death Valley and other parts of the Basin and Range region of the North American plate, crustal blocks between parallel, nearly vertical *normal faults* can drop down along those faults, forming basins or valleys called graben.

Because of the wet, tropical climate of much of the East African rift zone, Africa's largest freshwater lake, Lake Victoria, has formed within one of these graben regions. Bounded by Kenya, Tanzania, and Uganda, Lake Victoria is the largest tropical lake in the world, and the second-largest freshwater lake overall (second only to Lake Superior within the Great Lakes of North America). A leading hypothesis for the formation of Lake Victoria is that ongoing rifting starting about 400,000 years ago caused uplifted crustal blocks (*horst* ridges, to geologists) to dam rivers that had previously been freely flowing across the plains, leading to the accumulation of large volumes of water in the adjacent down-dropped graben valley. Today, only one river drains Lake Victoria: the mighty Nile.

There is ample fossil and geologic evidence that Lake Victoria has completely dried up multiple times since it formed, most likely because of a dwindling supply of snowmelt and precipitation during peak Ice Age periods. As the climate has warmed between the Ice Ages, however, the Lake has refilled—most recently after the end of the latest ice age around 11,000 years ago. Over that relatively short period of geologic time, Lake Victoria has been repopulated at an astonishing rate by a rich diversity of fish, reptiles, mammals, plants, and other species that have thrived in the tropical environment. Unfortunately, as with many inland lakes around the world, overfishing and poor environmental stewardship of the lake's ecosystem (pollution, development, and so forth) have led to the human-caused extinction of many of those species, as well as the proliferation of many non-native invasive species.

SEE ALSO East African Rift Zone (c. 30 Million BCE), The Dead Sea (c. 3 Million BCE), Death Valley (c. 2 Million BCE), End of the Last "Ice Age" (c. 10,000 BCE), The Great Lakes (c. 8000 BCE), Controlling the Nile (1902)

Fishermen heading out to work along the shores of Lake Victoria in Uganda.

Homo sapiens Emerges

Humans of the species *Homo sapiens* are relatively new on Earth. Our earliest appearance in the fossil record, judging by the oldest archaeological finds (in Africa), dates back only about 200,000 years. Fossil evidence shows that *Homo sapiens* coexisted for a time with our closely related subspecies, the Neanderthals, and that evidence for Neanderthal traits disappeared from the record around 30,000 years ago.

We *Homo sapiens* are a persistent lot, excelling at survival with tools, language, long memories, and hard-won experience. Our history and evolution reflect an inquisitiveness and a desire for more intangible nourishments of the soul, too, which may explain why music, dance, and art have apparently been such an important part of the human experience ever since it began. Indeed, visiting the 17,000-year-old Paleolithic cave paintings of the Dordogne region of France inspires amazement that our ancestors could make the time for art amid what must have been a constant struggle to stay alive. But they didn't just paint animals, plants, and other mundane things. Many archaeologists now believe that some of the dots, lines, and perhaps even animal figures were representations of constellations or other features of the night sky. If so, then not only are these the oldest paintings on Earth but they are also the oldest sky maps, painted by the world's first astronomers.

The emergence of modern humans, when combined with the emergence of technology, has had a profound influence on the history of planet Earth. The development of agriculture around the end of the last ice age and the ensuing formation and growth of cities has resulted in massive reworking of the distribution (or removal) of plants, and even in some places major changes in the geology and topography of the surface. Most recently, the Industrial Revolution and especially the development of the internal-combustion engine have also led to measurable human-induced changes in the composition of the atmosphere (especially the addition of significant CO_2) and rapid increase in the average temperature of the Earth's surface. Ensuing changes in climatic conditions from region to region, as well as a slow rise in sea level because of widespread glacial melting on land, will have impacts for many generations of *Homo sapiens* to come.

SEE ALSO Invention of Agriculture (c. 10,000 BCE), End of the Last "Ice Age" (c. 10,000 BCE), Industrial Revolution (c. 1830), Controlling the Nile (1902), Rising CO2 (2013)

Part of a reconstructed painting from the famous caves of Lascaux in southwestern France depicting a prehistoric horse and other symbols that some archeologists think might be representations of stars and constellations in the night sky.

The San People

Anthropologists don't know exactly when *Homo sapiens* and our precursor species' clans or societies began to adopt nomadic hunter-gatherer lifestyles, although there are various hypotheses suggesting that this transition was related to climate changes during the Stone Age that favored grasslands over tropical environments in regions such as eastern Africa, where many early hominid fossils have been found. One way to gain clues into the history of our species, then, would be to study any remaining examples of clans or societies that might still adhere to some of the traditions and lifestyles of our distant ancestors.

One such group is known as the San people (previously referred to as "Kalahari bushmen," even though only a subset of them dwell in the Kalahari Desert), a loose-knit society of around 100,000 indigenous people scattered throughout five countries spanning the southern third of the African continent. The San people have been nomadic hunter-gatherers in southern Africa for over 70,000 years, based on San artifacts such as stone tools and rock paintings that archaeologists have discovered that date back to that time. More recently, various government modernization programs have forced the San into a more sedentary farming lifestyle. Nonetheless, many take stewardship of their long culture and traditions seriously, and thus they (and their genetics) provide anthropologists with valuable information on the early evolution of human societies.

While there are differences among various San societies and cultures across the region, it is still perhaps not surprising that in general, the San people have been more closely attuned to the seasonal cycles of plants, animals, and weather than most other industrial-age human societies. Men primarily hunted for game and women primarily gathered food, but crossing these roles was not uncommon. Indeed, women appear to have a higher status in San society than in many "modern" societies, including in decision-making and ownership issues. Achieving consensus appears to have been a critical component of their society, and a spirit of equality and egalitarianism is evident in their traditions. They also have a rich and deep history of music, dance, art, and games.

SEE ALSO First Hominids (c. 10 Million BCE), The Stone Age (c. 3.4 Million BCE TO 3300 BCE), *Homo sapiens* Emerges (c. 200,000 BCE)

San (Bushmen) people crossing a sand dune in the Northern Cape region of South Africa.

Arizona Impact

We only need to look at the heavily cratered, ancient surface of our celestial neighbor the Moon to know that the Earth has been slammed by asteroid and comet impacts throughout its history. During the earliest part of solar system history, all of the planets and their moons experienced far more violent impacts than they experience today. Over time, as these early-formed asteroids, comets, and planetesimals destroyed one another through collisions, the number of catastrophic impact events has dramatically decreased.

Still, lots of rogue debris from early planetary formation remains in our solar system, from a relatively small number of large asteroids down to countless individual grains of dust. Dozens of tons of that dust impact the Earth every day; we call the slightly larger chunks of that debris "shooting stars" as they burn up from friction in our atmosphere. More rarely, somewhat larger but still relatively small objects (a few to tens of meters across) strike the Earth, producing spectacular fireballs and shock-wave explosions in the atmosphere. Much more rarely, even larger objects strike the Earth at extreme speeds and survive all the way to the ground, creating a large impact crater and potentially disrupting the climate and ecology of our planet.

The most recent impact event large enough to produce a sizable crater on Earth happened about 50,000 years ago, just east of what is now Flagstaff, Arizona. A small asteroid (about 160 feet, or 50 meters, across) made of iron and nickel slammed into the Colorado Plateau at a speed of around 40,000 miles per hour (16 kilometers per second). The resulting explosion had the force of about 10 megatons of TNT, or about 500 times more energy than the atomic bombs deployed in World War II, and created a hole in the ground about 3,900 feet (1,200 meters) across and 560 feet (170 meters) deep. Meteor Crater (also known as Barringer Crater or Canyon Diablo Crater) has slowly been eroding over time, but because the climate of Arizona has been so arid for much of the time since then, it remains one of the best-preserved and best-studied impact structures on Earth.

SEE ALSO Late Heavy Bombardment (c. 4.1 Billion BCE), Dinosaur-Killing Impact (c. 65 Million BCE), The US Geological Survey (1879), Hunting for Meteorites (1906), The Tunguska Explosion (1908), Understanding Impact Craters (1960), Torino Impact Hazard Scale (1999)

View from above the rim of Meteor Crater, a hole in the Arizona desert about ¾ mile (1.2 kilometers) wide, created about 50,000 years ago by the impact of a small iron-rich asteroid traveling more than 6 miles (10 kilometers) per second.

The First Mines

Economic geology is the study of the prospecting for, extraction, and use of Earth's surface and subsurface materials for economic or industrial purposes. Economic geology starts with the oldest available evidence for mining, dating back to more than 40,000 years ago. That oldest evidence has been found in the red-stained hills of Swaziland, in southeastern Africa, where archaeologists in the 1960s discovered small caves that had been excavated by prehistoric people using stone tools. Some of the tools in these oldest known mines were still there, providing the evidence that people had used stones to dig, chop, pick, and hammer their way as much as 42 feet (13 meters) into the hillsides to extract material.

But what were they mining? The hills are reddish because of the presence of abundant iron oxide minerals, and especially the fine-grained, bright red mineral known as hematite or sometimes red ochre. These kinds of minerals form in wet, tropical environments from the oxidation and weathering of darker iron-bearing volcanic rocks. But archaeologists and anthropologists were at first confused, because hematite is abundant right on the surface of these hills—why would these early miners have dug so deeply underground to extract what could be more easily just dug up on the surface? They eventually discovered the reason: deep within the caves, the miners were finding and extracting small pockets of a special kind of coarse-grained hematite called *specular hematite* or sometimes *specularite*. Black, shiny, and mirror-like, specularite was apparently prized by early clan leaders and shamans because it could be ground up into tiny pieces and pasted onto their bodies to make them glitter like gold during ceremonies and ritual practices. Those who could find and extract the glitter became important members of society.

The region where these caverns were discovered is now called the Ngwenya Mine, and it is still in operation as the world's oldest mine. In more recent times, red ochre has been the primary extracted ore, used as a pigment for rock paintings and other art by the San people of the region, as well as the smelting of iron ore, widely traded and exported across Africa.

SEE ALSO The San People (c. 70,000 BCE), The Bronze Age (c. 3300–1200 BCE), Magnetite (c. 2000 BCE), The Iron Age (c. 1200–500 BCE).

A small pond and terraced cliffs mark the location of the Ngwenya Mine in the Kingdom of Swaziland, the site of one of the earliest known iron mines (dating back to more than 40,000 years ago).

La Brea Tar Pits

Oil and natural gas form underground via the compression and heating (metamorphism) of organic matter over many millions of years. Most oil and gas deposits form very deep underground. By drilling deeply and/or relying on plate tectonics to uplift those materials more closely to the surface, economic geologists and oil and gas companies have been able to extract these materials as fossil fuels. Sometimes, however, the geologic histories of specific places have conspired to bring these deeply formed petroleum deposits right up to the current surface of the Earth, where they can start to flow as *oil seeps*.

One of the most famous and well-studied oil seeps in the world is the La Brea Tar Pits (*brea* means "tar" in Spanish), in urban Los Angeles. Gooey, black tar has been seeping out of the ground there for many tens of thousands of years. Native Americans used the tar as a water sealant for their canoes; in the 1800s, settlers in California used the tar as a roof sealant for their homes. But the most famous "use" of the tar at La Brea is as a preservative of the bones of animals that became fossilized within the hardened tar after the animals accidentally got stuck in it and died. Indeed, most of the pits at La Brea were human-excavated in the early twentieth century as a way to access hundreds of thousands of well-preserved specimens of ancient mammoths, bison, horses, sloths, wolves, lions, saber-toothed cats, birds, and many smaller animals.

The oldest specimens extracted from the La Brea Tar Pits are around 40,000 years old. But the tar itself is perhaps 10 million years old or more. Back then, the La Brea region was part of the deep seafloor just off the coast. Continental rivers and streams carried huge amounts of organic debris to the ocean, depositing them as sediments that, over millions of years, piled up to over a mile thick in some places, compressing the sand and turning the organic materials to oil. Subsequent complex interactions between the Pacific, Farallon, and North American tectonic plates has uplifted some of these tarry sands to the surface, where some of the liquid comes out as oil seeps.

SEE ALSO Plate Tectonics (c. 4–3 Billion BCE?), San Francisco Earthquake (1906)

In this 1913 textbook illustration, a saber-toothed cat and two dire wolves fight over a Columbian mammoth carcass in the La Brea Tar Pits.

Domestication of Animals

Most of us take for granted the many domesticated animal species that either live with us, work for us, or provide food resources. However, there was a time when domesticated species, or at least their non-domesticated precursor species, lived in the wild without such a close relationship to humans. *Domestication* of animals is defined by the US National Academy of Sciences as the establishment of "a mutual relationship between animals with the humans who have influence on their care and reproduction." Importantly, domestication is different from taming: the former is a genetic modification to members of a species via selective breeding, while the latter is a behavioral change carried out on one or more individuals from a wild species.

Fossils, ancient DNA, and other archaeological evidence indicate that dogs were the first domesticated animals, more than 30,000 years ago, around the height of the last major glaciation. Wolves are essentially dogs in the wild, and many hypotheses exist as to why and how some wolves were domesticated by prehistoric humans. Perhaps they helped hunters capture prey, maybe they helped defend people against other predators, or maybe our ancestors were just as susceptible as we are today to furry coats and floppy ears. Ecologists call this pathway to domestication *commensal* (co-beneficial).

With the invention of agriculture around 10,000 years ago and the need to work the soil, transport products, and support growing sedentary communities of people, most modern livestock species began to be domesticated. These include goats, sheep, pigs, cattle, fowl, and many other species raised for food resources, what ecologists call *prey* domestication. A final kind of pathway, known as *directed* domestication, was carried out in cases where animals were not necessarily either commensal or prey, but could still be useful to human societies for work, transportation, or other purposes. Examples include horses, donkeys, and camels.

Some species of animals are simply unable to be domesticated, at least not in large enough numbers to be generally useful to human societies. For example, none of the large herd animals in Africa (such as zebras and gazelles) can be domesticated, placing limits on the historical ability of African societies to develop agrarian and livestock-based lifestyles.

SEE ALSO Invention of Agriculture (c. 10,000 BCE), End of the Last "Ice Age" (c. 10,000 BCE), Genetic Engineering of Crops (1982)

Part of an Egyptian mural depicting people interacting with domesticated animals.

Invention of Agriculture

As the climate warmed after the end of the last Ice Age, around 12,000 years ago, and much of the ice retreated, large changes in human civilization began to occur as well. Perhaps most importantly, some clans of people, especially in certain special geographic regions, discovered that a nomadic lifestyle was no longer required to sustain the food needs of their group. For example, in much of the region just east of the Mediterranean Sea, flooding of major rivers such as the Nile, Tigris, and Euphrates began to bring new water and sediments to the surrounding flood plains in regular, predictable patterns. Edible wild grains flourished, and animals began to follow predictable migratory paths. It's no wonder that the region has come to be known as the "fertile crescent."

Within such lush river valleys, formerly nomadic peoples could instead settle down, banking on the reliable rains, flooding, and/or migrations to essentially bring food to them. It's not hard to imagine some entrepreneurial subset of people in such environments deciding to plant seeds in order to harvest grains, fruits, and vegetables in more easily accessible and centralized locations—the first farms. Reliable and often plentiful food sources supported growing populations in such regions, providing labor to tend even larger farms, spawning trade among groups cultivating different crops or livestock, and ultimately leading to the earliest permanent centralized architectural and political structures designed to organize and administer large groups of people—the first cities. The fertile crescent is often called "the cradle of civilization." Indeed, the Sumerians, who settled the fertile crescent from about 4500 to 1900 BCE, are widely regarded as the first civilization on Earth.

Similar scenarios played out around the world in the centuries that followed, with agricultural lifestyles and societies cropping up in China, Indonesia, sub-Saharan Africa, and the Americas. These transitions to agricultural societies may not have been as obviously driven by climate change as the earlier societies of the fertile crescent. Anthropologists continue to debate the importance of other factors besides climate, such as population pressure, plant and animal domestication, or even social pressures, in the emergence of agrarian lifestyles and the cultivation of crops.

SEE ALSO Domestication of Animals (c. 30,000 BCE), End of the Last "Ice Age" (c. 10,000 BCE), Population Growth (1798), Industrial Revolution (c. 1830), Controlling the Nile (1902), Genetic Engineering of Crops (1982), Large Animal Migrations (1997)

A lush valley within the "fertile crescent" in the Zagros Mountains near Dena, Iran.

End of the Last "Ice Age"

Milutin Milanković (1879–1958)

By definition, an ice age is a period of long-term (millions of years or longer) decreases in Earth's average surface and atmosphere temperature, which leads to the growth of continental glaciers and polar ice sheets. Within an individual long-term ice age, however, geologic, fossil, ice-core, and ocean sediment records show that there have been many separate shorter-duration (tens of thousands to hundreds of thousands of years) pulses of colder and warmer global average temperatures, referred to as "glacial" and "interglacial" periods. Long-term ice ages in Earth's history appear to have consisted of dozens, perhaps hundreds, of shorter-term glacial and interglacial cycles. In the early 1940s, Serbian astronomer and climatologist Milutin Milanković discovered that among the drivers for these shorter-duration cycles of climate change were slow variations in the Earth's tilt (obliquity) and the shape of Earth's orbit around the Sun (eccentricity), both of which influence the amount of solar heating that strikes the Earth's surface. In honor of his pioneering research in this area, these astronomical variations are called Milanković cycles.

We are all living within the most recent interglacial period during an ice age known as the *Quaternary glaciation*. This most recent of Earth's ice ages appears to have started about 2.6 million years ago, and our current interglacial period, which geologists call the *Holocene* epoch, began only about 12,000 years ago (coinciding with the dawn of modern civilization). The Quaternary glaciation has seen more than 60 glacial and interglacial cycles, triggered not only by astronomical forces, but also by long-term changes in greenhouse gases like CO_2, ocean currents, and the movement of the continents via plate tectonics. It is perhaps not obvious that we're still technically living in an ice age, but the persistence of the Antarctic and Greenland glaciers and ice sheets over the past few million years proves that to be the case. As the climate continues to warm at a historically unprecedented pace, however, we may be approaching the end of the Quaternary ice age.

SEE ALSO Snowball Earth? (c. 720–635 Million BCE), Domestication of Animals (c. 30,000 BCE), Invention of Agriculture (c. 10,000 BCE), Industrial Revolution (c. 1830), Discovering Ice Ages (1837), Rising CO_2 (2013)

A model of ice coverage in the northern hemisphere at the peak of the last glacial maximum about 20,000 years ago (top) versus modern ice coverage in the northern hemisphere (bottom).

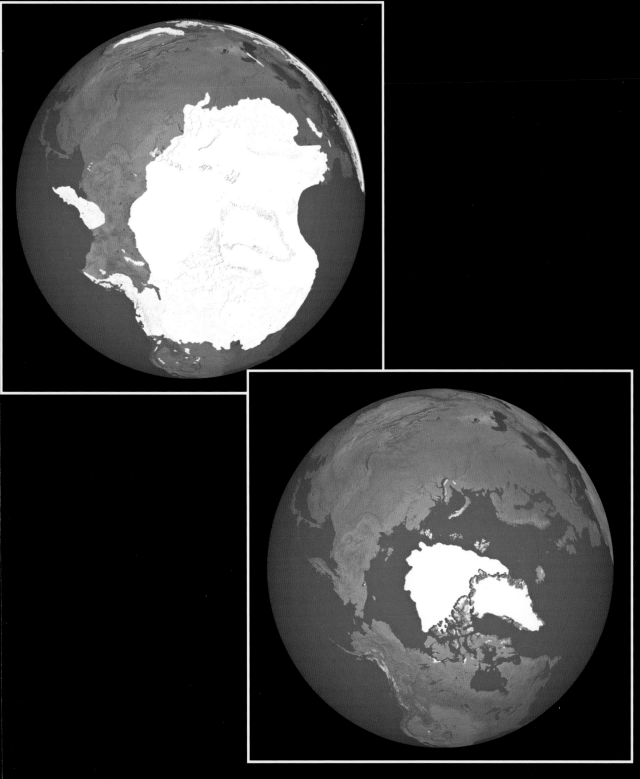

Beringia Land Bridge

While most of the Earth's continental land masses are above sea level, significant fractions of many continents extend below sea level as well, such as the continental "shelf" that extends underwater from the east coast of North America. Not all continental margins exhibit shelves; but where they do occur, they average about 50 miles (80 kilometers) in width, and about 500 feet (152 meters) below sea level.

The shallow nature of continental shelves means that historically, when sea level drops significantly, formerly underwater continental shelves can become dry land above sea level. This is precisely what occurred during the last glacial period around 15,000 to 25,000 years ago along one of the largest continental shelf regions on Earth. The Siberian shelf, in the Arctic Ocean, stretches for more than 900 miles (1,450 kilometers) along the northern margin of the Eurasian continent, where it eventually merges to the east with the Chukchi and Bering shelves along the Eurasian/North American boundary north of Kamchatka and Alaska. During the last glacial maximum period, sea level dropped by more than 160 feet (50 meters) as water accumulated in continental glaciers, exposing the Chukchi and Bering shelves and creating a "land bridge" between Eurasia and North America. Geologists refer to this past land bridge as "Beringia," because it is now underneath the Bering Strait.

Archaeologists broadly agree that modern humans have their roots in Africa and the Middle East, radiating outward from there into Europe and Asia. Prior to the last glacial maximum, however, there would have been no simple way for humans to get to North or South America because of the Atlantic Ocean to the west and the Pacific Ocean and Bering Sea to the east. Once Beringia was exposed, however, it became possible to simply walk from Eurasia to North America. A new era of human migration had begun.

As the continental glaciers began to melt at the beginning of the Holocene epoch, the sea level rose and covered the Beringia land bridge. While it was still possible to cross during iced-over periods, the closing of the Beringia land bridge would ultimately lead to the isolation and genetic divergence of the indigenous Eurasian and American populations.

SEE ALSO *Homo sapiens* Emerges (c. 200,000 BCE), Invention of Agriculture (c. 10,000 BCE), End of the Last "Ice Age" (c. 10,000 BCE)

Prior to about 11,000 years ago, the sea level was low enough that Alaska and Siberia were part of a continuous strip of land that geologists now refer to as "Beringia."

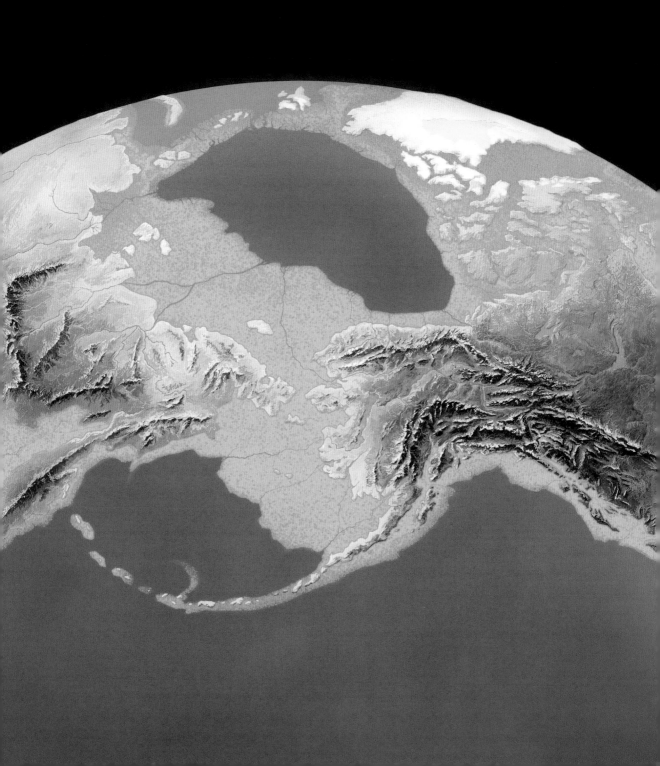

The Great Lakes

Lakes are critical storehouses of most of the world's freshwater supply. The world's single largest freshwater lake, Lake Baikal in Siberia, has about the same surface area as the relatively small U.S. state of Maryland but holds almost 20 percent of the world's freshwater supply by volume. A comparable volume of water is stored within the five large interconnected freshwater lakes in North America known collectively as "The Great Lakes." Lakes Huron, Ontario, Michigan, Erie, and Superior combined are the largest group of freshwater lakes by surface area in the world (comparable in area to the U.S. states of Pennsylvania and Ohio combined), and the second largest group in terms of volume.

Many geologists believe that the Great Lakes formed along zones of continental weakness associated with two ancient rifts in the North American plate; the first was a mid-continental rift that formed around 1 billion years ago, and the second was the St. Lawrence rift that formed around 570 million years ago. Neither rifting episode was extensive enough to split North America into separate continents, but they did create faults and deep valleys that became more susceptible to erosion. Indeed, many of these valleys were widened and deepened by repeated episodes of glaciation during the intervening ice ages. As the most recent continental glaciers began to retreat around 12,000 years ago, their meltwater began to fill these basins. By about 10,000 years ago, the remains of the last major North American glaciation took on the form of the Great Lakes as we see them today.

The Great Lakes are essentially inland seas, with rolling waves, strong currents, and lake-generated winds. They cover a large enough area to have significant local weather and climatic impacts on their surroundings, especially downwind. Among the most famous are winter "lake effect" snowstorms. Cold prevailing westerly winds cross the lakes and pick up warmer evaporating water vapor, and when that vapor passes over colder land to the east, it condenses out as snow, often in concentrated bands or streamers that dump epic amounts of snow (3 to 6 feet, or 1 to 2 meters, per day!) over small areas.

SEE ALSO The Caspian and Black Seas (c. 5.5 Million BCE), The Dead Sea (c. 3 Million BCE), Lake Victoria (c. 400,000 BCE), End of the Last "Ice Age" (c. 10,000 BCE), Beringia Land Bridge (c. 9000 BCE)

NOAA weather satellite mosaic of northeastern America, showing the five Great Lakes (left to right): Lake Superior, Lake Michigan, Lake Huron, Lake Erie, and Lake Ontario.

Fermentation of Beer and Wine

Louis Pasteur (1822–1895)

Fermentation is a natural metabolic process whereby microorganisms convert sugars like glucose into other organic molecules and gases, including alcohol. Fermentation within cells dates back to the age of anaerobic (oxygenless) respiration more than 3 billion years ago, but it also occurs in single-celled eukaryotes and more complex multicellular organisms that have evolved and thrived in Earth's more recent oxygen-rich environment.

It is perhaps not difficult to imagine that some late Neolithic (Stone Age) farmers or others who observed bubbles fizzing out of rotting (fermenting) stores of fruit or grain might have wondered what was going on and investigated further. Apocryphal stories abound in many cultures about the earliest discoveries that alcohol was produced by fermentation, and that it could be generally safe and potentially enjoyable to consume in limited amounts. Myths aside, starting around 9,000 to 10,000 years ago, people began manually controlling the process of fermentation to intentionally produce alcoholic beverages.

Beer, for example, is one of the oldest prepared beverages in human history. Evidence of people fermenting various grains like barley dates back to the earliest beginnings of organized agriculture. Beer was apparently so widely regarded in Mesopotamia and Egypt that it was used as partial payment for workers, including those involved in the building of the great pyramids at Giza. Fermentation of wine appears to have been developed around the same time, with organized wine cultivation developing across a wide swath of the Neolithic world from the Mediterranean to China. The development and advancement of pottery and other liquid-storage technologies around the same time period was critically important in the process of making both beer and wine.

The origin of fermentation in the production of alcohol was not understood until much more recently, however. Pioneering chemists like Louis Pasteur discovered that fermentation is caused by living microorganisms. By studying in detail the process of fermentation occurring in French breweries in the 1860s, he discovered the processes that would eventually lead to the process of pasteurization of milk, wine, and other foods and drinks, significantly extending their usable shelf lives. In the process, he and other late nineteenth-century scientists essentially invented the field of modern biochemistry.

SEE ALSO Life on Earth (c. 3.8 Billion BCE?), Photosynthesis (c. 3.4 Billion BCE), The Great Oxidation (c. 2.5 Billion BCE), Eukaryotes (c. 2 Billion BCE), Invention of Agriculture (c. 10,000 BCE), End of the Last "Ice Age" (c. 10,000 BCE), The Pyramids (c. 2500 BCE)

A pictorial brick depicting the making of wine, from the Eastern Han Dynasty of China, c. 25–220 CE.

Fertilizer

As more sedentary, agriculturally based ways of life began to be established as the climate warmed after the end of the last glacial episode, settlements, towns, and cities began to emerge. An increasing population needed an increasing food supply, which meant that even the earliest farmers had to rather quickly learn how to maximize the efficiency of their crops and land. One important tool in that quest for farm efficiency is the use of fertilizer—any substance that can act to enhance the growth of plants, and especially substances that help to provide plants with consistent levels of nitrogen, phosphorus, and potassium.

While many late Stone Age farmers may have been somewhat nomadic (for example, using slash-and-burn techniques to temporarily plant in one area, then moving on after sowing those crops), others appear to have realized the value of stable long-term farming of specific plots of land, and thus the need to replenish that land with nutrients over time. Indeed, evidence for the use of cattle manure as a fertilizing agent dates back to more than 8,000 years ago, based on chemical and isotopic analysis of preserved Neolithic farm sites. Ample additional archaeological evidence has been found about how the process of enhancing soil fertility using manure or minerals was managed and improved by Egyptian, Babylonian, Roman, and other societies since then.

The scientific study of fertilizers didn't really begin until the nineteenth century, however, when the chemical process of *nitrogen fixation* (pulling nitrogen out of the atmosphere and making it available to plants as ammonia or nitric acid) as a synthetic fertilizer was invented. Today the development and use of commercial fertilizers is a global industry worth tens of billions of dollars, with more than 100 million metric tons of nitrogen-based fertilizers being used annually. By some estimates, between 30 and 50 percent of today's crop yields are only made possible by the use of synthetic fertilizers. The challenge, of course, is balancing the obvious need for fertilizers to support the world's growing population, on the one hand, against the potentially devastating local or regional effects that many synthetic fertilizers have on the environment, including water pollution and soil acidification, on the other. Education and the adoption of sustainable practices will no doubt be the key.

SEE ALSO The Stone Age (c. 3.4 Million BCE TO 3300 BCE), Domestication of Animals (c. 30,000 BCE), End of the Last "Ice Age" (c. 10,000 BCE), Invention of Agriculture (c. 10,000 BCE), Population Growth (1798)

Fields being fertilized near Blythe, California.

The Bronze Age

The Bronze Age is the middle of the three technological and societal periods of prehistoric human history studied by anthropologists. The beginning of the Bronze Age is generally marked by the development of smelting (heating of ores to high-enough temperatures to extract metals), and specifically of the smelting of bronze—a hard metal alloy made mostly (85 percent or more) of copper but also containing smaller amounts of arsenic, tin, aluminum, and other metals. A softer alloy of bronze made using primarily copper and zinc is known as *brass*. Objects made out of copper-tin bronze were harder and more durable than objects made out of stone, copper, or brass, and thus societies that could forge bronze had a distinct technological advantage over their competitors in terms of tools and weapons.

Indeed, some of the world's earliest known dynasties were at least initially based on this technological advantage. The First and Second Dynasties of Egypt, for example, including the unification of Upper and Lower Egypt, roughly coincided with the first use of Bronze Age tools and weapons in that part of the world. Other concurrent examples include bronze-producing empires in southern Europe, Mesopotamia, and China. Trade among these groups helped to strengthen and enhance these empires. Some anthropologists have even hypothesized that the Bronze Age invention of writing, independently in these and other empires around the world, was made possible by the security that bronze weapons brought to scholars and leaders in those societies.

Production of the highest-quality, hardest bronze armor, helmets, weapons, and other implements required not only access to large supplies of copper-bearing ore, but also the separate discovery and smelting of tin-bearing ore. Copper-bearing and tin-bearing rocks are generally not found in the same places, thus making even more important the expanding global trade network in these and other metallic ores.

Today, hundreds of metric tons of copper and tin are extracted from mines globally each year to make bronze. Because of its hardness and durability (it does not "rust" in the traditional sense), bronze is still widely used in a variety of industrial and artistic applications. These include electrical circuitry, ball bearings, ship propellers, musical instruments, mirrors, sculptures, and coins.

SEE ALSO The Stone Age (c. 3.4 Million BCE TO 3300 BCE), The Iron Age (c. 1200–500 BCE)

A collection of middle Bronze Age tools (axes, chisels) and adornments (rings, necklaces, and pendants) found in England and dating back to 1300–1150 BCE.

Synthetic Pigments

A pigment is a material that can, requiring only small quantities, impart significant color changes to other materials. Naturally occurring pigments, like fine powders of iron-oxide–rich soils (ochre) or ground-up black powder from magnetic rocks (magnetite), have been widely used by people since prehistoric times for cave paintings, body art, pottery coloration, and other applications. It was only a matter of time until special "recipes" were developed from mixtures of naturally occurring materials to produce the first synthetic pigments.

Among the oldest synthetic pigments is a material called *Egyptian Blue*, which is a mixture of powdered silica (sand), limestone (carbonate), copper, and potash (an alkaline salt). Cooking this material at a high temperature results in an intensely blue powder with a color similar to that in many rare blue-colored gemstones like turquoise and lapis lazuli. The Egyptians appear to have perfected this recipe for synthetic blue pigment more than 5,000 years ago.

Because of the infrastructure and trade needed to acquire the ingredients and cook the materials, Egyptian Blue was highly prized among the elite in ancient Egypt, and was used widely as a coloring agent on tombs, coffins, and art objects for pharaohs and other members of the royalty. The Egyptians also appear to have created other synthetic materials for coloring or accenting their art and monuments, including the blue-green glassy material known as *faience*, which was widely used to color amulets and other forms of jewelry. Other societies from the Mediterranean to China appear to have created and used synthetic pigments like Egyptian Blue, although archaeologists debate whether these materials were independently discovered elsewhere or simply imported from Egypt.

The production of synthetic pigments and other materials provides evidence for an impressive grasp of chemistry and materials science among many Bronze Age societies. The methods used to make materials like Egyptian Blue and faience, as well as the contemporaneous bronze- and brass-making industries, would presage the development of the technology required to routinely make synthetic glass some 1,700 years later. Today, of course, the synthetic materials and glass industries are multi-billion-dollar global enterprises that have helped to fuel dramatic technological innovations and advances.

SEE ALSO *Homo sapiens Emerges* (c. 200,000 BCE), The First Mines (c. 40,000 BCE), Invention of Agriculture (c. 10,000 BCE), The Bronze Age (c. 3300–1200 BCE), Magnetite (c. 2000 BCE)

Main image: *Synthetic pigments were highly prized for use in royal pottery or artwork, like this hieroglyphic painting on the interior walls of an ancient Egyptian temple in Dendera.*
Inset: *A sample of Egyptian Blue, one of the earliest known human-made synthetic pigments.*

Oldest Living Trees

Based on fossil evidence, trees appear to have first appeared on Earth around the middle of the Devonian geologic time period, about 380 to 400 million years ago. Trees appear to have been part of a vast evolutionary experiment in the way that plants reproduce, as some of the earliest known species of trees, like conifers, were early examples of *gymnosperms*—seed-producing plants that required the physical exchange of pollen with other plants for fertilization. This reproductive path diverged from that of earlier land plants like ferns, which reproduce via spores, and requires the assistance of living pollinators (like insects) and/or natural pollinating processes like wind and fire. Flowering plants are a subset of seed-producing gymnosperm plants known as *angiosperms* that encase their seeds in fruit. During the middle Cretaceous geologic period (around 100 million years ago), tree forms of fruiting flowering plants evolved and eventually—along with the conifers—would cover the continents in vast temperate and tropical forests.

In the absence of other stressing factors, an average typical individual tree lives about 100 to 200 years. Some kinds of trees have been found to be much hardier, however, and can live for thousands of years. These include species like the tough bristlecone pine (*pinus longaeva*), which includes the oldest known single trees (as determined from counting tree rings as well as radiocarbon age dating), specimens in the White Mountains of central California (including one named Methuselah) that have been dated at about 5,000 years old. Other ancient living trees include specimens of giant sequoias, redwoods, juniper, and cypress.

Some trees and even some whole forest stands are part of vast *colonies* of interconnected tree-root systems that represent a single living organism. While the surface expression of these colonial organisms—individual trees—may only live for hundreds or rarely thousands of years, the subsurface root systems of some *clonal tree* colonies have been found to be ancient. Among the oldest yet identified are stands of temperate and boreal oaks, spruce, and pines estimated to be around 10,000 years old, and a remarkable colony of aspens in Utah with roots estimated to be from 80,000 to perhaps as much as an incredible 1 million years old.

SEE ALSO First Land Plants (c. 470 Million BCE), Flowers (c. 130 Million BCE), Deforestation (c. 1855–1870), Radioactivity (1896), Tropical Rain/Cloud Forests (1973), Temperate Rainforests (1976), Boreal Forests (1992), Temperate Deciduous Forests (2011)

Bristlecone pines, like these in Patriarch Grove in the White Mountains of California, are among the oldest living individual trees on Earth.

Stonehenge

While ancient peoples were clearly aware of the sky, it wasn't until the Bronze Age (about 3300 to 1200 BCE) that large-scale, often astronomically themed monuments began to appear. The most famous of these is the prehistoric monument of Stonehenge in southern England, which is just one of many ancient stone circles, burial mounds, and other earthwork structures found around the world with cultural, religious, and/or astronomical importance.

Stonehenge's construction is impressive—especially the 25-ton lintel stones somehow perched atop the 13-foot (4-meter), 50-ton standing post stones. Modern experiments and simulations have shown that building such structures was possible using Neolithic (late Stone Age) and Bronze Age tools and methods—neither magic nor architecturally gifted aliens were required. Still, it must have been near the limit of the available technology to build such previously unprecedented structures.

It also appears to have been an impressive feat of prehistoric astronomy. Detailed examination of the orientations of some of the site's various stones, postholes, pits, paths, and ridges has been interpreted by some archaeologists as evidence that Stonehenge was an astronomical observatory, designed in some ways as a giant sundial to mark the passing of the seasons and to reckon the specific dates of the winter and summer solstices. While the details of the monument's use as an observatory are the subject of active research and debate, there is broad consensus among both archaeologists and astronomers that the basic alignment of the structure was designed with the paths of the Sun and Moon in mind.

Other examples of prehistoric astronomical observatories include the Newgrange and Maeshowe burial mounds in Ireland and Scotland, respectively, aligned so that only the rising winter solstice Sun fills their inner tombs with light; solar-aligned trilithons and passage mounds in Portugal; and the stacked taula stones on the Spanish island of Minorca. The civilizations that built these remarkable monuments perhaps as far back as 5,000 years ago left no written records about themselves or their traditions and beliefs. They left lasting records of stone and earth, however, which reveal how much they must have valued their knowledge of the heavens.

SEE ALSO The Stone Age (c. 3.4 Million BCE TO 3300 BCE), The Bronze Age (c. 3300–1200 BCE), The Pyramids (c. 2500 BCE)

Aerial view of the prehistoric Stonehenge megalithic structure in southern England.

The Spice Trade

During the Bronze Age, a need developed for a diverse variety of natural resources (such as copper, tin, zinc, and iron) from different parts of the world, helping to establish the first truly global system of trade among different societies in Europe, Africa, and Asia. Trade in goods went beyond minerals and ores, however, and especially lucrative was the development of trade networks focused on highly coveted (and highly expensive) luxury goods, such as furs, ebony, pearls, and spices.

Trade among Europe, Africa, and Asia in spices such as cinnamon, ginger, cardamom, turmeric, nutmeg, and pepper, as well as other commodities, followed two distinctly different routes, one primarily overland (the "Silk Road") and the other primarily maritime. The spice trade connected what had been disparate civilizations in the East and West, creating lasting positive and negative political, economic, religious, and cultural interactions and amplifying the strategic importance of key cities and ports along the way, like Xi'an (China), Kerala (India), and Samarkand (Uzbekistan). These and other cities and ports became important hubs not only for traders and businessmen, but also for scholars, artists, and religious leaders. Battles for control of these centers of then-modern civilization were common.

Maritime control of the spice trade offered the possibility of avoiding many of the frequent turf wars and environmental challenges (such as deserts and rugged mountains) of the overland spice trade, and motivated advances in ships and seagoing technologies and significant additional investment in finding new trade routes by governments and private benefactors. Much of the impetus for Europe's so-called "Age of Discovery" (from roughly 1600 to 1900), for example, was to find shorter and safer maritime trade routes from western European ports to the spice islands of Indonesia and other parts of southeast Asia. The Americas got in the way for those sailing west, but the establishment of new spice-trading routes sailing east through the Red Sea or around the Cape of Good Hope helped to increase the economic and political reach of relatively small countries such as England, Portugal, and the Netherlands, and led to an extended and ultimately unsustainable era of colonialism dominated by those and other spice-trading nations.

SEE ALSO Invention of Agriculture (c. 10,000 BCE), Fertilizer (c. 6000 BCE), The Bronze Age (c. 3300–1200 BCE)

Common and exotic spices at a market in Kerala, India.

The Pyramids

The great pyramids of Giza are monuments to the technological prowess of the ancient Egyptian civilization. They are also testaments to the designers' astronomical skill, which figured prominently in Egyptian society and religion 4,500 years ago.

Because the Earth's spin axis slowly precesses (wobbles) like a spinning top, back in 2500 BCE the star Polaris was not the North Star. Indeed, much like the skies near our south celestial pole today, there was no bright star near the north celestial pole in those days. To the pharaohs, astrologers, and commoners, the sky at night appeared to spin around a vortex-like dark hole, thought to be a gateway to the heavens. In ancient Egypt, this gateway was located about 30 degrees above the northern horizon, and so the pyramids were carefully aligned to the north, with a small shaft leading from the pharaoh's main burial chamber to the outside, pointing directly into the center of the polar gateway. If the plan was to join the gods in the afterlife, why not go in through the main door?

Egyptian astrologers also played an important role in developing a rather sophisticated calendar system that was already well established by the time the pyramids were being built. A new year was defined by the first sighting of the brightest star in the sky, Sirius ("Sopdet" to the Egyptians), just before sunrise in midsummer. The year was divided into 12 months of 30 days each, with 5 extra days of worship or parties tacked on to the end for a 365-day year. They also knew from carefully observing and recording star positions on different dates that they needed to add an extra day every fourth year—what we call a leap day—to keep their calendar synced to the motions of the sky. The predawn rising times of a number of bright stars were tracked, in order to determine times for major religious festivals, as well as to plan for the annual floods of the Nile.

The pyramid shape itself may even represent a facet of ancient Egyptian cosmology, as some myths claim that the god of creation, Atum, lived within a pyramid that, along with the land, had emerged from the primordial ocean.

SEE ALSO Stonehenge (c. 3000 BCE), Controlling the Nile (1902)

The great pyramids of Giza, burial places of the pharaohs and astronomical pointers to the presumed gateway to the heavens at the north celestial pole. These were the largest human-made structures in the world for nearly 4,000 years.

Magnetite

Pliny the Elder (23–79)

Legend has it, according to the encyclopedia compiled by first-century Roman naturalist Pliny the Elder, that some 4,000 years ago a shepherd named Magnus from the northern Greek region of Magnesia noticed that, while tending his sheep in certain places, the nails in his shoes and the iron in his staff would mysteriously adhere to the ground. The source of the mystery turned out to be a rock called lodestone, which is dominantly made of a highly magnetic mineral called magnetite.

Magnetite is an iron oxide with the chemical formula Fe_3O_4. It has *ferrimagnetic* properties, meaning that it is attracted to other magnetic materials and it can be magnetized to become a permanent magnet itself. Early sailors and navigators used needle-shaped pieces of lodestone to fashion the first compasses, realizing for example that the needles, placed on a floating piece of straw, would align themselves north-south. Many iron meteorites are magnetic and were prized not only for their iron content but also for their "magical" ability to attract other pieces of iron. Living organisms can also produce magnetite, which appears to be used in some bacteria as well as some animals (including birds) to help navigate by sensing the direction of magnetic north.

Magnetite is a common mineral in typical terrestrial *basaltic* rocks erupted from mid-ocean ridges and mantle hotspots. As molten lava cools, magnetite grains in the basalt align themselves with Earth's magnetic field, recording both the direction and polarity of that field. Modern *paleomagnetic* geologists use this feature to understand the past history of the Earth's magnetic field, and patterns of magnetism observed in seafloor sediments in the 1960s were a key observation that led to the modern theory of plate tectonics.

Magnetite is magnetic because tiny positive and negative *domains* or *magnetic moments* in the Fe_2+ and Fe_3+ atoms that make up the mineral do not entirely cancel each other out, thus leaving a net magnetic force. The presence of such domains, and the fact that they can be flipped 180 degrees by applying the right kind of magnetic field, was used to develop magnetic tape-recording technology based on magnetite. The mineral is still widely used in other manufacturing, medical, and scientific instruments and processes today.

SEE ALSO Plate Tectonics (c. 4–3 Billion BCE?), Banded Iron Formations (c. 3–1.8 Billion BCE), Arizona Impact (c. 50,000 BCE), The First Mines (c. 40,000 BCE), The Bronze Age (c. 3300–1200 BCE), The Iron Age (c. 1200–500 BCE), Reversing Magnetic Polarity (1963), Magnetic Navigation (1975).

A hand sample of lodestone — rock containing the mineral magnetite — with small nails attracted to the surface by the rock's strong magnetism.

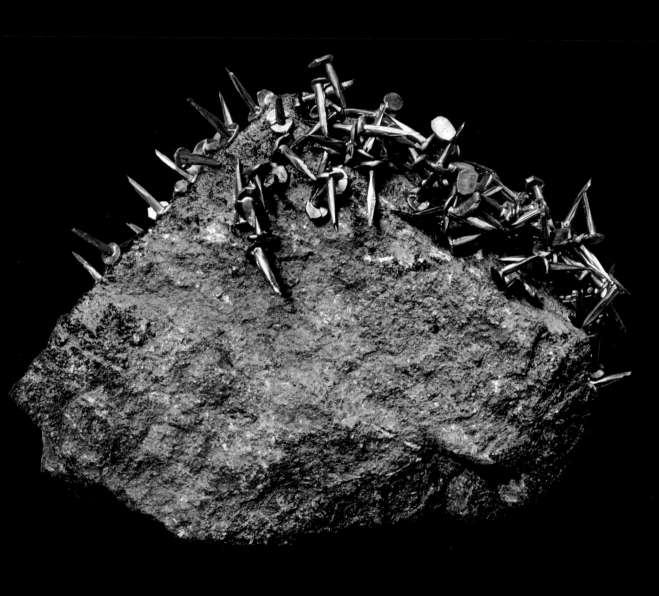

The Iron Age

Anthropologists have labeled the most recent of the three societal and technological periods of prehistoric human history the Iron Age. The transition from the Bronze Age to the Iron Age is generally marked by the production of weapons and tools by ironworking, and most specifically by the widespread use of carbon steel. Smelting of ores to extract iron is much more difficult than extraction of copper and tin, requiring specialized higher-temperature furnaces and precise mixtures of various alloys designed to yield high strength and protection from rust. As such, the invention and widespread dispersal of steel weapons and tools, with superior hardness and durability compared to their bronze equivalents, gave a significant technological advantage to those who could make them.

The earliest forms of ironworking involved smelting of mixtures of iron ore and charcoal in furnaces that used bellows to force air into the mix. At high enough temperatures, the carbon in the charcoal helped to convert the iron oxides in the ore to a mixture of stony minerals (*slag*) and metallic iron, which could be further worked by hand to remove the slag and leave behind the *wrought iron*. Late Bronze Age ironworkers (smiths) discovered that further heating wrought iron on beds of charcoal, followed by rapid cooling or quenching of the melt, yielded an even harder and more durable alloy—steel.

Archaeological evidence supports the earliest occurrences of mass-produced steel tools and weapons around 1200 BCE in the ancient Near East (Mesopotamia, Egypt, Iran, the Arabian Peninsula, and surroundings). From there, steel-smelting technology appears to have spread into the Mediterranean and eventually into Asia and Europe; routine steel production in northern Europe appears to have started sometime around 500 BCE, marking the approximate end of the prehistoric period known as the Iron Age. Not coincidentally, this is around the same time that routine, reliable historic records about individual and global civilizations began to appear.

Significant advances in technology have enabled modern steel alloys to have far superior strength and durability compared to that produced by our ancestors. Indeed, some scholars have cited the proliferation of massive skyscrapers, bridges, and other steel structures since the Industrial Revolution as evidence for a New Iron Age.

SEE ALSO The Stone Age (c. 3.4 Million BCE TO 3300 BCE), The Bronze Age (c. 3300–1200 BCE), Industrial Revolution (c. 1830)

Four Etruscan iron spearheads from around 550 BCE.

Aqueducts

The invention of agriculture and thus of settled, agrarian societies led inexorably to the growth of cities. Large population centers require large supplies of food and water to remain sustainable. As population continues to grow, however, local supplies of these resources might not prove adequate, meaning that food and water would have to be imported to sustain a city.

Bringing water into a city (or farming community) presents special additional engineering and logistical challenges compared to importing food. In many desert or plains environments, it was possible for people to dig canals to divert water from rivers into settlements, even back in prehistoric times. However, in areas of rugged topography, excavating canals might prove impractical or impossible. Hence the idea emerged, perhaps as long as 4,000 years ago according to some archaeological evidence, to connect canals and other waterways by water-bearing bridges—known as *aqueducts*— where the terrain required them. Perhaps the most famous of the early aqueducts known to historians was a 50-mile (80-kilometer) aqueduct made of limestone built by the Assyrian empire in Mesopotamia about 2,800 years ago, which included a water bridge about 30 feet high (10 meters) and about 1,000 feet long (300 meters) across a valley. Ruins of this monumental engineering achievement remain today in the region around Jerwan in northern Iraq.

Centuries later, the Romans would expand on these older designs and deploy a massive system of hundreds of canals and aqueducts to supply water to farms, towns, and cities across the Empire. Rome itself had eleven aqueducts by the third century, helping to sustain a population of over a million people. Some of the imported water fueled an extravagant water-rich lifestyle of fountains and public baths for the wealthy and noble; but just as importantly, some was also used to power water mills, and some to attempt to maintain higher standards of general public health by flushing sewers and drains.

Sections of many Roman aqueducts are still preserved today, and a few are still in partial use. More generally, aqueduct designs and building methods were widely mimicked in Roman and subsequent architecture, and the concept of an elevated water bridge is even in use today in some places where canals and lock systems encounter topographic obstacles.

SEE ALSO Invention of Agriculture (c. 10,000 BCE), Civil Engineering (c. 1500), Population Growth (1798)

A well-preserved ancient Roman aqueduct in Segovia, Spain, thought to date from the late first to early second centuries CE.

First World Maps

Maps have been a critical part of human history for as long as people have wondered "Where are we?" As the earliest settlements, cities, and societies began to develop after the end of the last glacial period, trade between these growing pockets of civilizations became more important and widespread. Especially important was the trading of the materials needed to support new Bronze Age and Iron Age technologies, like the smelting of bronze or steel. It was only natural for traders or sailors to want to know the best way to get their goods to and from the right markets.

Local or regional maps for city planning or to mark short trade routes date back more than 4,500 years, at least. However, the concept of creating an entire map of the known world is much newer. The Babylonians were among the first to attempt to make a global map, around 600 BCE. They depicted their world in a circular form, with the known lands surrounded by water. Their world map is important not only because of its antiquity, but because it also reveals that cartography (mapmaking) can include the intentional or implicit biases of the specific cartographers involved. In this case, the Babylonians knew about their Egyptian and Persian neighbors, but chose to ignore them in their depiction of the world.

More realistic spherical maps of the world would come to rely on knowledge of the size of the planet and the actual distances between different places. The Greek astronomer and mathematician Pythagoras was among the first to make convincing arguments that the Earth was spherical, around 500 BCE, and then some 250 years later the Egyptian astronomer and mathematician Eratosthenes made the first good estimate of the size of our planet (essentially inventing the field of geography). From there, it was up to generations of subsequent explorers, traders, and invaders to measure the actual distances involved, including the details of the shorelines, mountains, rivers, and new lands encountered. By the late eighteenth century, the basic shape of the world's lands and seas had been determined. Today, satellite technology allows us to globally map natural or human-made changes on our dynamic planet down to very small scales.

SEE ALSO Invention of Agriculture (c. 10,000 BCE), End of the Last "Ice Age" (c. 10,000 BCE), The Bronze Age (c. 3300–1200 BCE), The Spice Trade (c. 3000 BCE), The Iron Age (c. 1200–500 BCE), The Earth Is Round! (c. 500 BCE), Size of the Earth (c. 250 BCE), Global Positioning System (1973)

Top: *A more modern map of the known world as of around 1730, by Dutch engraver Daniël Stopendaal.*
Bottom: *A clay tablet with one of the first maps of the world, from sixth-century BCE Babylonia.*

The Earth Is Round!

Pythagoras (c. 570–495 BCE)

We take it for granted: the Earth is a beautiful, blue, spherical marble drifting in the blackness of space. But without the relatively recent benefit of being able to go out into space and look back, someone had to convince the world of the idea that the Earth might be round rather than flat, like it appears to anyone on the ground. By many accounts, that someone was Pythagoras, a sixth-century-BCE philosopher, mathematician, and part-time astronomer from Greece, also famous for his Pythagorean theorem in geometry.

The argument made by Pythagoras and his followers for a spherical Earth was an indirect one, based on a variety of observations. For example, sailors traveling south from Greece reported seeing southern constellations higher in the sky the farther south they went. Expeditions that departed for destinations along the African coast south of the equator, for example, reported that the Sun shone from the north rather than from the south (as it does in Greece). Another important piece of evidence came from observing lunar eclipses: when the full Moon passes directly behind the Earth relative to the Sun, the Earth's curved shadow is clearly visible as it eclipses the Moon.

It is a matter of some debate whether Pythagoras himself actually "discovered" that the Earth is spherical, or whether he was simply the most outspoken (and famous) advocate of what was becoming relatively common wisdom among educated people of the early Greek civilization. Regardless, the issue would be proven in another 250 years or so by the experiments of Eratosthenes; and nearly 2,500 years later, the first astronauts to leave Earth orbit, aboard the Apollo 8 mission, would share with the world their glorious photos of our beautiful, spherical, blue marble floating in space.

SEE ALSO First World Maps (c. 600 BCE), Size of the Earth (c. 250 BCE), Leaving Earth's Gravity (1968)

Above: *One piece of evidence that the Earth is round is the curved shadow of the Earth cast into the Moon during a lunar eclipse, like this one observed from Greece in 2008.*
Main image: *Engraving of the sixth-century BCE Greek philosopher and mathematician Pythagoras, among the earliest scientists to advocate that the Earth is a spherical body.*

PYTHAGORAS.

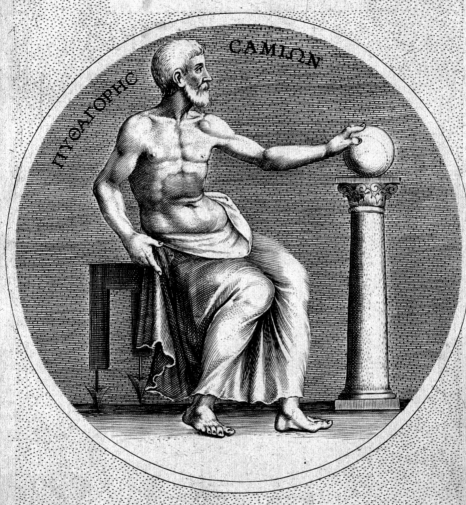

ΠΥΘΑΓΟΡΗΣ CAMIΩN

Apud Fuluium Vrſinum
in nomiſmate ӕreo.

Madagascar

While the specific details on exact dates and patterns of migration are the subject of ongoing archaeological and anthropological study, it's still clear to scholars that prehistoric humans very rapidly spread across the world, especially after the end of the most recent glacial maximum some 12,000 years ago, to fill nearly every habitable niche they could find. Based on evidence from fossils and tools, the last significantly sized landmass to be settled by humans was the large island of Madagascar, off the eastern coast of Africa.

Madagascar is the fourth-largest island in the world, having only a bit less surface area than France. Archaeological evidence indicates that people visited the island as long as 4,000 years ago (early sailors and traders, perhaps). Although the details are still the subject of much study, it appears that people didn't settle the island until much more recently, however, perhaps only around 2,500 years ago. What they found when they got there was a rich diversity of flora and fauna, including numerous large animal species like the hippopotamus, giant lemurs, giant mongoose-like animals known as fossa, and giant flightless birds like the elephant bird.

Madagascar's isolation from the continent of Africa began with the breakup of the Pangean supercontinent around 200 to 150 million years ago, and the island is thought to potentially have separated from India as its own continental microplate around 90 million years ago. Since that time, evolution and frequent tropical environmental conditions helped to create a megafauna population unique to the island. Human settlement changed the ecology of Madagascar dramatically, however. To support agriculture and a growing population (now nearly 25 million people), major fractions of the native forests have been cut down, disrupting animal habitats in the process. Human predation was also a major factor in driving many of the large animal species to extinction, leading to the proliferation of physically smaller species into those niches, including more than 100 known species and subspecies of lemurs.

Today, preserving some of Madagascar's unique biodiversity is a priority for the island's government, and ecotourism is a major economic driver. While extinct species cannot be recovered, efforts to preserve the island's existing unique plant and animal populations are widely supported by conservation organizations around the world.

SEE ALSO The Atlantic Ocean (c. 140 Million BCE), Primates (c. 60 Million BCE), Antarctica (c. 35 Million BCE), East African Rift Zone (c. 30 Million BCE), *Homo sapiens* Emerges (c. 200,000 BCE), The San People (c. 70,000 BCE), End of the Last "Ice Age" (c. 10,000 BCE), The Spice Trade (c. 3000 BCE)

Local people walking on the so-called "Avenue of the Baobab Trees" near Morondava, Madagascar.

Quartz

Theophrastus (c. 371–287 BCE)

Quartz (SiO_2) is one of the most common and most widely recognized minerals on Earth. It is the second-most common mineral in the Earth's continental crust (behind feldspar), and is the most common mineral in sedimentary and metamorphic rocks like granite, sandstone, and shale. Earth's beach sand is almost all quartz, because quartz-rich rocks can be mechanically broken down into smaller and smaller pieces, but quartz itself is difficult chemically to break down and weather away.

Quartz comes in a variety of colors (because of minor contamination by other elements: pure quartz is clear), and a variety of crystalline forms or *habits*. The most common habit for pure quartz crystals is a six-sided prism with pyramidal tips; this and other beautiful gem-like forms are often seen in rocks like geodes, where the crystals can grow unimpeded into empty void spaces. According to historians, some early scientists, such as the Greek naturalist and philosopher Theophrastus, thought that quartz was a mineral version of supercooled ice, partly because of their similarity in crystalline form.

Most quartz is formed within subterranean magma chambers as a result of *fractional crystallization* of cooling magma. As the temperature of the magma drops, minerals with lower silicon content (such as olivine) precipitate out first, followed by those with increasing silicon content. This process results in concentrated quartz-rich melt crystallizing out last, in quartz-rich caps at the tops of magma chambers, or as quartz-rich veins that fill fractures in the previously cooled surrounding rocks, or both. Further compression and heating of quartz results in the production of metamorphic minerals like quartzite and granite, which are highly resistant to erosion and thus often form major parts of exposed, uplifted mountain ranges like the Sierra Nevada in central California.

Quartz is used in many ways. Some gemstones as well as most ancient decorative glassware are primarily made of various kinds of quartz. Industrially, the fact that quartz crystals will vibrate at specific frequencies when subjected to a mechanical stress has made them critical for watches and other kinds of electronics (hence, "Silicon Valley"), where they are used essentially as tiny oscillating clocks. Today, quartz crystals can be synthetically created to a very high level of purity, relegating the quest for pure natural crystals primarily to jewelers and collectors.

SEE ALSO Continental Crust (c. 4 Billion BCE), Plate Tectonics (c. 4–3 Billion BCE?), The Sierra Nevada (c. 155 Million BCE), The First Mines (c. 40,000 BCE), Magnetite (c. 2000 BCE), Feldspar (1747), Olivine (1789)

A spectacular hand-sized specimen of a cluster of quartz crystals from Tibet.

Great Library of Alexandria

The invention of writing around 3000 BCE (in Mesopotamia, but independently at other times by other civilizations around the world) meant that, eventually, historical records could finally be kept of people, places, and events. Indeed, the boundary between "prehistoric" and more modern times is usually defined based on when written records began to be kept. A logical next step after recording written records is, of course, figuring out a place to store them for later access. Hence, the concept of the library was born.

Archaeologists have discovered evidence of the first libraries—collections of clay tablets bearing cuneiform characters—in Sumer, dating back to 2600 BCE. Clay proved to be an excellent material for preserving written records, because it was easy to mix from widely available raw ingredients like mud and limestone, and because it can be made hard and resistant to erosion by firing it in a kiln. Similar kinds of records were preserved on papyrus scrolls in ancient Egypt, which was a more difficult material to create but generally a better medium on which to write.

The most famous library of the ancient world was the Great Library of Alexandria, which was created sometime around 300 BCE as an intended repository for all of the world's scientific, engineering, cultural, and historical knowledge. Books, clay tablets, and scrolls from all over the world were purchased, borrowed, or stolen, and a small army of scribes copied their contents onto papyrus for the library's permanent collection. The library was a potent display of Egypt's power in the world, hosting dignitaries and visiting scholars from around the world. While there is much uncertainty, historians estimate that there may have been between 40,000 and 400,000 scrolls stored in the library in its heyday.

Sadly, most of the scrolls were lost when the library was sacked and burned during various sieges as the Roman Empire conquered Egypt between the first century BCE and the third century CE. This irretrievable loss of knowledge from the ancient world represents one of the greatest scholarly tragedies in all history. While some contents of the library were preserved in copies at other major libraries of the ancient world, we'll never really know what stories, poetry, myths, scientific and engineering insights, and other cultural information was lost when the Great Library of Alexandria was destroyed.

SEE ALSO The Pyramids (c. 2500 BCE), First World Maps (c. 600 BCE), Size of the Earth (c. 250 BCE)

A nineteenth-century engraving depicts an artist's rendering of a hall in the Great Library of Alexandria in ancient Egypt.

A Sun-Centered Cosmos

Aristarchus (c. 310–230 BCE)

The notion of the Earth as the center of the Universe permeated ancient Greek thinking about the cosmos by philosophers and mathematicians like Plato and Aristotle. Everyone could see that the Sun, Moon, and stars rotated around the Earth. Scholars added other supposedly irrefutable evidence: the Moon went through phases consistent with its orbiting our planet. If the Earth spun on its own axis, why was nothing flung off the surface? None of the stars showed any observed parallax, or shifting of position relative to other stars, that they would show if the Earth were moving in its own orbit.

There were doubters and skeptics, however. The earliest one on record was the astronomer and mathematician Aristarchus, from the Greek island of Samos, who made detailed naked-eye observations of the Sun and Moon and tried to interpret them in a geocentric (Earth-centered) context. His methods were limited by the acuity of the human eye, but nonetheless he was able to deduce from geometrical calculations that the Sun was at least 20 times farther away than the Moon (the actual value is 400). He then deduced that, because the Sun and Moon have about the same apparent angular diameter in the sky, the Sun's diameter must be at least 20 times larger than the Moon's diameter and 7 times Earth's diameter. Thus, according to his reasoning, the Sun's volume was more than 300 times the volume of the Earth (the actual value is about a million). It must have seemed foolish to him, then, that such a giant Sun would be indentured to such a relatively tiny planet like the Earth. Naturally, he advanced the idea that the Earth and other planets orbit the Sun and that the stars are so far away that no parallax could be observed. Aristarchus's universe was a much larger universe than anyone had described before.

Like most revolutionary ideas, Aristarchus's idea of a Sun-centered cosmos was met with ridicule by most of his colleagues; 250 years later, the idea was effectively crushed by the geocentric teachings and writings of the Egyptian astronomer Ptolemy during the Roman Empire. Aristarchus had planted a critical seed of doubt, but it would not germinate until the sixteenth century, when Copernicus and Kepler officially put the geocentric model to rest.

SEE ALSO The Earth Is Round! (c. 500 BCE), Size of the Earth (c. 250 BCE), Laws of Planetary Motion (1619), Gravity (1687)

Copy of a section of Aristarchus's original third-century BCE calculations of the relative sizes of the Sun, Earth, and Moon, supporting his then-radical notion of a heliocentric cosmos.

ad M B per-
pēdicularis.
parallela igi-
tur est CM ip-
si LX. est au-
tem & SX pa-
rallela ipsi M
R; ac propte-
rea triangu-
lum LXS si-
mile est trian-
gulo M R C.
ergo vt S X
ad MR, ita S
L ad RC. sed
S X ipsius M
R minor est,
quàm dupla;
quoniā & X
N est minor,
quàm dupla
ipsius MO. er-
go & SL ip-
sius CR mi-
nor erit, quā
dupla: &
R multo mi-
nor, quā du-
pla ipsius R
C. ex quibus
sequitur S C
ipsius CR mi-
norē esse, quā triplā. habebit igitur RC ad CS maio M

Size of the Earth

Plato (c. 427–347 BCE), **Aristotle** (c. 384–322 BCE), **Eratosthenes** (c. 276–195 BCE)

The Greeks had generally accepted the fact that the Earth is round at least as far back as the time of Pythagoras (6th century BCE), but estimates of the actual size of the Earth varied widely. Plato had guessed the Earth's circumference to be around 44,000 miles (70,000 kilometers), corresponding to a diameter of about 14,000 miles (22,000 kilometers), and Archimedes had estimated a circumference of about 34,000 miles (55,000 kilometers) and diameter of 109,000 miles (17,500 kilometers). To make a more accurate determination, Eratosthenes, a mathematician, astronomer, and the third chief librarian of Alexandria, devised a simple experiment that was akin to treating the Earth as a giant sundial.

Eratosthenes had learned that at noon on the summer solstice in the southern Egyptian city of Syene, the Sun was almost exactly overhead (at the zenith), so posts in the ground did not cast any shadows. He also knew that in his own city of Alexandria in the north of Egypt, posts in the ground did cast (small) shadows at noon on the summer solstice. He made some measurements and determined that the Sun was a little over 7 degrees south of the zenith in Alexandria. This corresponds to about 1/50 of the circumference of a circle, so he surmised that the circumference of the Earth was about 50 times the distance between Alexandria and Syene. With a distance of about 5,000 stadia (the stadium was an ancient Egyptian and Greek unit of measure) between Alexandria and Syene, he estimated the circumference of the Earth at about 250,000 stadia. Assuming that 1 stadium was about 175 yards (160 meters), this yields a circumference of about 25,000 miles (40,000 kilometers), which, given the various uncertainties and assumptions involved in the measurements, is essentially the correct answer.

Eratosthenes is widely regarded as the father of geography—indeed, he coined the word. It seems appropriate, then, that he was the first to accurately determine the size of the Earth. His method is also a fabulous example of the power of a simple, well-timed experiment. Archimedes had once quipped about levers: "Give me a place to stand, and I will move the Earth." Eratosthenes could easily have retorted: "Give me a few sticks and some shadows, and I will measure the Earth."

SEE ALSO First World Maps (c. 600 BCE), The Earth Is Round! (c. 500 BCE), Great Library of Alexandria (c. 300 BCE), A Sun-Centered Cosmos (c. 280 BCE).

A 1635 painting by artist Bernardo Strozzi of Eratosthenes (left) teaching a student how to estimate the size of the Earth from the lengths of shadows cast at different places at the same time.

Pompeii

Volcanoes release heat, gases, lava, and/or ash and fundamentally change the local and regional geology all over the world. In some places, like the Mediterranean, the confluence of civilization and plate tectonics has resulted in millions of people living within the eruption zones of active volcanoes. Sometimes the eruptions are relatively mild or remote (as has historically been the case for Mt. Etna on the island of Sicily, for example, or for Santorini, north of Crete). However, in other cases throughout history, major volcanoes have commingled with major population centers.

A famous example is the region around Naples, Italy. During the height of the Roman Empire, hundreds of thousands of people lived on the slopes of—or otherwise near—Mt. Vesuvius, one of the most active volcanoes in Europe. But past recorded Vesuvius eruptions were rare and generally not violent. All that changed in the summer of the year 79, however, when Vesuvius erupted violently, propelling a huge column of gas and ash up to nearly 100,000 feet (30 kilometers), and then raining that hot gas and ash down on surrounding cities and villages. The cities of Pompeii and Herculaneum were burned by hot *pyroclastic surges* from the collapsing cloud, and then buried by tens of meters of hot ash. It is estimated that 20,000 people or more were killed in the 2-day eruption.

Volcanoes continue to pose natural hazards to people all over the world. More than 4 million people now live around Mt. Vesuvius, for example, which erupted violently as recently as 1944. Why would they choose to live near an active volcano? One reason is that volcanic ash can develop into highly fertile soils, a fact exploited by farmers from ancient times to today. Regardless, millions of people are at potential risk around the world, and geologists actively monitor volcanoes and precursor seismic activity as lava or gases begin moving toward the surface, or as mountains start to bulge under the pressure of an impending eruption. In many cases, such as for the eruption of Mt. St. Helens in the U.S. Northwest in 1980, ample advance warning can help to inform evacuations and save lives.

SEE ALSO Plate Tectonics (4–3 Billion BCE?), The Mediterranean Sea (c. 6–5 Million BCE), Huaynaputina Eruption (1600), Krakatoa Eruption (1883), Mount St. Helens Eruption (1980), Mt. Pinatubo Eruption (1991)

Russian artist Karl Briullov's 1833 painting The Last Day of Pompeii *depicts both the geologic and human horrors of the volcano's massive ash eruption in the summer of the year 79.*

Polynesian Diaspora

People have been migrating to new places since prehistoric times, to follow prey, to seek new lands for farming, to escape persecution, and to explore. Whatever the reason, dispersion of populations from their indigenous homelands (known as a *diaspora*) has been common throughout human history. One of the most storied and dramatic migrations has been the dispersal of people from Southeast Asia out to the many hundreds of islands and atolls across the South Pacific, a 5,000-year-long human migration generally known as the Polynesian diaspora.

Archaeological, genetic, cultural, and linguistic clues are used by scientists to try to piece together the series of voyages and circumstances that led to the settlement of the South Pacific. Initial migration appears to have been from the areas around today's Taiwan and Indonesia first into Melanesia (islands immediately north and northeast of Australia), and then into Micronesia (islands north of Melanesia and east of the Philippines), over a few thousand years. Even though there is much debate and uncertainty about the details, over the next few thousand years additional voyages ultimately populated the so-called "Polynesian triangle," from New Zealand to Hawaii to Easter Island, with the farthest east of those settlements established between about 700 and 1200.

Sustained human settlement of numerous small and widely dispersed islands across the South Pacific is a testament to the clearly skilled and experienced shipbuilders, navigators, and sailors of those societies. Many new colonies may have been settled by hundreds of people at a time (instead of just a few boats), suggesting a sophisticated level of planning and logistical coordination of efforts. Trade among the island societies appears to have been vigorous, based on tracing the geochemical origins of archaeological artifacts. Wars were waged, too, within individual island communities and between major island clusters, often precipitated by famines or droughts that disrupted the fragile island ecosystems.

Western exploration and colonial expansion into the South Pacific beginning in the eighteenth century would ultimately dramatically disrupt (or destroy) the traditional kingdoms and many of the other indigenous political structures of the South Pacific islands. Today, descendants of the original Polynesian diaspora voyagers and settlers struggle to maintain their cultural and sociologic heritage, as well as their economic viability, in an ever-globalizing world.

SEE ALSO Beringia Land Bridge (c. 9000 BCE), The Spice Trade (c. 3000 BCE), Madagascar (c. 500 BCE), Transit of Venus (1769)

Drawing from around 1770 of Polynesian double canoes known as tipaerua, *based on encounters recorded during the South Pacific voyages of British explorer Captain James Cook.*

Mayan Astronomy

Prehistoric and Middle-Age (fifth to fifteenth century) astronomy was extensively studied and practiced by European, Arabic, Persian, and Asian scholars. In addition, a rich astronomical tradition had emerged in Mesoamerica going back to at least 2000 BCE in complex and advanced indigenous civilizations like the Maya, Olmec, Toltec, Mississippian, and other related cultures. Few written records remain from these earlier civilizations, however, partly because many were lost or destroyed during later European conquest.

For the Mayan civilization (peaking from c. 2000 BCE to 900 CE), only four surviving books are available to assess the level of scientific knowledge of this once-dominant Mesoamerican culture. One of those books, from the late period of Mayan history shortly before European contact, is called the *Dresden Codex* (after the location where it is currently archived); it provides fascinating and revealing evidence that Mayan astronomy had reached a level of advancement and sophistication comparable to that of the Greeks, Arabs, and other early societies.

The Dresden Codex is part history and part mythology, but it is mostly a series of detailed astronomical tables for charting and predicting the motions of the Sun, Moon, Venus, and the other known planets. After deciphering the glyphs and numeric symbols, archaeoastronomers determined that the 74 pages of illustrated tables track the cycles of Venus (which repeats its pattern of rising and setting every 584 days) and the Moon (857 full Moons repeat every 25,377 days). The tables could also be used to predict eclipses, as the Mayans recognized the various repeating eclipse cycles to a much higher level of precision than their earlier Babylonian and Greek counterparts. They could also apparently predict lunar and planetary alignments or conjunctions with great accuracy. Knowledge of these periodicities in the heavens to such a high level of precision must have required centuries of careful, detailed observations and sophisticated naked-eye instruments. Once the Mayans discovered the cycles, these tables could be used essentially forever to predict the heavens.

What did the Mayans use this information for? Much remains a mystery, but historians have identified many potential religious, agricultural, social, and even military events and traditions tied to their astronomically derived calendar system.

SEE ALSO The Pyramids (c. 2500 BCE), Laws of Planetary Motion (1619)

Part of page 49 of the Dresden Codex, *one of three known surviving books from the Mayans, depicting part of the cycle of appearances and disappearances of Venus and the Moon goddess Ixchel.*

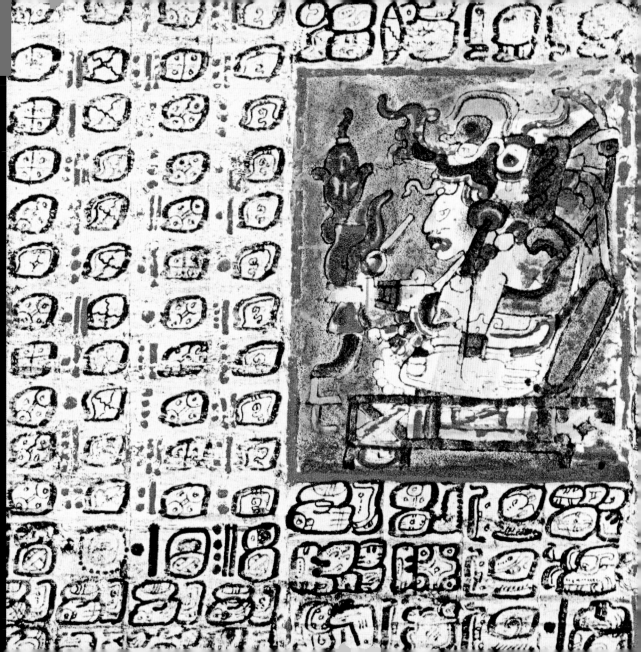

Great Wall of China

Since prehistoric times, humans have been modifying the Earth's surface. Clearing of forests and tilling of soil for farming as well as digging of irrigation ditches and canals are a few pragmatic low-tech examples. Other perhaps less pragmatic but still impressive higher-tech ancient examples include earthen and stone monuments like Stonehenge and the Great Pyramids at Giza, and sophisticated aqueducts in Mesopotamia and the Roman Empire. By far the grandest and arguably most pragmatic example of such large-scale ancient architectural and civil engineering projects, however, is the Great Wall of China, which runs east to west roughly along the border between China and Mongolia for more than 5500 miles (8850 kilometers).

Archaeologists trace the creation of the earliest segments of the Great Wall back to the seventh century BCE. That relatively primitive wall (parts of which were apparently just fortifications made of dirt and rubble) was rebuilt in the third century BCE by China's first unifying emperor, Qin Shi Huang, and essentially the entire wall has been updated and rebuilt multiple times since then. Most of the modern Great Wall of China is from the Ming Dynasty, which ruled from the late fourteenth to mid-seventeenth centuries. Typical segments are more than 25 feet (8 meters) high and 16.5 feet (5 meters) wide, and some sections are built across treacherous mountain terrain with rugged topography. It is truly a remarkable and historic engineering achievement, especially given the limited tools and techniques for civil engineering available at those times for work on that kind of scale.

As the orientation along the Mongolian border suggests, the primary purpose of the wall was defensive—large and well-armed fortifications could help keep various nomadic invaders from the north from getting into the country. However, several key branches of the Wall also served as a series of immigration control and customs/trade control stations along the Silk Road between eastern Europe and the markets of inland and coastal China. Now a major tourist destination for visitors to China, the Great Wall is often credited by historians as being a key part of the strategy to maintain and defend a unified China.

SEE ALSO Stonehenge (c. 3000 BCE), The Spice Trade (c. 3000 BCE), The Pyramids (c. 2500 BCE), Aqueducts (c. 800 BCE), Civil Engineering (c. 1500), Hydroelectric Power (1994)

Part of the Great Wall of China, near Jinshanling, northeast of Beijing.

Native American Creation Stories

We tend to think of cosmology as a modern pursuit—the scientific study of the origin and evolution of the Universe. But people have been thinking about cosmology long before the advent of science, attempting to answer fundamental questions at the heart of existence: Where do we come from? What will the future bring? Are we alone?

Anthropologists have found that many early human societies expressed and shared their concepts of cosmology through song, dance, art, and/or oral creation stories, attempting to make sense of the observed mysteries of life, the Earth, the sky, and everything else that they encountered and to place themselves into a broader context within the Universe. Among the most detailed and revered such stories are those shared by oral tradition among the tribes of various indigenous Native American societies.

In a story from the Iroquois of Upstate New York, for example, the Earth started as entirely ocean, and land was created by frogs and other animals piling mud on the back of a large sea turtle. The Spirit of the Sky World then created the people to live on and care for this land. Plants, animals, and all-powerful spirits permeate many other Native American creation stories, consistent with those societies' general spirit of reverence for the natural world. These kinds of stories were passed down from generation to generation from prehistoric times through the very late 1400s, when European explorers and then settlers began to interact with, and ultimately decimate through disease and war, these indigenous populations.

In a sense, the modern hypothesis that the observable Universe was created at a specific time some 13.8 billion years ago in a gigantic energetic expansion that we call the Big Bang is also a creation story. While we study the origin and subsequent evolution of our Universe using modern technologies and the scientific method, there are aspects to the Big-Bang story that in many ways echo the creation stories of precursor societies. For example, much of what modern science thinks happened during (or even "before") the creation of our Universe could ultimately end up being scientifically untestable, at least with any kind of physics that we currently understand, and thus would ultimately be a matter only of faith and belief.

SEE ALSO The Pyramids (c. 2500 BCE), Mayan Astronomy (c. 1000), Many Earths? (1600)

A modern artist's interpretation of the Iroquois creation myth of the world originating on the back of a large sea turtle.

The Little Ice Age

Weather is the day-to-day and place-to-place variation in temperature, humidity, wind, and other environmental parameters. *Climate*, in contrast, is the long-term average of the weather, typically measured over decades to centuries and potentially described on a global scale. Evidence for climate excursions such as ice ages comes from a variety of sources, including the fossil record, ice cores, tree rings, or—in the last few centuries—records of direct temperature measurements and other meteorological data. Indeed, such direct measurements, supplemented by anecdotal historical accounts, reveal that much of the Northern Hemisphere experienced a significant drop in average annual temperatures, and a significant increase in glaciation, during much of the fourteenth to nineteenth centuries, compared to mid-twentieth century averages. The cause is uncertain, but probably related to slight variations in the Sun's energy output and the cooling effects of atmospheric ash and dust from volcanic eruptions.

This span of time is often informally called "The Little Ice Age." Historical accounts tell of significantly colder winters in Europe and North America during this time period, including unusual (compared to previous centuries) freezing-over of rivers, damage to mountain villages due to advancing glaciers, and widespread closure of ports and normal sailing routes because of extensive sea ice. Climate data spanning this time period show a decrease of only about 0.5°C to 1.0°C in average global temperature—a small change, to be sure, but large enough to cause major climatic, economic, and human impacts. Colder, longer winters led to shorter growing seasons and widespread famine, drought, and loss of life across Europe, for example. European explorers and early settlers reported similar extreme conditions and food shortages in North America as well.

The end of The Little Ice Age in the climate data corresponds approximately (and probably coincidentally) to the beginning of the Industrial Revolution in the mid-nineteenth century. Since that time, global average temperatures have increased by about 1°C (to about 0.5°C *above* mid-twentieth century averages) and mountain glaciers and polar ice sheets have retreated dramatically. As the climate continues to warm over the coming decades and perhaps centuries, we should be prepared for a very different environment than that experienced by our pre–Industrial-Revolution ancestors.

SEE ALSO Snowball Earth? (c. 720–635 Million BCE), End of the Last "Ice Age" (c. 10,000 BCE), Industrial Revolution (c. 1830), Discovering Ice Ages (1837), Rising CO_2 (2013)

Spanish artist Francisco de Goya's The Snowstorm *was painted in the late 1780s, during a colder-than-average span of several centuries informally known as the Little Ice Age.*

Civil Engineering

Leonardo da Vinci (1452–1519)

Engineers are problem-solvers, and thus their expertise has long been coveted within human societies. To solve the kinds of problems that society asks them to solve (in, for example, transportation, irrigation, or the construction of buildings, roads, and bridges), engineers also have to be polymaths, comfortable with many of the basic principles of physics, mathematics, materials science, earth science, and perhaps even project management. There was apparently no shortage of people with such skills in the ancient world, as evidenced by remarkable engineering achievements like Stonehenge, the Egyptian Pyramids, the Parthenon, the Roman aqueduct system, the cities of the Mayan empire, the Great Wall of China, and many others.

One of the most famous engineers in human history was the fifteenth- to sixteenth-century Italian Renaissance polymath Leonardo da Vinci. Widely known not just as an engineer but also as an inventor, artist, mathematician, scientist, historian, architect, and musician, among other vocations, his work as an engineer focused on both military and civilian applications to pragmatic problems, like defending cities, designing new weapons, and building bridges. He was a prolific inventor who also had an imaginative, if not impractical, fascination with machines of all sort, but especially with flying machines. Probably to his frustration, many of the machines that he imagined (like helicopters) were technologically infeasible to manufacture during his lifetime. Like many of the best engineers, his ingenuity outpaced the technology of his time.

Civil engineering has emerged as a critical and highly respected field of expertise for the development of a technologically advanced, sustainable, and global civilization. Like Leonardo, today's engineers must know a little bit of everything but also be able to focus on solving problems in increasingly specialized subfields. In addition to construction and transportation, modern engineers also focus on and aim for dedicated advanced degrees in materials science, thermal control, software, electronics, power systems, environmental issues, forensics, city planning, water-resource management, security, communications, mechanisms, and dozens of other specialized areas. Military engineering overlaps significantly with these fields, but of course includes additional applications related specifically to defense, advanced weaponry, and other related areas. Engineers continue to be highly valued in a society that has no shortage of problems to solve.

SEE ALSO Stonehenge (c. 3000 BCE), The Pyramids (c. 2500 BCE), Aqueducts (c. 800 BCE), Great Wall of China (c. 1370–1640), Hydroelectric Power (1994)

A drawing from around 1480 by Leonardo da Vinci of a design for a lightweight, transportable parabolic swing bridge for military use.

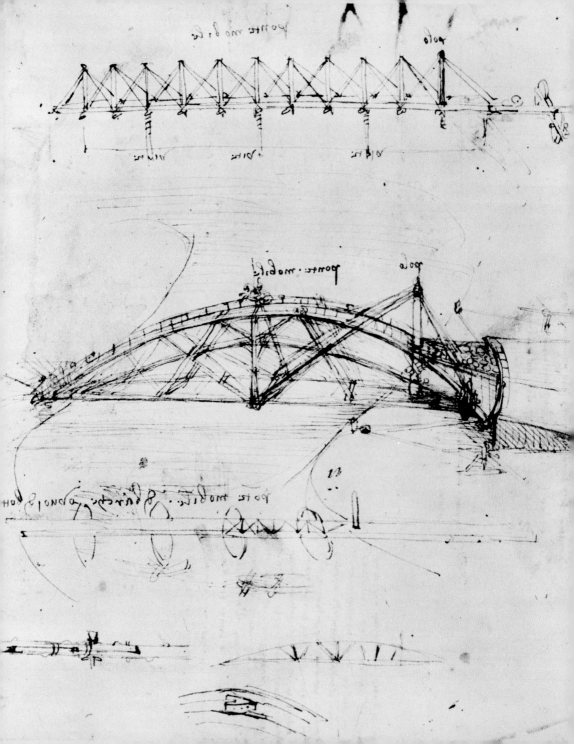

Circumnavigating the Globe

Ferdinand Magellan (1480–1521)

Early sailors and shipbuilders had displayed remarkable prowess in maritime navigation, trade, and warfare dating back to prehistoric times. In many ways, the successes and advancements of knowledge of ancient societies of the Mediterranean, western Europe, Scandinavia, and other regions bordering the sea were based on their ability to transport goods and people so reliably. Many seafarers were also explorers as well, helping to chart and inventory new (to them) lands.

Much of the so-called Golden Age of European Exploration in the late fifteenth through mid-sixteenth centuries was fueled by the desire to find shorter, safer routes to the Spice Islands of Southeast Asia. Columbus's expedition in 1492 set out west from Europe to Asia, but ran into an obstacle called North America. In 1498 and 1502 he sailed farther south, but this time South America, and then Central America, thwarted his efforts to continue west.

It would take until 1520 for a crew to finally make it past these obstacles and become the first Europeans to sail into the South Pacific. That voyage, which departed Spain in 1519, consisted of a fleet of five Spanish ships funded by King Charles V and under the command of Portuguese explorer Ferdinand Magellan. Three of the ships eventually made it into what Magellan called the Peaceful (Pacific) Ocean around the southern tip of South America, two made it across to the Philippines, and eventually just one—*Victoria*—made it all the way back to Spain, in 1522, almost exactly three years after departing.

The voyage was historic and successful for the Spanish crown, but that success came at an enormous cost in human lives. Of the original 270 crewmembers who set out on the voyage, 232 died along the way, including Magellan himself, who was killed in a skirmish with locals in the Philippines. The long distance required to get to Asia by sailing west, including the vast span of the Pacific Ocean, made it clear to subsequent investors and explorers that the preferred route for spice trading was indeed eastward. Europe's westward voyages would soon focus more on exploitation and colonization of the new lands "discovered" by the early explorers, and the riches and glory that they could provide.

SEE ALSO The Spice Trade (c. 3000 BCE), Polynesian Diaspora (c. 700–1200)

A map of the Pacific Ocean from 1589 depicting Ferdinand Magellan's ship Victoria *entering the South Pacific in 1520.*

Amazon River

Rivers are the primary conduits for freshwater—rain and snowmelt—to get from the land to the sea. More than 80 rivers longer than 1000 miles (1600 kilometers) flow across Earth's continents, discharging a combined total of more than a million cubic meters of water per second into the world's oceans. Nearly 20 percent of that discharge comes from just one river, the Amazon in South America. The Amazon is the first- or second-longest river on Earth (depending on where you start counting), but it carries more than five times more water than any other river on the planet.

The first European explorers to sail into the Amazon in the early 1500s discovered millions of indigenous peoples living along the river. The Amazon drains a huge part of South America, with source tributaries (first recorded in 1541 by Spanish explorers) in Brazil, Colombia, Ecuador, Peru, and Bolivia all feeding the main trunk of the river that runs nearly entirely east–west across the South American continent in northern Brazil. Part of the reason that the Amazon carries so much water is because it courses through some of the densest, highest-rainfall tropical forests of the world.

Geologists speculate that back before the formation of the Atlantic Ocean, the South American Amazon and African Congo rivers formed a great drainage basin within the supercontinent of Gondwana, with the Amazon draining to the west and the Congo to the east. After the supercontinent broke up, the Amazon probably continued to drain South America to the west until around 20–10 million years ago, when the Andes Mountains began to be uplifted by the oceanic–continental plate collision along the western margin of the continent. Flow to the ocean was blocked, and an enormous inland sea formed in what is modern-day Brazil. While substantial flooding still occurs in the Amazon basin, changes in climate conditions over time have led to most of that sea being drained through the vast tributary network that is the modern Amazon River.

The Amazon flows through and sustains the largest and biologically most diverse tropical rainforest in the world; indeed, it is estimated that more than half of the world's remaining rainforest is in the Amazon basin.

SEE ALSO The Atlantic Ocean (c. 140 Million BCE), The Andes (c. 10 Million BCE), Deforestation (c. 1855–1870), Tropical Rain/Cloud Forests (1973)

NASA satellite topographic image of the Amazon River and its many branches and tributaries.

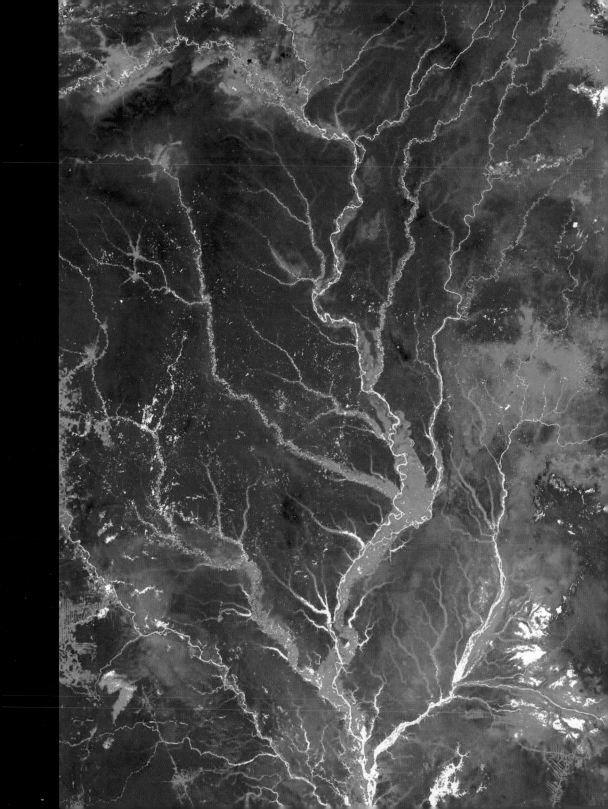

Many Earths?

1600

Giordano Bruno (1548–1600)

The heliocentric (Sun-centered) view of the solar system promoted by pioneering Polish astronomer Nicolaus Copernicus in 1543 was not widely accepted by his sixteenth-century peers. Although the idea that the Earth was not the center of the Universe was inconsistent with the scriptures of the sixteenth-century Roman Catholic Church, Copernicus, a church canon, was, ironically, never the focus of much controversy regarding his views. Others like Galileo would soon inherit that controversy, however.

One of Copernicanism's earliest and most vocal advocates was the late–sixteenth-century Italian philosopher, astronomer, and Dominican friar Giordano Bruno. Bruno appears to have been an outspoken advocate of a number of unorthodox and controversial views about science, religion, and natural philosophy. While not known for any particular observations, skills, or discoveries, Bruno eventually came to believe in a form of non-geocentrism far more extreme than even Copernicus had espoused.

In his 1584 book, *De l'Infinito, Universo e Mondi (On the Infinite Universe and Worlds)*, Bruno postulated that Earth was just one of an infinite number of inhabited planets orbiting an infinite number of stars, which are just suns like our own. To the Church, advocating such a plurality of worlds was mildly heretical; Bruno made it wholesale heresy with other brash demotions of the central tenets of Christian theology, such as the non-centrality of even God in his infinite Universe. He escaped persecution by the Inquisition for more than 15 years but was eventually arrested, tried, convicted, and burned at the stake in Rome in 1600.

It is tempting to romanticize Bruno simply as a scientific martyr, fighting for the truth against a dogmatic regime, especially because some of his ideas about cosmology and the plurality of worlds have turned out to be right. But others before him had held views at odds with the Church, as did others of his contemporaries, without suffering as drastic a fate (most famously, Galileo, who was only placed under house arrest for his heliocentric view of the cosmos). Bruno's demise may not have been so much about his Copernicanism as it was about his confrontational style and his passion for outspoken criticism of authority and the so-called common wisdom.

SEE ALSO Laws of Planetary Motion (1619), Earthlike Exoplanets (1995)

Part of a bronze relief by Italian sculptor Ettore Ferrari (1845–1929) depicting the trial of Giordano Bruno by the Roman Inquisition in 1600.

Huaynaputina Eruption

When it comes to volcanoes, looks can be deceiving. Some of the largest and most destructive volcanoes in geologic history—so-called "supervolcanoes"—reveal themselves today not as tall, iconic volcanic mountains like Vesuvius or the volcanoes of the Cascades Range, but instead as inconspicuous, broad circular depressions. Such eroded craters are all that remain of past supervolcanoes near Yellowstone National Park or Crater Lake in Oregon, although these dormant geologic giants could once again waken in the future.

Truly colossal volcanic eruptions are rare. The most recent isolated extreme eruption anywhere on Earth was the 1991 eruption of Mt. Pinatubo. A larger concentration of geologically recent extreme eruptions has occurred within the Andes Mountains of South America, however, including the eruption of the Huaynaputina volcano in southern Peru in 1600.

Residents in cities near the volcano began reporting feeling earthquakes and seeing small steam eruptions come from the summit starting about four or five days before the actual eruption, which occurred on the morning of February 19. An enormous plume of ash and smoke (now called a Plinian plume, after the chronicler of the Vesuvius eruption's plume in the year 79) jetted high into the stratosphere and influenced the global climate for decades. Soon after the eruption, ash began to blanket the surrounding terrain. The ash and pyroclastic rocks fell into nearby rivers, creating a rapidly flowing slurry of water, mud, and volcanic debris called *lahars* that laid waste to forests, fields, and towns along the riverbanks all the way to the ocean, some 75 miles (120 kilometers) away. Nearly a dozen villages were buried by ash, and an estimated 1500 people were killed during the initial eruption as well as subsequent intermittent eruptions over the next month.

The Huaynaputina eruption is just the most recent of five colossal volcanic eruptions that have occurred in the Andes over the last 700 years or so. Will there be others along that highly active collisional plate boundary? Absolutely. Can scientists provide adequate warnings about them to help save lives and property? Most likely, based on the eyewitness accounts of previous supervolcano eruptions there and elsewhere in the world. But that means continuing to be diligent, and to make volcano monitoring a priority.

SEE ALSO Plate Tectonics (4–3 Billion BCE?), Cascade Volcanoes (c. 30–10 Million BCE), Pompeii (79), Krakatoa Eruption (1883), Mount St. Helens Eruption (1980), Volcanic Explosivity Index (1982), Mount Pinatubo Eruption (1991), Yellowstone Supervolcano (~100,000)

The Huaynaputina volcanic crater in southern Peru, which represents the relatively inconspicuous remains of the largest historical volcanic eruption in South America.

Laws of Planetary Motion

Johannes Kepler (1571–1630)

While there is significant overlap, astronomers today can be generally characterized as either observationalists, those who primarily collect data from telescopes or space missions, or theorists, those who primarily try to develop models or theories to explain existing observations. Most astronomers (and astrologers) from antiquity through the Middle Ages were observationalists who dabbled in theory. Theoretical astronomy had been primarily considered to be the realm of philosophers, not physicists.

The Renaissance German mathematician, astrologer, and astronomer Johannes Kepler changed that paradigm and arguably became the world's first theoretical astrophysicist. Kepler worked with data from others—most notably from the Danish astronomer Tycho Brahe and of course from the famous Galileo Galilei—in his quest to develop a single unifying model of the cosmos. A deeply religious man, Kepler believed that God had designed the Universe in an elegant geometric plan, and that the plan could be reasoned out through careful observations.

Kepler believed in Copernicus's heliocentric (Sun-centered) view of the cosmos, and also believed that a Sun-centered solar system was entirely consistent with biblical writings. Kepler's book *Astronomia Nova* (*New Astronomy*, 1609) described the orbits of Mars and the other planets as elliptical, not circular (his first law of planetary motion), and asserted that the planets change speed in a way that allows them to sweep out equal areas in equal time as they orbit (second law). Later, in *Harmonices Mundi* (*Harmony of the Worlds*, 1619) he showed that a planet's orbital period squared is proportional to its average distance from the Sun cubed ($P^2 \propto a^3$; third law). Through Kepler's patience and persistence, the harmony that he sought among the worlds was finally revealed.

Kepler's laws were not widely appreciated until observationalists had verified their precise timing predictions during rare eclipses and transits of planets across the disk of the Sun (Kepler's predictions were right). Soon, Isaac Newton would find (in 1687) that Kepler had discovered the natural consequences of a universal law of gravitation.

SEE ALSO Gravity (1687), Transit of Venus (1769), Earthlike Exoplanets (1995)

Main image: *Johannes Kepler struggled to find divine perfection in the orbits of the known planets by trying to match them with the shapes of the so-called perfect solids in this illustration from his book* Mysterium Cosmographicum (1596). **Inset:** *A 1610 portrait of Kepler by an unknown artist.*

TABVLA III ORBIVM PLANETARVM DIMENSIONES, ET DISTANTIAS PER QVINQVE
REGVLARIA CORPORA GEOMETRICA EXHIBENS.

miraris opus, SPECTATOR, olympi
quæ nunquam Vita figura tibj
Planetarum distantia quanta sit inter
; Euclidis Corpora quinque docent.
enè conuenit quod dogma COPERNICVS olim
dit. Autoris nunc tibi mon strat opus.
exhibuit tanto se munere gratum
TECCIACO non sine laude DVCJ.

Christophorus Leibfried ff.
Tubing: 1597.

∝. Sphæra ♄.
β. Cubus primum Corpus regulare Geometricum
    distantiam ab orbe ♄ usq ad ♃ exhibens
γ. Sphæra ♃.
δ. Tetraedron siue pyramis ♃ exterius Sphæ-
    ram ♃ attingens, interius ♂ maximam
    inter planetas distantias Causans
ε. Sphæra ♂.
ζ. Dodecaedron, ♀ corpus à Sphæra ♂ usq ad
    Magnum orbem tellurem cum Luna fe-
    rentem repraesentans distantiam
η. Orbe Magno
θ. Icosaedron ab orbe Magno ad Speram ♀ Ve-
    ram distantiam ẏ dicans
ι. Sphæra ♀
κ. Octaedron à Sphæra ♀ ad ☿ orbem exhibens
    distantiam
λ. Sphæra ☿.
μ. Sol Medium siue Centrum Vniuersj
    immobile.

Ponatur tabula ad
    pagin. 26.

Foundations of Geology

Nicolas Steno (1638–1686)

To a geologist, layered rocks are like the pages of a book, just waiting to be read. But many of the basic principles underlying the scientific study of layered rocks—the field of study known as *stratigraphy*—had to be painstakingly developed and advocated by early practitioners of geology. Among the first to formalize what have essentially become the bedrock principles underlying all of modern geology was the Danish scientist (and, later, bishop) Nicolas Steno.

Steno was trained in anatomy and medical sciences, and was a consummate observer of features and structures in biological specimens as well as the natural world. After dissecting the head of a shark, he noted striking similarities between its teeth and certain shark-tooth–shaped objects embedded within rocky geologic formations. He soon became a leading advocate of the idea that these and other bone-like structures embedded in rocks were the fossilized remains of once-living organisms.

Steno's observations of geologic formations took him further, and in 1669 he developed and published a book called *Dissertationis Prodromus* that laid out what are still the fundamental principles of stratigraphy today: the Law of Superposition (in layered rocks, younger strata are above older strata), the Principle of Original Horizontality (strata in layered rocks were originally formed horizontally), the Principle of Lateral Continuity (layers are continuous across large distances), and the Principle of Cross-Cutting Relationships (layers that cut across or intrude into other layers must be younger). These principles are still taught in introductory geology classes today.

Steno's stratigraphic principles might seem relatively obvious to us (or to beginning geology students) today, but they were both innovative and controversial in his time. The concept of coherent solid objects like fossils occurring deep within other coherent solid objects such as layered rocks was difficult to reconcile with the classical, Aristotelian view of these features merely being an inherent characteristic of the Earth. Steno's principles of stratigraphy also hinted at the potential for truly ancient ages of rock formations (indeed, of the entire Earth) and enormous spans of time across geologic layers. However, neither the technology nor the social and religious climate would exist to realize that potential for several more centuries.

SEE ALSO Unconformities (1788), The Age of the Earth (1862), Radioactivity (1896)

Main image: *Spectacular folded, layered rocks in the Hajar Mountains, Oman.*
Inset: *Cover of* Dissertationis Prodromus *(1669) by Nicolas Steno.*

NICOLAI STENONIS
DE SOLIDO
INTRA SOLIDVM NATVRALITER CONTENTO
DISSERTATIONIS PRODROMVS.
A D
SERENISSIMVM
FERDINANDVM II.
MAGNVM ETRVRIÆ DVCEM.

FLORENTIÆ
Ex Typographia sub signo STELLÆ MDCLXIX.
SVPERIORVM PERMISSV.

Tides

Isaac Newton (1643–1727)

Coastal communities and seafaring civilizations have been tuned in to the twice-daily rise and fall of the sea—the tides—throughout human history. Babylonian and Greek astronomers recognized the relationship of the height of the tides with the position of the Moon in its orbit, and thought they were connected through the same almost-spiritual forces that governed the motions of the planets. Early Arabic astronomers thought the tides were mediated by sea-temperature changes. Seeking support for a Sun-centered view of the cosmos, Galileo proposed that the tides come from the sloshing of the oceans during the Earth's motion around the Sun.

The first person to propose the correct origin of the tides—linking the Earth, Moon, and Sun—was the English mathematician, physicist, and astronomer Isaac Newton. Among other things, Newton had been working on a generalized theory to explain Kepler's Laws of Planetary Motion, and by 1686 he had developed the basic outline for new theories of universal gravitation and general laws of motion. Newton hypothesized that both the Moon and the Sun exert a strong gravitational attraction to the Earth (and vice versa). His breakthrough discovery, since verified and enhanced by space-age observations, was that gravitational forces alone (not the Earth's spin or orbital motion), acting on the Earth's thin fluid ocean "shell," are almost entirely responsible for the tides.

The Moon's gravitational attraction raises a deep-water ocean tide of about 20 inches (50 centimeters); the Sun's tidal effect is about half that. Tidal heights can be up to ten times larger in shallow water, and the tides at any particular coastal location are a strong function of the positions of the Sun and Moon, as well as of the loc al seafloor depth and coastline shape. The solid parts of the Earth and Moon bulge in response to gravitational tidal attraction as well, with an amplitude typically about half that of the oceanic tides. Solid and liquid deformation dissipate energy in the Earth–Moon system through what is called tidal friction. As a result, the Earth's spin is slowing down by a few milliseconds per century, and the Moon is slowly receding from the Earth by about 13 feet (4 meters) per century.

SEE ALSO Laws of Planetary Motion (1619), Gravity (1687), Earth's Spin Slows Over Time (1999)

The Moon, Earth, and Sun are all connected gravitationally, a link made by Sir Isaac Newton. Each exerts strong gravitational attractive forces on the others in proportion to their masses and inversely proportional to the square of their distances. The fluid nature of Earth's oceans allows these forces to be expressed as tides.

Gravity

Isaac Newton (1643–1727)

The scientific revolution begun by Aristarchus, who proposed removing the Earth from its cosmological position of centrality, was carried on for 2,000 years by other like-minded scientific rebels such as the Indian mathematicians Aryabhata and Nilakantha Somayaji, the Persian scholar Al-Biruni, and the European scholars Copernicus, Tycho, Kepler, and Galileo. This revolution culminated—decisively—in the work of the Englishman Isaac Newton. Newton was a mathematician, physicist, astronomer, philosopher, and theologian, and is widely regarded as one of the most influential scientists in all of human history.

Newton developed new concepts and tools in optics, including the first astronomical telescope to use mirrors instead of lenses, a design that bears his name. In the theoretical realm—using basic principles of then-modern physics and essentially inventing the new mathematical field of calculus as he went along—Newton discovered that Kepler's

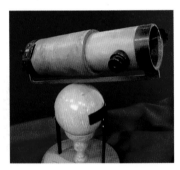

Laws of Planetary Motion were the natural consequence of a force that exists between any two masses and that decreases as the square of the distance between them (known as $1/r^2$ behavior). He called this force *gravitas* (Latin for "weight"). We now call it gravity, and the $1/r^2$ behavior is known as Newton's Law of Universal Gravity.

Newton built on that foundation to derive his three famous laws of motion: (1) Bodies at rest or in motion remain at rest or in motion unless acted on by an external force; (2) A body of mass (m) subjected to a force (F) will accelerate (a) at a rate according to $F = ma$; and (3) The mutual forces of action and reaction between two bodies are equal and opposite. Newton published these transformational theories in 1687 in a book called *Philosophiae Naturalis Principia Mathematica*, known and revered today simply as the *Principia*. Newton's laws of gravity and motion destroyed any remaining shreds of geocentrism and were the definitive solutions to planetary orbits for more than 200 years, until Albert Einstein showed them to be a subset of an even bigger theory known as General Relativity. In one of the most famously quoted examples of scientific humility, Newton once wrote: "If I have seen further, it is by standing on the shoulders of giants."

SEE ALSO A Sun-Centered Cosmos (c. 280 BCE), Laws of Planetary Motion (1619), Tides (1686), Proof that the Earth Spins (1851), Leaving Earth's Gravity (1968), Earth's Spin Slows Over Time (1999)

Above: *A replica of one of Isaac Newton's reflecting telescopes, built in 1672.*
Main image: *A vintage engraving of Newton from 1856.*

Feldspar

Johan Gottschalk Wallerius (1709–1785)

The surface of the Earth consists of about 60 percent oceanic crust and 40 percent continental crust (including the shallow continental shelves just offshore of the main continents). The oceanic crust is primarily composed of *mafic* (which means magnesium- and iron-rich) basaltic minerals erupted from the upper mantle via mid-ocean ridge or hotspot volcanoes. In contrast, continental crust is composed of less dense, higher silicon content *felsic* minerals. Felsic means "feldspar-rich," because the dominant kind of mineral within the assemblage of rocks that make up continental crust is called feldspar.

Feldspars are a class of "framework silicate" minerals built around tetrahedra of SiO_2. Besides about 50 percent silicon and oxygen, feldspars mostly also consist of calcium, potassium, sodium, and aluminum. With little to no iron or magnesium, feldspars have lower density than basalts, which is a primary reason why oceanic crust sinks (subducts) under continental crust when the two rock types collide at plate boundaries. Feldspars lack significant iron and magnesium because they form during the crystallization of cooling subsurface magma chambers *after* higher-density mafic minerals such as olivine and pyroxene have crystallized out. With heavy minerals sinking to the bottom of the magma chamber, lower-density minerals like feldspars (and, eventually, quartz) accumulate at the top as well as within the fractures of surrounding higher-density rocks.

Swedish chemist and mineralogist Johan Gottschalk Wallerius studied this mineral class extensively and gave feldspar its name (which is a contraction of the German words for "field" and "ore-free rock") in 1747. Feldspars occur in several major subgroups, including *the alkali feldspars* (dominated by potassium and sodium) and the *plagioclase feldspars* (containing sodium and various degrees of calcium). Many feldspar minerals are economically valuable because of their aluminum, potassium, and sodium content, and feldspars are important raw ingredients in ceramics and glassmaking.

When exposed to water and humidity at the Earth's surface or in the shallow subsurface, feldspars weather chemically into clay minerals, which can form impermeable layers in the crust that help to confine surface drainage and subsurface aquifers. Isotopes of potassium and other elements in clays or unweathered feldspars are also important for radioactively dating sedimentary layers and fossilized organisms.

SEE ALSO Continental Crust (c. 4 Billion BCE), Plate Tectonics (c. 4–3 Billion BCE?), Radioactivity (1896)

A piece of granite from Victoria, Australia, dominated by alkali and plagioclase feldspar minerals (pinkish-red and grayish in color, respectively), and also containing quartz (shiny) and mica (black).

Transit of Venus

Captain James Cook (1728–1779)

The exploration of the so-called "new world" by European explorers in the sixteenth and seventeenth centuries was motivated by economics (the desire to find shorter and safer routes to the profitable Spice Islands in southeast Asia) as well as national glory (the conquest and settlement of new lands and the acquisition of new riches). It wasn't until the eighteenth century that the desire to advance science or to learn more about the natural world became part of the motivation for voyages of exploration.

Among the first such voyages was the expedition of the British ship HMS *Endeavour*, commanded by Captain James Cook, to explore the Pacific and observe a rare transit (crossing) of the planet Venus across the disk of the Sun for the British Royal Society, one of the world's oldest scientific organizations. Cook was sent to observe the 1769 transit from Tahiti, half a world away from Europe, so that the difference in perspective (parallax) and timing from such distant observations could help to determine the Astronomical Unit—the average distance between the Earth and the Sun.

Transits of Venus are rare, occurring only about once per century in pairs about eight years apart. Prior to Cook's voyage, the previous pair of transits had occurred in 1631 and 1639, and only the latter was observed—and that only from a single site in England using a fairly crude early telescope. It was thought that Cook's observations, and those of others using more modern instruments back in Europe and elsewhere, would significantly advance knowledge of the scale of the solar system as well as the size and nature of Venus.

The coordinated observations of the Venus transit of 1769 represented the world's first large-scale international scientific collaboration. Despite ongoing wars and the hardships of long-distance travel, more than 120 observers at more than sixty separate observing sites around the world participated. The collective results of those observations established the Astronomical Unit to within about 1 percent of its modern accepted value (making this an amazingly accurate observation), and led to the widespread hypothesis that Venus might have a thick atmosphere like the Earth's.

SEE ALSO The Spice Trade (c. 3000 BCE), First World Maps (c. 600 BCE), Circumnavigating the Globe (1519), Charting North America (1804), Natural Selection (1858–1859)

Main image: *A 1776 portrait of Captain James Cook by British artist Nathaniel Dance-Holland.*
Inset: *Time-lapse image from NASA's Solar Dynamics Observatory satellite capturing the June 5, 2012 transit of Venus.*

Unconformities

James Hutton (1726–1797)

Beginning in the late seventeenth century, pioneering geologists like Nicolas Steno and others noticed that, in some places, the layers in rocks are not parallel, instead abutting each other at an angle. Steno and others initially interpreted these formations as having formed that way deep within the Earth.

Scottish geologist and naturalist James Hutton wasn't convinced that such sharp and angled changes in rock layers were so-called "primary" features, formed that way within the earth. He set out to examine a variety of these features across the countryside and came to a remarkable conclusion: sudden changes in the angles or other features among layered rocks represent potentially large gaps in time during which the lower (older) layers were being eroded before the upper (younger) layers began to be deposited. These visible gaps in time in the geologic record have come to be known as *unconformities*.

The specific examples first studied by Hutton are called angular unconformities because the older and younger layers are not parallel. Hutton was among the first to realize that beds of layered rocks, originally deposited horizontally, could later be uplifted, tilted, and eroded over vast spans of time before new horizontal layers would start to form. Further, he realized that such sequences of sedimentary layers appear to have occurred in cycles over geologic time, likely representing numerous advances and retreats of ancient seas and oceans. He published these ideas in 1788 in his book *Theory of the Earth*, which was controversial but spurred significant additional research and study among geologists, naturalists, and philosophers of the time. Hutton introduced the concept of "deep time"—millions or even billions of years of ancient history—into the common lingo of earth science. His ideas, along with those of the nineteenth century Scottish geologist Charles Lyell, would form the foundation of the concept of uniformitarianism in the Earth's history. That is, processes working to change the Earth's surface today are the same as those at work long ago. The present is the key to the past.

SEE ALSO Foundations of Geology (1669), Modern Geologic Maps (1815), Uniformitarianism (1830), Basin and Range (1982).

Main image: *Layered rocks at Siccar Point, Scotland, an example of one of the earliest-recognized unconformities in the geologic record.*
Inset: *A sketch from a 1920 geology textbook of Scottish geologist James Hutton.*

Olivine

The most common elements that make up the Earth are silicon and oxygen, which together make up more than 75 percent of our planet. It is not surprising, then, that the most common class of minerals that make up the Earth are the silicates, and especially the "framework silicate" minerals built around tetrahedra of SiO_2. The Earth's crust consists of lower-density quartz as well as potassium, calcium, aluminum, and/or sodium-bearing feldspar, which were more buoyant than denser minerals during our planet's early "magma ocean" phase, and which have been subsequently enriched in near-surface magma chambers during the formation of continental crust. In contrast, the Earth's mantle consists of denser silicates, dominated by magnesium and iron-bearing minerals such as olivine and pyroxene.

Olivine is a magnesium–iron silicate named by mineralogists in 1789 for its olive-green color; it is the most abundant mineral in the Earth's mantle. Geologists call olivine a "solid solution" mineral because its chemistry can vary from one extreme containing only magnesium (called forsterite) to the other extreme containing only iron (called fayalite), and can include any relative combination of magnesium and iron in between. Olivine changes its crystalline structure under high-pressure conditions like those deep within the Earth's mantle. These different phases of olivine are called *polymorphs* because they have the same chemical formula but different crystalline structures. Geophysicists can determine the depths in the mantle where these different phases occur, and thus in the structure of the mantle itself, by monitoring changes in the velocities of seismic waves from earthquakes.

Because the mantle represents most (85 percent) of the volume of the Earth, olivine is the most abundant mineral on (or in) our planet. Olivine can also be found on the Earth's surface, for example in places where volcanic rocks from the deep crust or upper mantle have erupted. However, it is very susceptible to chemical weathering, and quickly transforms to clay minerals and iron oxides in the presence of water. Olivine has also been discovered within rocks from the Moon and Mars, as well as within a class of meteorites known as pallasites, which also contain high abundances of iron-nickel minerals and are thought to come from the regions near the core–mantle boundary of early-formed protoplanets that were smashed to pieces in collisions during the solar system's violent early history.

SEE ALSO Earth's Mantle and Magma Ocean (c. 4.5 Billion BCE), Continental Crust (c. 4 Billion BCE), Magnetite (c. 2000 BCE), Quartz (c. 300 BCE), Feldspar (1747)

Green sand—grains of the mineral olivine—on a volcanic beach on the island of Hawaii.

Desalination

Thomas Jefferson (1743–1826)

Fresh (non-salty) water represents less than 3 percent of the world's water supply, and more than two thirds of that is locked up in glaciers, polar caps, and deep crustal groundwater. This leaves only a very tiny fraction of Earth's water (mostly in lakes, rivers, and shallow underground aquifers) potentially available for plants, animals, and human consumption and farming. With a growing global population and an increasing fraction of that small supply suffering from pollution and/or dwindling from overuse, the world is currently in a freshwater crisis. The United Nations estimates, for example, that 14 percent of the world's population—more than a billion people—will encounter water scarcity by 2025.

Beyond recycling and conservation, another part of the solution to the world's freshwater problems could be the removal of salt—*desalination*—from seawater, which is abundant on our planet. Desalination is an ancient concept and practice. The ancient Greeks, for example, knew that boiled seawater would condense as freshwater. Sailors in the Roman Empire would boil seawater onboard and capture and drink the condensed steam using sponges. Thomas Jefferson, George Washington's secretary of state in 1791, researched and experimented on a variety of ways for sailors and other "sea-faring citizens" to desalinate seawater for drinking, and arranged for descriptions of the then-known desalination methods to be printed on the backs of all permits issued to ships departing from US ports.

Historical methods of desalination work reasonably well for small groups of people with modest freshwater needs, like sailors on a ship. They become problematic in terms of energy consumption and infrastructure, however, when the demand for freshwater becomes large, like that needed to sustain a farm or a city. Nonetheless, significant advances in desalination technology and efficiency have been made, especially in the last few decades, and especially in countries in arid regions such as the Middle East and Australia. These include methods such as reverse osmosis filtration, which uses electrically charged membranes without needing to heat/boil seawater—now the most commonly used and economical desalination method. Use of solar, wind, and even wave energy for electricity generation in those kinds of facilities is helping to make them less polluting and even more sustainable and economical.

SEE ALSO The Mediterranean Sea (c. 6–5 Million BCE), The Dead Sea (c. 3 Million BCE), Civil Engineering (c. 1500)

1964 photo of the view of the Zarchin desalination plant on the Red Sea in Eilat, Israel.

Rocks from Space

Ernst Chladni (1756–1827), Jean-Baptiste Biot (1774–1862)

We take it for granted nowadays that rocks sometimes fall from the sky; but for much of human history, such an idea was regarded as simply crazy. Many ancient and indigenous cultures were aware of special stones with unique magnetic properties or high concentrations of iron metal, but the fact that these were extraterrestrial samples coming from asteroids from near-Earth space—and sometimes from as far as the Main Belt between Mars and Jupiter—was not deduced until the late eighteenth and early nineteenth centuries.

The German physicist Ernst Chladni proposed in 1794 that some of these special rocks, including a large metal-rich sample that he studied called the Pallas Iron, found in 1772 near the Russian city of Krasnoyarsk, were debris from outer space. The idea was dubbed preposterous, as many scientists believed these rocks to be volcanic or produced by lightning strikes. By the early nineteenth century, it was possible to make detailed laboratory measurements of these rocks. In 1803, the French physicist and mathematician Jean-Baptiste Biot proved Chladni's hypothesis by showing that the chemical compositions of rocks found by the thousands near the French town of L'Aigle shortly after a spectacular shower of shooting stars were unlike any known Earth rocks. Such rocks became known as meteorites, and the scientific field of meteoritics was born.

Scientists have now collected more than 40,000 meteorites from around the world—many from desolate deserts or Antarctic snowfields, where it is relatively easy to notice an odd rocky interloper that fell from above. The vast majority of rocks that fall to Earth (86 percent) are made of simple silicate minerals and tiny spherical grains called chondrules, thought to be some of the first materials condensed from the solar nebula and to be the building blocks of asteroids and eventually planets. About 8 percent more are silicates without chondrules—samples of igneous rocks from formerly geologically active crusts of large asteroids, the Moon, and Mars—and only about 5 percent are made of iron and iron-nickel (like the rocks originally studied by Chladni and Biot) and are pieces of the cores of ancient now-shattered asteroids and planetesimals that had grown large enough to differentiate into core, mantle, and crust before being destroyed by impacts in the solar system's violent early history.

SEE ALSO Arizona Impact (c. 50,000 BCE), Olivine (1789), The US Geological Survey (1879), Hunting for Meteorites (1906), Understanding Impact Craters (1960), Meteorites and Life (1970)

A small (14-ounce [408-gram]) ordinary chondrite meteorite that was found in 2008 on the hard pebbly surface of the Rub 'al-Khali desert near Ash-Sharqı̄yah, Saudi Arabia. The black surface is a thin fusion crust formed during the rock's short, fiery passage through Earth's atmosphere.

Population Growth

Thomas Robert Malthus (1766–1834)

The invention of agriculture after the end of the last glacial period some 12,000 years ago spurred the development and expansion of cities and a slow growth of the population (from a post-glacial estimate of about 1 to 10 million people worldwide at that time). The scientific and technological advances of the Enlightenment helped the world's population exceed 1 billion people by around 1800. Since then, the rapid expansion of agriculture and transportation infrastructure associated with the Industrial Revolution, as well as the rapidly accelerating pace of technological (especially medical) advances, have further spurred dramatic population growth. The world's population was about 1.8 billion by 1900, but it then grew rapidly to more than 6 billion by 2000 and is expected to surpass 7.5 billion people by 2020.

The "Malthusian trap," named after English political-science scholar Thomas Robert Malthus, suggests that advancements in the standard of living due to scientific or technological innovation that lead to increased food production will essentially be cancelled out by concurrently large increases in population. Malthus focused much of his research on the economics and demographics of population growth. The "trap" that he noted in his famous 1798 essay on population growth, *An Essay on the Principle of Population*, was that population grows dramatically faster than food supply, causing the poor to suffer disproportionately from famine and disease, and increasing the pressure on societies to further increase food production, reinforcing the cycle. Malthus was a pessimist about the future of human society and human happiness in the face of continued scientific and technological advancement—in stark contrast to many of the Utopian writers and philosophers of his day, who saw those advancements as ultimately improving the human condition.

Malthus's views were controversial in his day, and they remain controversial today. The following are just a few of the hot topics that are under severe scrutiny around the world today: Must human societies inevitably grow during times of abundance, as Malthus hypothesized? Is it the responsibility of individuals to help keep population growth in check? Should governments incentivize (or force) limits on family size? These and other related questions will continue to be debated as the world's population potentially swells to 10 billion people in the next few decades.

SEE ALSO Invention of Agriculture (c. 10,000 BCE), Industrial Revolution (c. 1830).

Main image: *A crowded subway station in Brazil, 2017.* Inset: *1834 engraved portrait of Thomas Robert Malthus.*

Platinum Group Metals

Precious metals are those that occur relatively rarely in nature and thus are the most expensive to find, mine, and extract from their associated ores. The precious metals that immediately come to mind include gold, silver, and copper because of their widespread use in coinage, jewelry, art, architecture, and so forth. However, there are many other precious metals on the periodic table. Among the most useful but least well known are the so-called platinum group metals—six elements that comprise platinum plus rhodium, palladium, osmium, iridium, and ruthenium. Platinum had been known since antiquity; rhodium, palladium, osmium, and iridium were discovered between 1802 and 1805, and ruthenium in 1844. These six elements have similar physical and chemical properties and are often found together in the same ore deposits.

Platinum group metals are among the most important precious metals used industrially because of some of their specific chemical properties. For example, platinum and rhodium can be used to oxidize ammonia, producing nitric oxide for fertilizer as a byproduct. Another important use of platinum and its related metals is as a catalyst in the reduction of fumes or emissions from the burning of fossil fuels. Most modern cars and trucks, for example, use catalytic converters containing platinum and other platinum group metals to convert toxic hydrocarbon exhaust fumes to less-toxic phases such as carbon dioxide and water.

Like most metals, platinum group metals are dense, which is why they are so rare in the Earth's crust. When Earth *differentiated* into core, mantle, and crust, heavier elements such as iron, nickel, and the platinum group metals preferentially sank into the core and lower mantle, compared to the lighter elements. Upwelling mantle plumes and deep-seated volcanism can dredge up these heavier elements closer to the surface, where they can be extracted and mined from metal-rich ore bodies. Another source of crustal enrichment of platinum group and other metals is the impact of metallic asteroids—cores of ancient, disrupted protoplanets. Indeed, a thin global layer of 65-million-year-old sediments enriched in the platinum-group element iridium was one of the key discoveries in the hypothesis that the impact of a large metallic asteroid had ended the Cretaceous period and the age of the dinosaurs.

SEE ALSO Earth's Core Forms (c. 4.54 Billion BCE), Earth's Mantle and Magma Ocean (c. 4.5 Billion BCE), Dinosaur-Killing Impact (c. 65 Million BCE), Fertilizer (c. 6000 BCE), Rocks from Space (1794)

Laboratory-grown crystals of pure platinum, about 1 inch (2.5 cm) across.

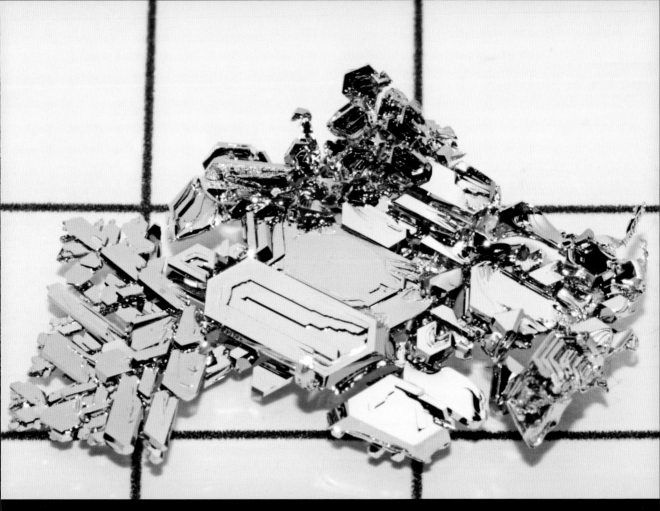

Charting North America

Meriwether Lewis (1774–1809), William Clark (1770–1838)

The era of government-sponsored voyages of exploration that also made meaningful scientific discoveries arguably began with the seafaring expedition of British Captain James Cook and the crew of the HMS *Endeavour*, sent to Tahiti to observe and record the transit of Venus in 1769. In overland travel, among the earliest and most noted such voyages was the nearly 8,000-mile (13,000-kilometer) trek of the specially commissioned (by President Jefferson) Corps of Discovery expedition between 1804 and 1806, led by Meriwether Lewis and William Clark.

President Thomas Jefferson had just completed making the Louisiana Purchase from France and was keen to see that new territory surveyed and US sovereignty declared over its native inhabitants. He also wanted to stake a claim to the Pacific Northwest territories that had not, as of yet, been claimed by any European powers. Thus, Jefferson dispatched Lewis, Clark, and thirty others on a Congressionally funded journey from St. Louis to the Pacific Ocean. While the purpose of the expedition was not scientific discovery per se, the Corps was charged with accurately cataloguing the flora, fauna, climate, and geography along their route, and was thus equipped with a variety of scientific instruments and was trained in some aspects of the relevant sciences.

Following the Missouri River upstream to its headwaters in the Rocky Mountains, the expedition encountered, traded with, and learned from several Native American tribes, avoiding conflict for the most part, and making careful observations and records of what they encountered along the way. The Rockies proved to be an obstacle to Jefferson's hope for a continuous river route to the Pacific, and the Corps struggled to traverse the tall, snowy peaks. Guidance and material help from Native American tribes along the way proved critical to the survival of the group. Eventually, the Snake and Columbia rivers provided their path to the Pacific, where they arrived in late 1805. They returned to St. Louis in 1806.

The expedition resulted in the discovery and description of more than 200 new species of plants and animals, and the establishment of peaceful contact (at least for the time being) with more than 70 Native American tribes.

SEE ALSO Circumnavigating the Globe (1519), Transit of Venus (1769), Modern Geologic Maps (1815), Exploring the Grand Canyon (1869), National Parks (1872).

Mid-nineteenth century painting by artist Thomas Burnham of Lewis and Clark scouting ahead during their 1804 expedition to the American Northwest.

Reading the Fossil Record

Mary Anning (1799–1847)

Nicolas Steno's work on the foundational principles of stratigraphy and geology in the seventeenth century and James Hutton's eighteenth-century idea of "deep time" were slow to be embraced by the world's scientific community. Significantly more evidence was needed to support the hypothesis that the Earth was truly ancient and that countless different climates and species had come and gone long before the arrival of humans. The primary source of that evidence would prove to be fossils, and fossil hunters with the skill and experience to find, extract, and identify important new specimens would be critical to our modern understanding of the history of our planet.

Among the most accomplished early fossil hunters was the nineteenth-century self-taught British fossil collector and paleontologist Mary Anning. Anning grew up in a popular English seaside resort town and worked with her father and brother to find and sell fossils ("curios") that they extracted from the layered, sedimentary cliffs along the ocean. While the family fossil business helped put food on the table, it also fueled Mary's intense scientific curiosity about the nature and origin of the strange and wonderful forms being dug out of the cliffs.

Landslides after winter storms would erode those cliffs and expose new fossils, which Mary would collect and catalog (often in dangerous circumstances, as the cliffs were quite unstable). She made her first critical find in 1811, when she was just 12 years old, discovering the skeleton of an ancient large marine reptile that would later be known as an *ichthyosaur*. Later, she discovered the first complete plesiosaurus specimen, another large extinct marine reptile, and then the first British example of the flying reptiles known as *pterosaurs*. Her reputation grew as she met and sold specimens to noted geologists and other fossil collectors of the day, and she familiarized herself with the scientific literature hypothesizing the origin and evolution of these fascinating ancient animals.

Despite being more knowledgeable and experienced than her scientifically trained male contemporaries, as a woman Anning was not accepted into the academic world, and often not given credit for her scientific contributions. Today, however, she is recognized as having made major contributions to the discovery of extinctions and the great age of our planet.

SEE ALSO Foundations of Geology (1669), Unconformities (1788), Modern Geologic Maps (1815), Discovering Ice Ages (1837)

Main image: *1823 letter and sketch by Mary Anning announcing the discovery of* Plesiosaurus *fossils.*
Inset: *Portrait of nineteenth-century geologist Mary Anning, with her rock hammer, her sample bag, and her dog Tray.*

Scale One Inch to each Foot

Sir

I have endeavoured by a rough sketch to give you some idea of what it is like. Sir you understood me right in thinking that I said it was the supposed plesiosaurus, but its remarkable long neck and small head, shows that it does not in the least answer their congeners; in its analogy to the Ichthyosaurus, it is large and heavy, but of one thing I may venture to assure you it is the first and only one discovered in Europe. Colonel Birch offered one hundred guineas for it unseen, but your letter came one days post before

Sunlight Deciphered

Isaac Newton (1643–1727), **William Hyde Wollaston** (1766–1828),
Joseph von Fraunhofer (1787–1826)

In 1672, experiments by Isaac Newton showed that sunlight is not white or yellow but is instead composed of many colors of light that can be separated into a spectrum because they refract slightly differently when passing through an object, such as a prism. Newton's experiments were widely repeated and expanded by others, including fellow English scientist William Hyde Wollaston, who, in 1802, was the first to observe that some parts of the Sun's spectrum showed mysterious dark lines.

Scientists needed a tool, a method, for understanding these dark lines and deciphering exactly what they meant in the solar spectrum. In 1814, the German optician Joseph von Fraunhofer developed that tool, called a spectroscope, a specially designed prism that could be used to measure the positions or wavelengths of the lines in an experimental technique known as spectroscopy. He observed more than five hundred narrow dark lines in the solar spectrum with his spectroscope—astronomers still call these Fraunhofer lines. In 1821, he constructed a higher-resolution spectroscope using a diffraction grating instead of a prism, and founded stellar spectroscopy by discovering that bright stars such as Sirius also have spectral lines, and they are different from the Sun's.

By the mid-nineteenth century, physicists and astronomers were able to reproduce these kinds of lines in the laboratory by filtering light through various gases, thus discovering that the lines are caused by different kinds of atomic elements absorbing different, very narrow and specific, wavelengths of light. Spectroscopy instantly became the primary way to measure the atomic and molecular composition of distant light sources such as the Sun, planetary atmospheres (including Earth's), stars, or nebulae without having to touch the object directly: all that was needed was a telescope and some kind of spectral-line measuring device, or spectrometer. Indeed, spectroscopy from ground- and space-based telescopes and from orbiting and landed space missions studying Earth and other worlds continues to be an important part of modern astronomy, Earth science, and planetary exploration.

SEE ALSO Gravity (1687), Solar Flares and Space Weather (1859), The Greenhouse Effect (1896), The Ozone Layer (1913), Earthlike Exoplanets (1995), North American Solar Eclipse (2017)

A series of high-resolution visible-light spectra of the Sun showing Fraunhofer lines, from the McMath-Pierce Solar Facility at Kitt Peak National Observatory in Arizona. Wavelengths increase from bottom to top in each row, starting with violet at bottom and ending with red at top.

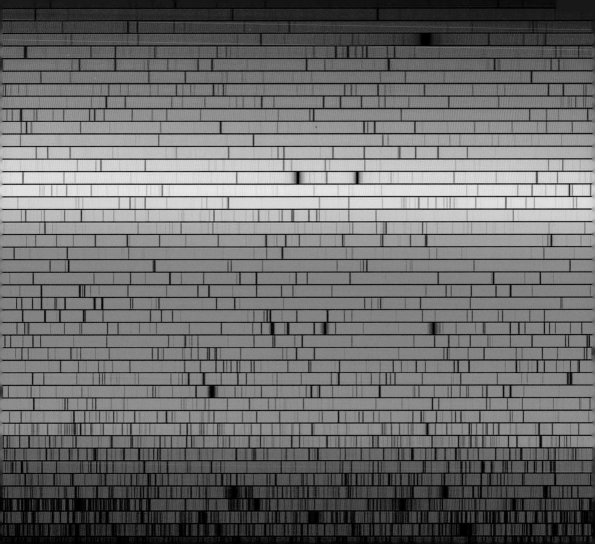

Mount Tambora Eruption

The so-called Pacific Ring of Fire is a hemispheric-sized zone of extensive earthquake and volcanic activity extending along tectonic plate boundaries from South America north to the Aleutian Islands, then back south past Japan and Indonesia, and then eventually to New Zealand. Geologic activity along the Ring of Fire is spurred by collisions among the Pacific plate and several others, and is especially intense where oceanic crust is subducting and melting underneath continental plates. Among the most active collisional zones is a long, curving stretch of crust extending from Myanmar to Papua New Guinea, where the Indian and Australian plates are rapidly subducting under part of the Eurasian plate.

Indeed, the most powerful volcanic eruption yet recorded in modern history occurred in this region at Mount Tambora, on April 5–11, 1815. Tambora was a 14,000-foot (4,300-meter) mountain prior to the eruption, but was reduced by a third to its current height of 9,350 feet (2,850 meters). The eruption is classified as having had a Volcanic Explosivity Index of 7 (out of 8), the kind of event that only occurs once every 500 to 1,000 years. The sound of the explosions was heard more than 1,500 miles (2,400 kilometers) away, and ash covered the ground as far away as 800 miles (1300 kilometers).

The eruption of Mt. Tambora instantly killed about 10,000 people and led to the indirect deaths of perhaps 100,000 more because of subsequent famine and disease in the region. The volcanic plume injected enormous amounts of ash and volcanic gases into the stratosphere, which had a rapid cooling effect on the global climate (1816 was known as "the year without a summer" for much of the world). Crops failed for several years in many regions, causing more loss of life because of famine and disease, including outbreaks of typhus and cholera.

Tambora's 1815 event is the most deadly volcanic eruption so far in recorded human history. The volcano experienced another minor eruption around 1880 and is still active today. Diligent monitoring and prohibitions by the Indonesian government on development close to the summit are ongoing, in an effort to try to protect the lives of the nearly 10 million people who now live within the broader zone affected by the 1815 eruption.

SEE ALSO Plate Tectonics (c. 4–3 Billion BCE?), Cascade Volcanoes (c. 30–10 Million BCE), The Andes (c. 10 Million BCE), Pompeii (79), Krakatoa Eruption (1883), Mount St. Helens Eruption (1980), Volcanic Explosivity Index (1982), Mt. Pinatubo Eruption (1991), Sumatran Earthquake and Tsunami (2004)

A 2005 photo of the volcanic island and caldera of Mount Tambora, Indonesia, from the International Space Station. In April 1815, Mount Tambora was the site of the most powerful volcanic explosion in recorded history.

Modern Geologic Maps

William Smith (1769–1839)

Geologists are not sedentary creatures. Much of their research comes from time spent in the field, cracking open rocks and taking samples out among the sediments, mountains, or volcanoes. Critical to this kind of activity is the accurate recording and eventual mapping of geologic features observed in the field. Resulting geologic maps describe the relationship of the different kinds of rocks and formations to one another, and place the area being mapped into a regional or even global context.

Among the earliest of modern geologic mappers was the English geologist William Smith. Smith started his career as a surveyor, a job that took him out in the field often and allowed him to become extremely familiar with the topography and geology of England, Wales, and Scotland. In particular, surveying work that he did in and around coal mines—as well as around layered rocks exposed in the walls of excavated canals—allowed him an opportunity to see the landscape in three dimensions. Over time, his careful observations and notes allowed him to connect the dots from place to place, and in 1815 he created the first modern geologic map of Britain.

Smith's map was a tour de force of geology, covering a larger area in more detail than any previous geologic map, based on his extensive travels and observations across the region. He used colors to delineate different geologic *units* on the map; he developed the idea of using the similarity of fossils within distinct layers to help connect far-flung units; and he included a geologic cross-section with the map to describe the three-dimensional nature of the geology exposed at the surface. Smith's map is remarkably similar to modern maps of Britain.

Sadly, because he was a self-taught "outsider," Smith and his mapping work were largely ignored by the academic community. He self-funded the publication and distribution of his maps, ultimately landing him in debtor's prison. It would be more than fifteen years until Smith's map was recognized as a major achievement in geology, and he was ultimately exonerated by academics as "the father of English geology." Smith's innovative observational and presentation techniques are now commonplace among modern geologic maps.

SEE ALSO First World Maps (c. 600 BCE), Charting North America (1804), Reading the Fossil Record (1811), Uniformitarianism (1830), Discovering Ice Ages (1837).

Main image: *The first geologic map of Britain, published by William Smith in 1815.*
Inset: *1837 portrait of English geologist William Smith.*

Uniformitarianism

Charles Lyell (1797–1875)

It is difficult to imagine that the Earth has a history spanning more than 4.5 billion years, because such timescales are far beyond human experience. But this fact was teased out of observations and measurements of rocks and rocky layers by early pioneering geologists well before the advent of modern radioactive dating methods. Indeed, modern geology traces its roots to the fundamental principles of stratigraphy (layering) worked out by Nicolas Steno in 1669, and the remarkable insights on unconformities (gaps in time preserved in the rock record) worked out by James Hutton in 1788. Putting these concepts together into a unifying model of earth science and Earth's history was the job of nineteenth-century Scottish geologist Charles Lyell, who, in his 1830 book *Principles of Geology*, expanded and explained in detail Hutton's earlier concept of *uniformitarianism*.

Uniformitarianism is the idea that the present is the key to the past—in other words, that the detailed study and understanding of processes at work on the surface and in the interior of the Earth today would provide the ability of geologists to understand and reconstruct the past history of the Earth from evidence preserved in the geologic record. Furthermore, the principle of uniformitarianism could also allow geologists to predict the kinds of geologic processes and events that would likely occur in the future. An underlying assumption that Lyell embraced was that the Earth changed gradually, over long expanses of time, and in predictable ways.

Lyell's strong advocacy of uniformitarianism, also known as *gradualism* by some geologists and philosophers of the time, was met with fierce resistance by other scientists, in addition to religious leaders and philosophers, who all instead advocated the concept of *catastrophism*—the idea that changes in the Earth's history and geologic record were instead caused by sudden, violent events. Examples of such events were thought to include Noah's Biblical flood (and other major floods), sudden upthrusting of mountains via earthquakes, and major volcanic eruptions. A major assumption of many advocates of catastrophism was that the Earth was relatively young, thus only rapidly operating processes could have caused Earth's observed features. Today, modern geologists accept the idea that Earth's history has been a combination of *both* gradual *and* catastrophic processes.

SEE ALSO Foundations of Geology (1669), Unconformities (1788), Modern Geologic Maps (1815), The Age of the Earth (1862), Radioactivity (1896), Basin and Range (1982)

Top: *1865 engraving of geologist Sir Charles Lyell.* Bottom: *Illustration from Lyell's* Elements of Geology, *showing an idealized cross-section of igneous, metamorphic, and sedimentary rocks in the geologic record.*

SIR CHARLES LYELL, BART., D.C.L. AND F.R.S.

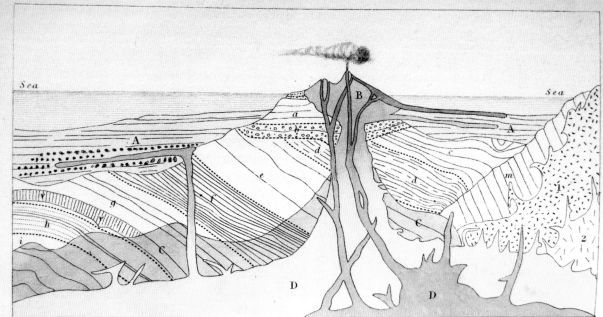

IDEAL SECTION of part of the Earth's crust explaining the theory of the contemporaneous origin of the four great classes of rocks — see Chap.1.

| A | Aqueous | B | Volcanic | C | Metamorphic (Gneiss, mica-schist &c.) | D | Plutonic (Granite &c.) |

All the rocks older than A,B,C,D are left uncoloured.

Industrial Revolution

Anthropologists use advancements in the technology of tools, weapons, architecture, transportation, or other areas to define important developmental milestones in human history. The timing of these milestones has often been set by the development of a single key idea or methodology, such as the domestication of animals, the invention of agriculture, or the transition to bronze or iron toolmaking. Among the more recent such milestones in human history was the transition in the developed world from the production of things by hand to the mass production of things by machines. This transition, which is called the *Industrial Revolution*, started in Britain around 1760 and was largely completed worldwide by around 1830.

Key aspects of the Industrial Revolution included the widespread use of steam power, the development of factories, and general (slow) increases in the standard of living as well as the size of the world's population. The first business sector to take advantage of widespread mechanization was the textile industry, although advances in the mass production of chemicals and in the production of advanced alloys of iron and steel were also important early byproducts.

The Industrial Revolution has had a dramatic effect not only on the world's population, but also on the world itself. Large increases in the capacity to manufacture goods drove large increases in the need for both natural resources, such as water; and raw materials, such as iron and other metal ores, wood, rubber, and so forth. An increasing population required farm and grazing land for increased food production, as well as additional housing, spawning widespread deforestation. Greater supplies of goods drove a more expansive transportation network to get those goods to market, leading to new roads, bridges, canals, and railroads. Keeping the machines running around the clock drove the development of massive piping networks for natural gas lighting. Mechanization was applied to mining to extract gas, metallic ores, and other raw building materials more quickly.

Despite serious issues with pollution and poor working conditions (including slavery and child labor), the Industrial Revolution remains a major net positive milestone in human history, helping to enable modern technological advancements as well as a more modern view of the appropriate relationship between human activities and the natural environment.

SEE ALSO Stone Age (c. 3.4 Million BCE to 3300 BCE), Domestication of Animals (c. 30,000 BCE), Invention of Agriculture (c. 10,000 BCE), The Bronze Age (c. 3300–1200 BCE), The Iron Age (c. 1200–500 BCE), Population Growth (1798), Deforestation (c. 1855–1870), The Anthropocene (c. 1870)

An 1868 engraving of a textile factory in Chemnitz, Germany.

Discovering Ice Ages

Louis Agassiz (1807–1873)

Geologists trying to piece together the past history of the Earth are a bit like forensic detectives trying to figure out what happened at a crime scene. Evidence is scattered about, some glaringly obvious, some subtle and hidden. The scene has to be observed carefully for clues, some of which may require detailed laboratory analysis to understand. In geology, the rocks and fossils are the only surviving witnesses, and they must be examined and cross-examined carefully and thoughtfully to tease out the truths that only they can reveal.

Among the most challenging detective work geologists faced in the nineteenth century was explaining what appears to be evidence for the erosive force of massive glaciers and sheets of ice in landscapes that are now ice-free. For example, enormous linear striations—scratches like those formed from rocks dragged along by high-latitude glaciers—were carved into ancient granite bedrock through mid-latitude regions in Europe, Asia, and North America. Massive piles of unconsolidated rocky debris remarkably similar to moraines—glacially formed rocky debris piles—were also found far south of current glaciers on these continents. Isolated rocks and boulders with clear origins from specific mountain ranges or layered sedimentary deposits were found to have been somehow transported far to the south of where they had formed.

Late eighteenth- and early nineteenth-century geologists concluded that many of the then-known glaciers had once been more extensive. One geologist in particular, the Swiss-American Louis Agassiz, took the evidence a step further, concluding along with several colleagues in 1837 that much of the northern hemisphere had been covered in a thick sheet of ice—thicker than the tallest mountain ranges and extending perhaps halfway or more in latitude down to the equator. They coined the term "ice age" for such periods of extensive glaciation. The idea would take decades to be widely accepted.

Agassiz and colleagues found geologic evidence for multiple cycles of extensive ice ("glacials") followed by periods of ice retreat ("interglacials"). Later geologists would add clues from chemistry and paleontology (fossils) to learn that Earth appears to have experienced at least five major ice ages over the past 2.5 billion years, each one consisting of hundreds of cycles of longer glacials and shorter interglacials—the most recent of which began about 12,000 years ago.

SEE ALSO Snowball Earth? (c. 720–635 Million BCE), End of the Last "Ice Age" (c. 10,000 BCE), The Little Ice Age (c. 1500), Next Ice Age? (~50,000)

Main image: *An 1870 photograph of geologist and biologist Louis Agassiz.* **Inset:** *Illustration by J. Bettanier of the Matterhorn glacier, near Zermatt, Switzerland, from Louis Agassiz's 1840 book,* Études sur les Glaciers.

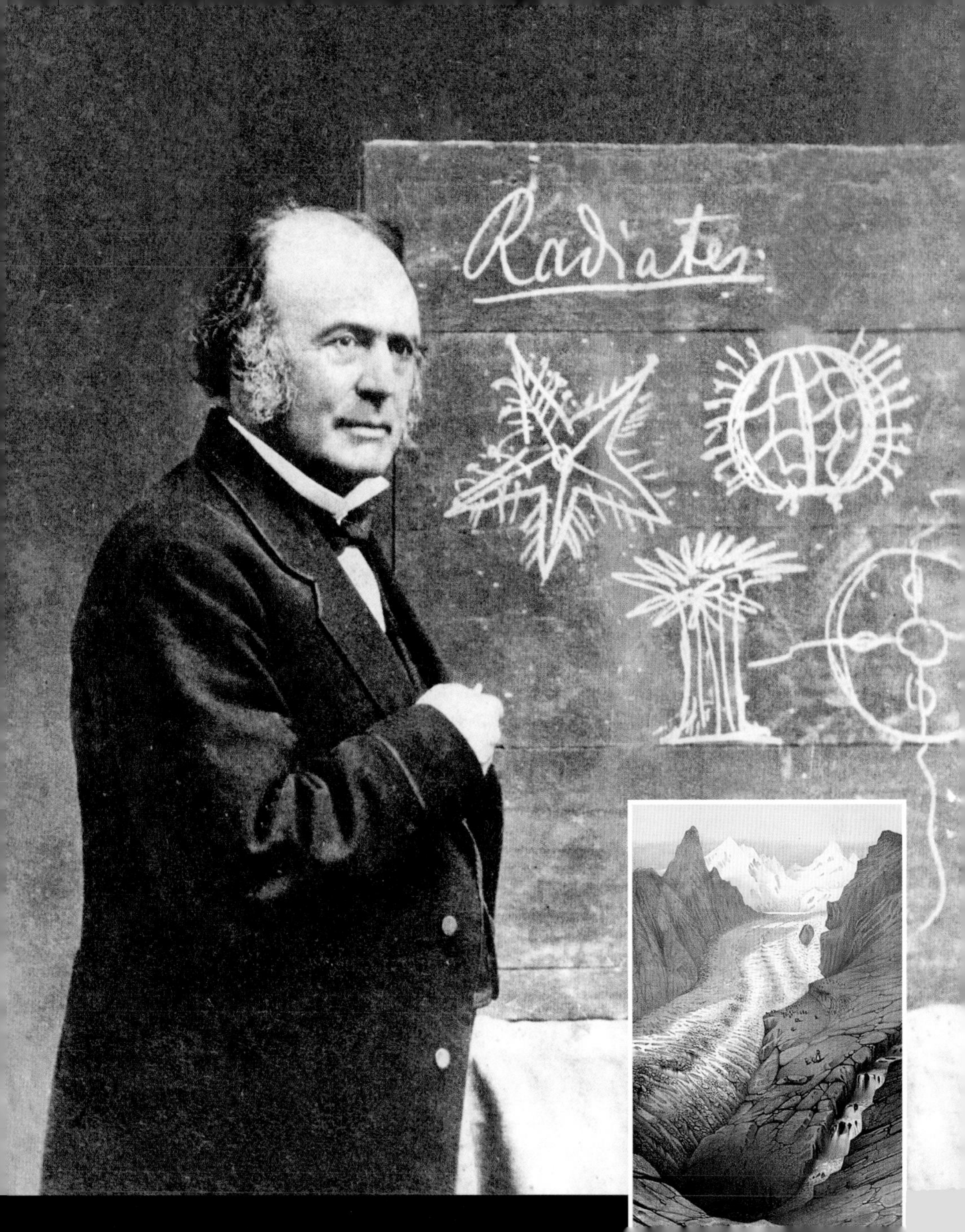
Radiates

Birth of Environmentalism

Alexander von Humboldt (1769–1859)

The history of science and exploration is replete with individuals who, working mostly on their own, have made significant advances in our understanding of the world and humanity's place within it. Such people include Eratosthenes, Galileo, Copernicus, Newton, Darwin, Einstein, Hubble, Hawking, and many others who have been largely forgotten despite their work's continuing to have a lasting influence on the world today. That is the unfortunate legacy of the late eighteenth-/early nineteenth-century naturalist and explorer Alexander von Humboldt, who arguably gave birth to what we call the environmental movement today.

Humboldt was a polymath (good at everything he tried) from a wealthy family, with an early passion for botany, anatomy, and other sciences that eventually led him to a degree in geology and a government job inspecting mines. His work cataloguing plants, minerals, and fossils drove a wanderlust to explore, and in 1799 he set out on his own, using his family money and with the blessing of the Spanish crown, to explore and catalog the uncharted flora and fauna in new Spanish territories in today's Venezuela, Colombia, Ecuador, and Peru.

Humboldt brought sophisticated scientific instruments into the field and collected enormous numbers of plant, animal, and fossil samples during a series of such expeditions in the early 1800s. He made extensive, systematic observations of the physical geography, plant life, and meteorology of as-yet undocumented regions from the Amazon to the Andes. He was among the first to record and interpret the interrelationships among plants, animals, people, climate, and geology and to establish a holistic view of nature. His five-volume book *Kosmos* (first published in 1845) would establish an entirely new worldview: the Earth as an interconnected set of ecosystems; nature experienced through emotions; science as the path to understanding the physical world. Humboldt's work and writings were inspirational to Charles Darwin, Henry David Thoreau, John Muir, and others who would be so critical to an emerging sense of environmental responsibility in society. According to biographer Andrea Wulf, "Humboldt gave us our concept of nature itself. The irony is that Humboldt's views have become so self-evident that we have largely forgotten the man behind them."

SEE ALSO Size of the Earth (c. 250 BCE), Transit of Venus (1769), Charting North America (1804), Discovering Ice Ages (1837), Natural Selection (1858–1859), Exploring the Grand Canyon (1869), The Sierra Club (1892)

Top: *1806 painting of Alexander von Humboldt (standing, left) collecting flora and fauna samples near Chimborazo volcano in Ecuador.* **Bottom:** *Humboldt's drawings comparing different ecosystems at different mountain elevations on Chimborazo and on Mont Blanc in the Alps.*

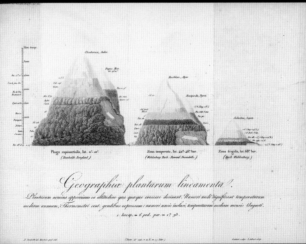

Geographiæ plantarum lineamenta.

Plantarum nomina apposuimus eis altitudine quas quæque crescere desinunt. Numeri multi significant temperaturam
mediam annuam. Thermometri cent. gradibus expressam: numeri uncis inclusi, temperaturam mediam mensis Augusti.

Proof that the Earth Spins

Jean Bernard Léon Foucault (1819–1868)

Our space-age perspective on our planet now makes the fact that the Earth spins a matter of common knowledge. But imagine for a moment going back in time to an era when there were no satellites or space probes or fancy computerized planetarium programs, and try to convince someone that the Earth is actually spinning. It's not intuitive—the Sun and sky appear to move, not the Earth! If the Earth were spinning as fast as it would need to be spinning to rotate once per day (about 1,000 miles [1,600 kilometers] per hour at the equator), wouldn't we all get flung off into outer space? Even today, it's hard to prove to someone that the Earth spins. What is needed is a simple and repeatable experiment that provides a physical demonstration of the Earth's rotation.

Although a number of such experiments have been proposed and conducted, by far the most famous is the one first performed in 1851 by the French physicist Jean Bernard Léon Foucault. Foucault (pronounced foo-KOH), like any good physicist, understood Newton's laws and exploited the first law (bodies at rest or in motion remain at rest or in motion unless acted on by an external force) for his experiment. He constructed a long, heavy, stable pendulum using a ball (or bob) made from lead-coated brass that was suspended on a wire 220 feet (67 meters) long from the ceiling to the floor of the Panthéon in Paris. Foucault knew that in the absence of any other forces, once he started the pendulum swinging, it would continue to swing in that same plane—that is, it would remain in the same inertial reference frame relative to the "fixed" stars, rather than to the Earth. By setting up hour markers like those on a sundial (or small obstacles for the bob to knock over), and by compensating for friction in the wire or from the bob's motion through the air, it became easy to demonstrate that the room (indeed, the whole Earth) was slowly rotating relative to the plane of the pendulum's swing. The following year, Foucault perfected a gyroscope (a device for measuring velocity and orientation) based on similar principles.

Foucault's pendulum became a nineteenth-century sensation because of its simplicity; hundreds can still be found around the world in universities, museums, and science centers.

SEE ALSO Size of the Earth (c. 250 BCE), Gravity (1687)

A large Foucault's pendulum from the Príncipe Felipe Science Museum of the City of Arts and Sciences in Valencia, Spain. In this setup, a little ball-and-stick model is tipped over roughly every 30 minutes because the Earth is spinning underneath the inertially fixed plane of the pendulum's swing.

Deforestation

Trees and the wood that they produce have been an important natural resource since prehistoric times. Trees take up significant arable land, however, and so after the invention of agriculture and the development of sedentary city-based lifestyles, large stands of trees began to be routinely cut or burned down to make room for crops, homes, and grazing land (and as fuel). In places where the supply of trees was highly limited—like Easter Island, for example—such deforestation was complete, leading to a dramatic change in the economics and social structure of such societies.

With the advent of the Industrial Revolution around 1830, and especially with the early reliance on steam to power a growing number of factories, the global rate of deforestation began to increase dramatically around 1855, not coincidentally around the same time that global population also began to increase dramatically. Efforts to replant forest stands in unforested areas (*afforestation*) were typically relatively minimal in the nineteenth and much of the twentieth centuries, and, as a result, more than 60 percent of the area of the Earth that had been forested in prehistoric times has since been deforested.

Deforestation enables increased food production and helps to support a continually growing human population. However, there are also a number of adverse short- and long-term effects that make unchecked deforestation ultimately unsustainable. Removal of trees and their root systems results in increased rates of soil erosion and landslides, for example. Trees provide habitats for countless species of animals and insects, and so deforestation leads to decreases in hunting yield for some societies as well as to loss of species diversity (or even to extinction) for boreal inhabitants. Trees shade the surface and help keep moisture in the ground, and thus deforestation can lead to dramatic soil aridity. Trees also sequester a significant amount of atmospheric CO_2, and thus deforestation can only work to increase the relative abundance of that strong greenhouse gas. While cutting down forests may seem like the right short-term solution for some people and societies, the long-term effects of unmitigated deforestation could be quite negative.

Efforts to promote sustainable deforestation and afforestation are currently focused heavily on rainforests like the Amazon, which continue to be cut down at a dramatic rate. Over the past fifty years, more than half of the world's tropical rainforests have been cut down—a rate that would see rainforests completely disappear by the middle of this century without preventive action.

SEE ALSO Invention of Agriculture (c. 10,000 BCE), Population Growth (1798), Industrial Revolution (c. 1830), Birth of Environmentalism (1845), Big Burn Wildfire (1910), Tropical Rain/Cloud Forests (1973), Temperate Rainforests (1976), Boreal Forests (1992), Temperate Deciduous Forests (2011)

Clear-cut and deforested jungle in the Amazon River basin, Brazil.

Natural Selection

Charles Darwin (1809–1882), Alfred Russel Wallace (1823–1913)

The discovery, in the eighteenth and nineteenth centuries, of fossils of now-extinct plants and animals, and the realization that many of those fossils are ancient, fueled an innate curiosity among many biologists and geologists about how and why life on Earth has changed over time. One such scientist, the young British naturalist Charles Darwin (then only twenty-two), took it upon himself (and his family's wealth) to board a Royal Navy ship set to conduct geographic and botanical surveys of South America. The HMS *Beagle* set out from England for its five-year expedition around the world in 1831.

Darwin made careful observations and collected numerous plant, animal, and fossil specimens. His time on the *Beagle* opened his eyes to the incredible diversity of life on Earth, and to the relatively incremental changes among similar species that had become geographically isolated, for example among finches on the Galápagos Islands. Some of what Darwin saw was consistent with the idea of uniformitarianism (gradual changes in geology and life), but other observations hinted at more complex and rapid forces driving changes among species.

After returning to England in 1836, Darwin set about attempting to make sense of the observations he had made and the samples he had collected. His conclusions led him to believe that differences in an individual organism's physical and behavioral characteristics (what biologists today call *phenotype*) can lead to differential success (or not) in survival and reproduction of those individuals—a process that he called *natural selection*. In 1858, he realized that fellow British naturalist Alfred Russel Wallace was independently coming to the same conclusion, and they presented their theory to the scientific community jointly. Darwin's more detailed ideas on natural selection and its role in evolution subsequently appeared in his landmark 1859 book, *On the Origin of Species*.

Darwin's book and later work were extremely popular and controversial among academics and theologians, especially regarding its implications that humans, too, have evolved over the millennia partially in response to natural selection. Today, natural selection and evolution are almost universally accepted as the prevailing theory of human (and other species) origins, and Darwin is widely hailed as one of the most influential scientists in human history.

SEE ALSO The Origin of Sex (c. 1.2 Billion BCE), Primates (c. 60 Million BCE), First Hominids (c. 10 Million BCE), Galápagos Islands (c. 5 Million BCE), *Homo sapiens* Emerges (c. 200,000 BCE), Reading the Fossil Record (1811), Uniformitarianism (1830), Endosymbiosis (1966)

Top: *Photograph of Charles Darwin from around 1870.* Bottom: *An 1845 drawing by the ornithologist John Gould of the variety of finch beaks encountered by Darwin during his travels through the Galápagos Islands in 1835.*

1. Geospiza magnirostris.
2. Geospiza fortis.
3. Geospiza parvula.
4. Certhidea olivasea.

Airborne Remote Sensing

Gaspard-Félix Tournachon (Nadar) (1820–1910)

The idea of creating a map of a large region—a city, a country, the world—required early mappers or cartographers to change their perspective, from an Earth-bound one to an imagined airborne or even extraterrestrial one high above the surface. Some early mappers succeeded more than others in moving themselves to this perspective; but, regardless, most early large-scale maps still suffered from a lack of geographic accuracy and/or scale.

That would all change with the advent of *remote sensing*—the ability to determine information about a place or object without needing to be physically in contact with it. Astronomy and spectroscopy of the Sun and other stars and planets through a telescope, for example, is a kind of remote sensing, as is charting a coastline from a ship using a spyglass or other surveying tools. For the mapping of the Earth, a major advance came with the development of airborne remote sensing, the ability to use information like photographs to make accurate maps of large areas.

The first examples of airborne remote sensing came from pioneering balloonists like the French photographer Gaspard-Félix Tournachon, who went by the pseudonym Nadar. In 1858, Nadar became the first person to obtain aerial photographs, from a balloon floating above Paris. While none of his earlier photographs has survived, numerous examples of airborne photos that he acquired in the decades after he perfected the process of shooting and developing photos in the challenging environment of a balloon basket attest to his skill. Nadar's photos were used for city and government surveys as well as for tourist advertising.

Soon thereafter, kites began to be used to take aerial remote sensing photos, and eventually the first airplanes began to be used for aerial reconnaissance. During World War I, aerial photos could provide unique new information on enemy defenses and troop movements. After the war, commercial photographic remote-sensing companies in Europe and the US began providing mapping and surveying services from the air for government, industrial, and academic clients. Beginning in the 1960s, the inevitable start of remote sensing from space-based satellites would ultimately lead to today's modern era of sophisticated spy satellites, as well as scientific satellites focused on imaging and studying our planet at ever-finer resolution and at ever-higher cadences.

SEE ALSO First World Maps (c. 600 BCE), Sunlight Deciphered (1814), Modern Geologic Maps (1815), (Geo) Science Fiction (1864), Structure of the Atmosphere (1896), The Ozone Layer (1913), Exploration by Aviation (1926), Geosynchronous Satellites (1945), Weather Satellites (1960), Earth Selfies (1966)

A series of 1868 photographs of the Arc de Triomphe and Place de l'Étoile in Paris, obtained from the balloon "Le Géant" by the French photographer Nadar.

Solar Flares and Space Weather

Richard Carrington (1826–1875)

The Sun is the most massive, energetic, and important (to us, at least) non-Earth object in the solar system, and so it should not be surprising that many astronomers have chosen to train their increasingly powerful telescopes on our nearby star in order to study its inner workings. By using proper filters or projecting the solar disk onto a wall or screen, astronomers could measure and monitor features like sunspots on the Sun's visible "surface," the photosphere. Sunspots had been studied for centuries, with telescopic observations going back to the early seventeenth century, and naked-eye (filtered) observations spanning an even longer history. Improvements in telescopes and observing methods over time have allowed sunspots to be studied in ever more exquisite detail.

One of the most noted and prolific observers of sunspots was the English amateur astronomer Richard Carrington. On September 1, 1859, Carrington observed an intense brightening on the Sun near a particularly dense cluster of sunspots. The event lasted only a few minutes. The next day, however, saw reports from around the world of intense auroral activity and major disruptions to telegraphs and other electrical systems.

What Carrington had witnessed was the first recorded example of a solar flare—an enormous explosion in the Sun's atmosphere that can hurl high-energy particles at enormous speeds out into the solar system. The dramatic effects from this solar "wind" crashing into the Earth's protective magnetic field are known as a solar storm. Many such flares and storms have been observed since then, but visual records and ice-core data indicate that the 1859 event was not only the first but also the largest in recorded history—perhaps a once-in-a-millennium mega-flare.

Carrington's scientific observations established a connection between the Sun's activity and the Earth's environment and led to intense interest in the study of space weather—the interaction of the solar wind with all the planets. Today's armada of Earth-orbiting communications, weather, and remote-sensing satellites, in particular, represents billions of dollars of technology and infrastructure that is highly vulnerable to disruption by solar flares and their ensuing storms. This is just one reason why NASA and other space agencies are highly motivated to continue Carrington's important work of predicting, monitoring, and understanding the effects of space weather.

SEE ALSO Airborne Remote Sensing (1858), Reversing Magnetic Polarity (1963), Magnetic Navigation (1975), The Oscillating Magnetosphere (1984)

A spectacular solar-prominence eruption captured in time-lapse frames on March 30, 2010, in the extreme ultraviolet light of ionized helium by the NASA Solar Dynamics Observatory satellite. For scale, hundreds of Earths would fit into the loop in the top frame.

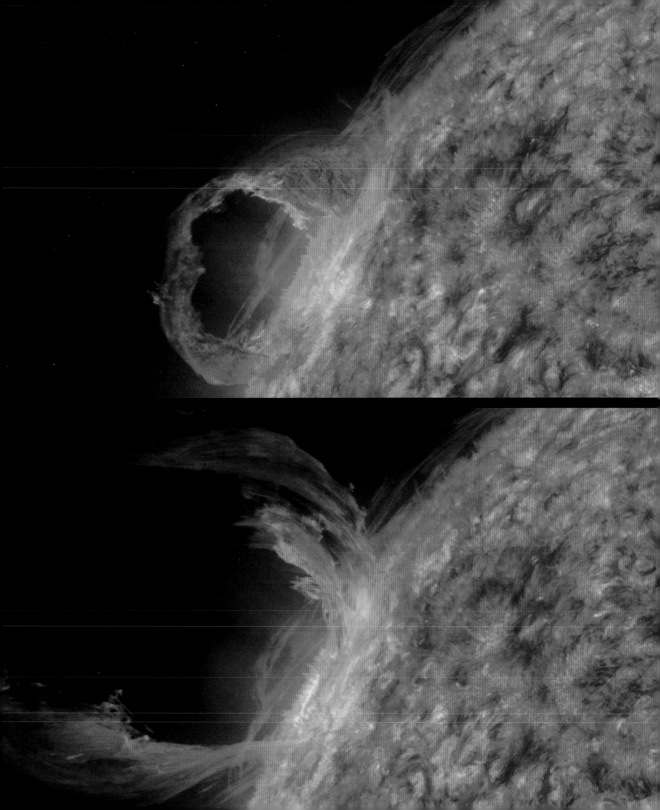

The Age of the Earth

William Thomson (Lord Kelvin) (1824–1907)

How old is the Earth? Among the earliest reputable scientists to estimate the age of our planet was the Scottish-Irish physicist and engineer William Thomson. Thomson was an expert in *thermodynamics*—the study of the relationships between different kinds of energy—and he had become the first British scientist elevated to the House of Lords because of his discoveries. Subsequently known as Lord Kelvin, he tried to estimate the age of our planet by assuming that it had started out completely molten and simply cooled off over time to its current surface and interior temperatures, with no additional sources of internal heat. In that model, Kelvin estimated in 1862 that the Earth was somewhere between 20 and 400 million years old, an estimate that he later refined to between 20 and 40 million years old.

Kelvin's age estimate instigated significant debate among nineteenth-century scientists. While it seemed too old for some to fathom (compared to, for example, an age of only about 6,000 years based on some religious interpretations of the Bible), it seemed too *young* for many geologists to grasp, especially those who had the most experience studying layered sedimentary rocks and unconformities. To proponents of uniformitarianism like Charles Lyell, a few tens of millions of years was not enough time to explain the extensive history of layered rocks and climate cycles recorded in the geologic record. To proponents of natural selection like Charles Darwin, it was also not enough time to explain the evidence for the slow species changes seen in the fossil record.

The discovery of radioactivity in 1896 and the subsequent development of radioactive age-dating techniques revealed that one of Kelvin's fundamental assumptions was incorrect: there was an additional source of internal heat—decay of radioactive elements such as uranium. Modifying Kelvin's calculations to take that factor into account resulted in a *substantially* older age estimate, up to several billion years. Since then, using radioactive age-dating techniques on meteorites—the oldest known materials in the solar system and the building blocks of the planets— scientists of the twentieth and twenty-first centuries now estimate the age of the Earth at the staggeringly precise and ancient value of 4.567 billion years, plus or minus a few million.

SEE ALSO Foundations of Geology (1669), Gravity (1687), Unconformities (1788), Sunlight Deciphered (1814), Uniformitarianism (1830), Natural Selection (1858–1859), The Greenhouse Effect (1896), Radioactivity (1896)

1910 photo of the physicist William Thomson (Lord Kelvin) making compass measurements.

(Geo) Science Fiction

Jules Verne (1828–1905)

The pace of scientific discoveries about the Earth and the natural world in general accelerated dramatically in the nineteenth century. Scientists exploring and studying the Earth and its history, like Alexander von Humboldt, Charles Darwin, Charles Lyell, and William Thomson (Lord Kelvin) became the celebrities and pop stars of the time, delivering lectures and writing popular science books for an increasingly scientifically literate general public. It is perhaps not surprising, then, that fiction writers would pick up the mantle of popular science and create memorable stories and characters that paralleled exciting advances in science.

Among the earliest writers to widely popularize the relatively new genre of science fiction was the French author Jules Verne. Verne began his literary career in the 1850s writing popular magazine articles and short fictional pieces focusing on popular topics in the science and technology of the day, and especially on his own personal fascination with geography and exploration. He was particularly inspired by the work of Charles Lyell and other leading geologists of the time, coupled with the relatively new idea that the Earth had gone through numerous cycles of change over vast expanses of time, and that the history of those changes was preserved in the geology of the surface and subsurface. Verne's characters would descend deep into that subsurface realm, attempting to read the record and follow the clues to reach the elusive core. Verne even seized upon the new idea that perhaps life on Earth was ancient, and in his imagined voyage deep underground his characters encountered numerous examples of prehistoric creatures that readers could imagine may have once dwelled on the surface.

Some of the best science fiction since (and including) Verne is popular, educational, and entertaining precisely because many authors tap into the same new technological advancements and scientific discoveries that are being widely discussed and debated at the same time among cutting-edge scientists and engineers themselves. Indeed, recurrent science fiction themes revolving around archaeology, biology/genetics, polar and deep sea exploration, aviation, robotics, space travel, and many other areas of science and technology often include real science and engineering (or sometimes fanciful extrapolations of them) as an effective "hook" to bring in motivated readers and viewers.

SEE ALSO First World Maps (c. 600 BCE), Native American Creation Stories (c. 1400), Industrial Revolution (c. 1830), Birth of Environmentalism (1845), Natural Selection (1858–1859), The Age of the Earth (1862), Geosynchronous Satellites (1945)

An illustration from Jules Verne's 1864 science fiction story A Journey to the Center of the Earth.

Exploring the Grand Canyon

John Wesley Powell (1834–1902)

The Grand Canyon of the Colorado River is among the top natural wonders of the world and is a veritable mecca for geologists. The Grand Canyon is a 277-mile (446-kilometer) section of the river in northern Arizona where it has carved deeply and dramatically into the layered sedimentary rocks, exposing a cross-section spanning more than 2 billion years of Earth's geologic history. In some places, the canyon is more than 6,100 feet (1,850 meters) deep but less than 1,000 feet (300 meters) wide.

While several groups of Spaniards were known to have visited the Canyon as early as 1540, no one had yet documented its full extent until the first formal scientific exploration of the Grand Canyon was initiated in 1869 by the geologist, explorer, and soldier Major John Wesley Powell. He and nine other men set out in four boats to chart the course of the river and to document the flora, fauna, and meteorology of this little-known (to European settlers, at least) region. It was a dangerous voyage through arid, uncharted territory that involved substantial portage of boats and supplies around waterfalls and intense rapids. Three members of Powell's crew would perish during the three-month voyage.

Powell's drawings, measurements, books, and presentations, from both his 1869 trip as well as a second, more ambitious, 1871–1872 expedition down the Canyon (which included a photographer), caused a sensation among the public as well as the scientific community. In a single cross-section down to the bottom of the canyon, changes in rock types (igneous, metamorphic, and sedimentary), texture, color, and fossil content provide textbook evidence for the basic principles of geology and stratigraphy, including one of the most famous unconformities on the planet, named after Powell and representing more than a billion years of missing geologic time. Few other places on Earth would provide such a dramatic window into the planet's geologic history, which is one reason why the Grand Canyon became a National Park in 1919. As Powell himself wrote in 1895, "The elements that unite to make the Grand Canyon the most sublime spectacle in nature are multifarious and exceedingly diverse."

SEE ALSO Foundations of Geology (1669), Unconformities (1788), Charting North America (1804), Modern Geologic Maps (1815), Uniformitarianism (1830), National Parks (1872), The US Geological Survey (1879)

Main image: *1872 photo of boats moored along Marble Canyon.*
Inset: *Photograph of John Wesley Powell and a Native American named Tau-gu.*
Both photos are from Powell's second expedition down the Grand Canyon.

The Anthropocene

Geologists divide the history of the Earth into successively smaller chunks of time, from the longest, called *eons*, down through *eras*, *periods*, *epochs*, and finally the shortest-duration time periods, ages. The major boundaries of geologic time are typically marked by major geologic or paleontological events (like mass extinctions). However, the boundaries of more recent ages and epochs have been defined based on major climatic events, such as glacial and interglacial episodes. For example, we are currently in the *Quaternary* geologic period, which started at the beginning of the current ice age, about 2.6 million years ago. Within that period, we are currently in the *Holocene* epoch, which is defined to have begun at the start of the current interglacial warming period, around 10,000 BCE.

Some geologists and climatologists think that we have recently entered a new geologic epoch, however, an epoch in which humans have become a major (perhaps the major) agent of change to the Earth's surface, atmosphere, and biosphere. This proposed new epoch is being called the Anthropocene, deriving its first three syllables from the Greek word for "human."

Evidence cited for the Anthropocene's being a distinct geologic epoch from the Holocene dates back to the 1870s, and includes the impact of the Industrial Revolution, and especially the development of internal combustion engines, on increasing the amount of pollution and greenhouse gases (such as CO_2) in the Earth's atmosphere. Some cite the dramatic increase in deforestation and resulting changes in surface drainage patterns that began in the mid-nineteenth century. Others cite the dramatic human impact on biodiversity over the past century, based on numerous direct examples of species extinctions correlated with human activities. Still others with a more philosophical perspective believe that the dawn of the nuclear age in the 1940s represents a new epoch for our planet, as there is now a single species that could not only destroy itself but could also dramatically and fundamentally wreak havoc on the entire biosphere.

Does the concept of the Anthropocene portend a depressing, dystopian view of the current and future state of the Earth? Or might the Anthropocene represent an epoch that began in unchecked ignorance of humanity's potential effect on our world but then became a time of awakening of environmental responsibility? It's up to us to find out.

SEE ALSO The Stone Age (c. 3.4 Million BCE TO 3300 BCE), End of the Last "Ice Age" (c. 10,000 BCE), The Bronze Age (c. 3300–1200 BCE), The Iron Age (c. 1200–500 BCE), Industrial Revolution (c. 1830), Discovering Ice Ages (1837), Deforestation (c. 1855–1870), The Greenhouse Effect (1896), Earth Day (1970)

Has humanity's "footprint" on planet Earth resulted in the beginning of a new geologic era, which some call the Anthropocene? Some scientists think so.

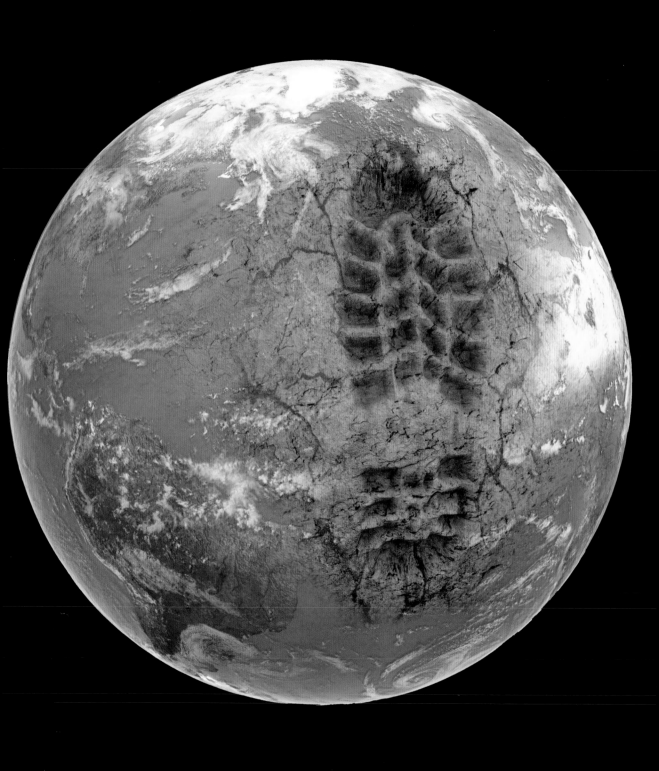

Soil Science

Vasily Dokuchaev (1846–1903)

Soils are the fine-grained physical and chemical weathering products of rocks exposed to Earth's surface conditions. Wind, water, ice, tectonics, volcanic explosions, and even impact cratering are some of the agents that can break down or alter bedrock into soil. And as every farmer knows, soil is the critical carrier of nutrients and the storage medium for water that makes plant growth possible. Without soils, plants would struggle to survive, large-scale agriculture would be essentially impossible, and life on Earth would be dramatically different.

The study of soils as a distinct research area only began formally around 1870, spurred on by the Russian geologist Vasily Dokuchaev, who established many of the foundational principles upon which modern soil science is based. For example, Dokuchaev is credited with introducing the idea that variations in soils are correlated with variations in their starting bedrock composition, local climate conditions, local topography and drainage conditions, nature and influence of organisms that soils host and support, and the amount of time that soil formation processes have operated. He conducted extensive field studies of soils and developed one of the earliest soil classification systems (based on grain size, color, organic content, and other factors), some aspects of which are still used to classify soils today.

The study of the formation, chemistry, morphology, and classification of soils is known as *pedology*, and it encompasses a diverse range of academic and social disciplines, including geology, chemistry, mineralogy, ecology, microbiology, agronomy, archaeology, engineering, and even city and regional planning. In both academic and commercial realms, soil is rightly treated as a natural resource that, like other resources, deserves careful monitoring and proper management.

Because microscopic and macroscopic organisms play such an important role in the formation and modification of soils on Earth, it may be fair to ask, then: Are there soils on other moons or planets? Recently, the Soil Science Society of America decided that the answer is yes, changing their formal definition of "soil" to potentially include soils on other worlds by defining it as "the layer(s) of generally loose mineral and/or organic material that are affected by physical, chemical, and/or biological processes at or near a planet's surface, and that usually hold liquids, gases, and biota, and support plants."

SEE ALSO Invention of Agriculture (c. 10,000 BCE), Civil Engineering (c. 1500), Controlling the Nile (1902)

Soils come in a variety of colors, textures, and compositions, as seen in this representative cross-section of soils beneath an asphalt road.

National Parks

Population growth and associated deforestation, urbanization, and industrialization all require resources that, over time, can dramatically alter the natural world. As these forces began to increase at an unprecedented rate in the nineteenth century, some naturalists, conservationists, and government officials began to realize that many of the most beautiful and fragile environments and ecosystems could be destroyed unless they were granted special government protections. In the US, this led to the creation of state and national parks intended to preserve some of the country's most special and scenic natural wonders.

Yellowstone National Park became the first such designated special area in 1872, when Congress and President Ulysses S. Grant set aside more than 2 million acres of pristine mountains, canyons, forests, rivers, and hot springs covering parts of what would later become the states of Wyoming, Montana, and Idaho. Sequoia National Park, in California, famous home to the giant redwood trees, was the next region designated for federal protection, in 1890. To administer these and a growing number of additional parklands set aside for preservation, the National Park Service was established by Congress and President Woodrow Wilson in 1916, "to conserve the scenery and the natural and historic objects and wildlife therein, and to provide for the enjoyment of the same in such manner and by such means as will leave them unimpaired for the enjoyment of future generations."

Today there are 59 national parks in the US, as well as hundreds of other specially protected areas such as memorials, monuments, seashores, lakeshores, trails, and recreational areas. Added up across the country, an area nearly equivalent to the size of the state of Montana has been set aside as national parklands. Similar parks have been created in countries across the world, and many of them (including many in the US) are also designated as World Heritage Sites by the United Nations Educational, Scientific and Cultural Organization (UNESCO). Overall, UNESCO estimates that, as of 2016, just under 15 percent of the Earth's land area consists of totally or partially protected areas, including scientific reserves with limited public access, national parks, natural monuments, nature reserves or wildlife sanctuaries, protected landscapes, and areas managed mainly for sustainable use.

SEE ALSO The Sierra Nevada (c. 155 Million BCE), Oldest Living Trees (c. 3000 BCE), Exploring the Grand Canyon (1869), The Sierra Club (1892), Extremophiles (1967)

Spectacularly colored mineral hot springs deposits at Lion Geyser and Heart Spring in Yellowstone National Park.

The US Geological Survey

G. K. Gilbert (1843–1918)

The dramatic increase in the size of the United States as a result of first the Louisiana Purchase in 1803 and then the Mexican-American War in 1848 drove the desire for the federal government to have these new lands surveyed and their flora, fauna, and geography inventoried. The job was passed on to a new federal agency, the United States Geological Survey (USGS), established by Congress in 1879 and charged with "the classification of the public lands, and examination of the geological structure, mineral resources, and products of the national domain."

Among the earliest leading USGS geologists was Grove Karl Gilbert, who studied the origin and evolution of a variety of geologic landforms (*geomorphology*) and was a primary assistant of John Wesley Powell's 1874 expedition to explore the Rocky Mountains. He eventually became embroiled in a controversial debate among geologists on the origin of a mysterious, 3,900-foot-wide (1,200 meters), 560-foot-deep (170 meters) circular hole in the ground near Winslow, Arizona. Three hypotheses for its origin were being debated: Was it a traditional volcanic crater, a special kind of volcanic crater called a maar formed from the explosive interaction of hot magma and groundwater, or a crater caused by the impact of an asteroid or comet? Gilbert concluded in 1895 that it was a maar, formed by a powerful volcanic steam explosion.

However, Gilbert would be proven wrong nearly sixty-five years later by another prominent USGS geologist, Eugene Shoemaker, who in 1960 conclusively showed that what is now called Meteor Crater was indeed formed by the high-velocity crash to the ground of a small metallic asteroid. To his credit, Gilbert also studied similar circular features on the Moon, and advocated (correctly) that the circular depressions on the Moon are indeed the result of such cosmic collisions.

Partly because of Gilbert's extraterrestrial studies, as well as those of Shoemaker and others, impact cratering is now considered a major force for change on a planetary surface. USGS scientists today are deeply involved in the mapping and exploration of not just the Earth's surface, but also the surfaces of many other worlds in our solar system.

SEE ALSO Arizona Impact (c. 50,000 BCE), Charting North America (1804), Exploring the Grand Canyon (1869), Hunting for Meteorites (1906), The Tunguska Explosion (1908), Understanding Impact Craters (1960), Torino Impact Hazard Scale (1999)

Main image: *USGS geologist G. K. Gilbert climbing an outcrop near Berkeley, California, c. 1906.*
Inset: *Photo from around 1891 of experiments by Gilbert that demonstrated the formation of impact craters by throwing balls of hard clay into soft clay.*

Krakatoa Eruption

Indonesia and its immediate surroundings are among the most geologically complex and active areas on Earth. Two continental plates (the Eurasian and Australian) and two oceanic plates (the Philippine Sea and Pacific) all converge in the area around Indonesia, producing a zone of subduction, faulting, folding, and uplift greater than 3,000 miles (5,000 kilometers) wide. Some of the largest recorded earthquakes and volcanic eruptions in history have occurred in this region. Among the most colossal of those events was the August 26–27, 1883 eruption of the island volcano Krakatoa.

Krakatoa, located between the islands of Sumatra and Java, is part of the volcanic island arc created by the subduction of the (higher-density) Indian Ocean plate underneath the (lower-density) Eurasian continental plate. As the oceanic plate melts and subducts, less-dense molten magma rises up into the overlying continental plate, melting some of it and producing extensive earthquakes and volcanic activity.

What sets Krakatoa apart from other earlier and later volcanic eruptions is the level of destruction caused by the enormous amount of energy released in the event. By the late nineteenth century, European explorers, settlers, missionaries, and merchants had established many colonies and outposts across southeast Asia, most of them sustained by trading and commerce associated with rare and exotic spices found on many of the islands in the region. While accounts were known of prehistoric and earlier violent eruptions of the volcano (most recently in 1680), little was known in the 1880s about the potential for future explosions. More than 36,000 people near and around Krakatoa were directly killed by the eruption and subsequent ash fall and tsunamis. The main explosion is estimated to have had the force of about 200 megatons of TNT—more than four times the force of the world's largest atomic bomb. Two thirds of the island of Krakatoa itself was obliterated in the eruption.

Like other major volcanic eruptions throughout history, Krakatoa had a dramatic and lasting effect on the Earth's weather and climate. Skies were darkened, sunsets were more colorful, and global average temperatures were a degree or more lower for five years after the eruption because of all the ash and aerosols injected into the stratosphere. Since the 1883 eruption, Krakatoa has started to rebuild, and numerous smaller eruptions continue to occur.

SEE ALSO The Spice Trade (c. 3000 BCE), Pompeii (79), Huaynaputina Eruption (1600), Mount Tambora Eruption (1815), Island Arcs (1949), Mount St. Helens Eruption (1980), Volcanic Explosivity Index (1982), Mount Pinatubo Eruption (1991), Sumatran Earthquake and Tsunami (2004), Eyjafjallajökull Eruption (2010), Yellowstone Supervolcano (~100,000)

An 1888 lithograph of the 1883 eruption of Krakatoa.

View of Krakatoa during the Earlier Stage of the Eruption.
from a Photograph taken on Sunday the 27th of May, 1883.

The Sierra Club

John Muir (1838–1914)

The mid-to-late nineteenth century witnessed a dramatic increase in global environmental awareness and activism, partly motivated by publicly engaging voyages, writings, and presentations by explorers and naturalists, but also partly by the seemingly boundless growth of cities caused by the Industrial Revolution.

A key player in the early history of environmental activism was the Scottish-American naturalist, philosopher, scientist, and author John Muir. Muir studied geology and botany in college but never graduated, instead succumbing to a spirit of wanderlust that took him on numerous personal voyages of exploration and itinerant work across the US and Canada. Making his way to California in the late 1860s, Muir eventually settled in Yosemite Valley, a spectacularly scenic region of the Sierra Nevada mountain range, where he lived, wrote, and studied the geology. Muir hiked, climbed, and interpreted the landscape; his ideas about the prominent role of glaciers in forming Yosemite were controversial at the time, but were consistent with the growing evidence of past ice ages propounded by Louis Agassiz and others.

Muir recognized that unchecked development, logging, and mining could threaten pristine natural lands like Yosemite and the Giant Sequoia Forest, and so he helped to advocate for state and federal protection of these areas. His influence helped persuade Congress to make Sequoia the nation's second National Park, and to place Yosemite Valley under California state protection, in 1890. Recognizing the potential power and political influence of like-minded citizens in such advocacy efforts, Muir founded the Sierra Club in 1892 and became its first president. Among the Club's earliest legislative victories under Muir's guidance were the creation of Glacier and Mount Rainier National Parks, the official transfer of Yosemite Park to federal control, and the establishment of the National Park Service.

Today, the Sierra Club boasts over three million members, and its mission is "to explore, enjoy, and protect the planet." As development and commercial activity continue to expand and population continues to grow across the planet, the role of citizen-based environmental advocacy and watchdog groups like the Sierra Club continues to be critical.

SEE ALSO The Sierra Nevada (c. 155 Million BCE), Charting North America (1804), Industrial Revolution (c. 1830), Discovering Ice Ages (1837), Birth of Environmentalism (1845), Deforestation (c. 1855–1870), Natural Selection (1858–1859), Exploring the Grand Canyon (1869), National Parks (1872), Earth Day (1970)

A 1903 photograph of American conservationist John Muir, founder of the Sierra Club, with US President Theodore Roosevelt on Glacier Point in Yosemite Valley, California.

The Greenhouse Effect

Joseph Fourier (1768–1830), **Svante Arrhenius** (1859–1927)

We often think of our home planet as a natural "Goldilocks" world, not the hellish inferno of closer-to-the-Sun Venus or the frozen ice world of farther-from-the-Sun Mars. It wasn't until the end of the nineteenth century, however, that scientists realized that Earth is a habitable oceanic world only because of the influence of two relatively minor, but critically important, atmospheric gases: water vapor (H_2O) and carbon dioxide (CO_2). Without them, Earth's oceans would freeze solid, and life on our planet would definitely be very different, if indeed any life existed at all.

In the 1820s, the French mathematician Joseph Fourier (pronounced FOO-yay) was the first scientist to realize that the Earth's equilibrium temperature—how warm the surface would be if it were heated by sunlight alone—was actually well below freezing. So why are the oceans liquid? Fourier speculated that the atmosphere might act as an insulator, perhaps trapping heat like the panes of glass in a greenhouse. But Fourier wasn't sure.

It was the Swedish physicist and chemist Svante Arrhenius who provided the answer, showing that gases in our atmosphere are indeed warming the surface, by more than 30 degrees Celsius (by more than 50 degrees Fahrenheit)—thus keeping our planet above freezing. The specific gases responsible—H_2O and CO_2—are transparent and thus let sunlight reach the surface, but they absorb a large part of the outgoing infrared heat energy emitted by our planet, thereby warming the atmosphere. Even though the cause of the warming is different from that in a closed glass room like a greenhouse, it's still called the "greenhouse effect," partly because of the earlier ideas and experiments discussed by Fourier.

Arrhenius knew that greenhouse warming was a simple—and fortunate—consequence of Earth's natural abundance of H_2O and CO_2, and he speculated that past decreases in carbon dioxide, especially, could explain the ice ages. He was the first to further speculate that future burning of fossil fuels could enhance the carbon dioxide abundance and lead to global warming. The Earth's climate is more complex than Arrhenius envisioned, but still his concern over the role people might have on changing the Earth's environment has turned out to be prescient.

SEE ALSO Life on Earth (c. 3.8 Billion BCE?), Cambrian Explosion (c. 550 Million BCE), Dinosaur-Killing Impact (c. 65 Million BCE), End of the Last "Ice Age" (c. 10,000 BCE), The Little Ice Age (c. 1500), Discovering Ice Ages (1837), Rising CO_2 (2013)

The Earth's atmosphere acts somewhat—though not exactly—like a transparent greenhouse surrounding our planet, letting in sunlight, which warms the surface, but then on its way out trapping heat, which warms the atmosphere.

Radioactivity

Wilhelm Röntgen (1845–1923), **Henri Becquerel** (1852–1908),
Pierre Curie (1859–1906), **Marie Skłodowska Curie** (1867–1934)

Late-nineteenth-century physics labs in Europe and America were quite literally abuzz with new discoveries related to electricity and magnetism. The newly created ability to generate and store large voltages and currents for use in a variety of experiments often led to surprises, such as the German physicist Wilhelm Röntgen's 1895 study of high-voltage cathode-ray tubes that generated a mysterious new form of radiation, which he dubbed X-rays.

The French physicist Henri Becquerel suspected that the ability of some natural materials to phosphoresce (glow in the dark) might be related to these X-rays. He conducted a series of experiments in 1896 to determine if these materials emitted X-rays when exposed to sunlight, and accidentally discovered that one of the materials, uranium salts, spontaneously emitted radiation on its own. He had discovered radioactivity, something altogether different from X-rays.

Becquerel went on to collaborate with fellow French physicists Pierre Curie and Marie Skłodowska Curie, who were also interested in the strange and exotic behavior of this newly discovered spontaneous radiation. Marie Curie's studies of uranium, in particular, led her to the discovery of two new radioactive elements: polonium (named for her native Poland), and radium. In recognition of their fundamental discoveries, Becquerel and the Curies were awarded the 1903 Nobel Prize in Physics. Marie Curie was the first woman to win a Nobel Prize; she would win another in Chemistry in 1911 and is still the only scientist to have won the prize in two different fields, and the only woman to have won two.

Over the past century, radioactivity has been exploited as a natural "clock" because radioactive elements release energy and decay into other elements at a predictable rate. Radioactivity has been used to accurately determine the ages of the Earth, the Moon, meteorites, and, by extension, the entire solar system and beyond, using the age and evolution of the Sun as a guide. Because of radioactivity and the pioneering work of scientists like Becquerel and the Curies, we now know, with astonishing precision, that the Earth is 4.54 billion years old, and that it formed very soon after the Sun itself formed 4.568 billion years ago.

SEE ALSO Earth Is Born (c. 4.54 Billion BCE), Birth of the Moon (c. 4.5 Billion BCE), Foundations of Geology (1669), The Age of the Earth (1862)

Main image: *Pierre and Marie Curie, at work studying radioactivity in their laboratory in Paris, sometime before 1907.* **Inset:** *A 1918 photo of Henri Becquerel, discoverer of radioactivity.*

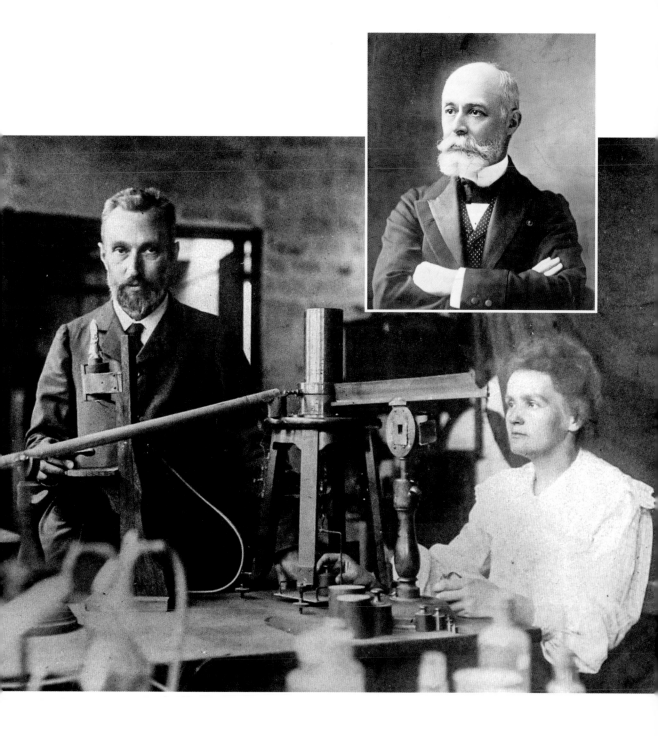

Structure of the Atmosphere

Léon Teisserenc de Bort (1855–1913), **Richard Aßmann** (1845–1918)

The advent of balloon-borne remote sensing in the mid-nineteenth century and improvements in meteorological instruments and the balloons themselves enabled researchers to make direct measurements of the temperature and pressure of the Earth's atmosphere up to very high altitudes. The pioneers in this new field of *aerology* (study of the structure of the atmosphere) were the French meteorologist Léon Teisserenc de Bort and the German meteorologist Richard Aßmann, contemporaries who were independently using high-altitude hydrogen balloons to measure the structure of the Earth's atmosphere. Starting in 1896, for example, de Bort began launching hundreds of instrumented, unmanned weather balloons high into the atmosphere. Balloons at the time could operate up to altitudes of about 56,000 feet (17,000 meters).

These early pioneers discovered that the atmosphere is divided into at least two distinct layers. In the lowest layer close to the surface, which de Bort called the *troposphere* (from the Greek "turning"), the temperature slowly decreases with altitude up to around 40,000 feet (12,000 meters). Above that in the higher layer, however, the temperatures remained relatively constant for as high as their balloons could fly. De Bort called this upper zone the *stratosphere* (from the Greek "spreading out"), and the boundary between the layers the *tropopause*.

Subsequent weather balloons, high-altitude aircraft, suborbital sounding rockets, and eventually orbital satellites have significantly enhanced our understanding of the structure of our planet's atmosphere. Rather than just two main layers, Earth's atmosphere actually has five. From bottom to top, these are: (1) the troposphere, going from the surface to about 7 miles (12 kilometers) altitude and containing about 80 percent of the mass of the atmosphere; (2) the stratosphere, which was eventually discovered to extend up to 31 miles (50 kilometers) and to get significantly warmer above the limit of early balloons because of heating by atmospheric ozone; (3) the *mesosphere*, extending up to 50 miles (80 kilometers), and where temperatures once again decrease with altitude; (4) the *thermosphere*, extending up to 440 miles (700 kilometers), and where the solar wind and Earth's magnetic fields can ionize this rarefied air and produce auroral displays; and finally (5) the *exosphere*, extending up to about 6,200 miles (10,000 kilometers), and where atoms and molecules easily escape and return to our atmosphere along those same magnetic field lines.

SEE ALSO Airborne Remote Sensing (1858), The Greenhouse Effect (1896), The Ozone Layer (1913), Weather Radar (1947), Weather Satellites (1960), The Oscillating Magnetosphere (1984)

Photo of the hand-guided release of a typical US Weather Service atmospheric sounding balloon, sometime between 1909 and 1920.

Women in Earth Science

Florence Bascom (1862–1945)

The pursuit of modern science has been a male-dominated field for millennia. Academically as well as socially, women were discouraged from pursuing careers in scientific or technical fields, were actively prohibited from admission to leading academic institutions, and were barred from membership in and even recognition by leading scientific societies. While there are rare exceptions documented throughout history, for the most part science has been a boys' club for most of its history.

In the earth sciences, at least, this situation began to slowly change in the nineteenth and twentieth centuries. The British fossil collector Mary Anning (1799–1847), for example, made important contributions to paleontology in the early nineteenth century despite receiving no formal education in the field and little formal academic recognition at the time. As another example, geologist and paleontologist Mary Holmes (1850–1906) became the first American woman to be awarded a PhD in an Earth science field, in 1888.

Another pioneering late-nineteenth-century earth scientist who significantly advanced the cause of women in science was Florence Bascom, an American geologist who was the second woman to receive a PhD in the field (in 1893), and the first woman to be hired by the United States Geologic Survey, in 1896. Bascom's area of research was the composition and origin of volcanic rocks near current and former continental margins. She was an expert in mineralogy, crystallography, and petrology (the study of the formation of rocks), and, like many modern geologists, she combined field and laboratory observations. She developed and tested novel hypotheses for the origin and evolution of the Appalachian piedmont (foothills) region along the US eastern seaboard. While continuing to work for the USGS, she founded Bryn Mawr College's Department of Geology in 1901, teaching and advising students there until 1928.

While still facing numerous sexist obstacles and inadequate recognition during their careers, Anning, Holmes, Bascom, and a handful of others began slowly to open the academic doors for women. Those doors are still not fully open today, however: women now earn 50 percent of science PhDs, but account for less than 25 percent of university science faculty positions. As science educator Bill Nye "The Science Guy" says, "Half the world are women and girls, so let's have half the scientists and engineers be women and girls."

SEE ALSO The Appalachians (c. 480 Million BCE), Reading the Fossil Record (1811), The US Geological Survey (1879), Geology of Corals (1934), The Inner Core (1936), Mapping the Seafloor (1957), Seafloor Spreading (1973), The Oscillating Magnetosphere (1984)

Undated Smith College archive photo of geologist Florence Bascom, holding her geologic compass.

Galveston Hurricane

Hurricanes, also known as tropical cyclones, are rapidly rotating low-pressure storm systems characterized by strong winds, thunderstorms, and heavy rains. The storms are "tropical" because they form in warm open waters near the equator. Hurricanes are fueled by the energy released when intense tropical sunlight evaporates ocean water and that moist water rises, cools, and re-condenses as clouds and rain. Hurricane winds blow counter-clockwise in the northern hemisphere, and clockwise in the southern hemisphere. These storms generally move from east to west within about 30° latitude from the equator, but can veer dramatically north or south, or even back to the east, when they encounter continents or large islands. In Asia and the Western Pacific, hurricanes are called *typhoons.*

The deadliest hurricane to strike US soil (indeed, the deadliest natural disaster in American history) occurred on September 8–9, 1900, when a tropical cyclone slammed into Galveston, Texas, then a city of around 40,000 people on the Gulf of Mexico just south of Houston. Weather records from the Caribbean indicate that the storm passed over the islands of Hispaniola and Cuba five to six days earlier and then intensified dramatically once it entered the warm waters of the Gulf, damaging telegraph lines along coastal states and impairing communications. The storm hit Galveston with maximum sustained winds of 145 miles per hour (230 kilometers per hour) and caused massive destruction of homes, buildings, and other infrastructure. As many as 12,000 people may have been killed by the storm, devastating the city socially as well as economically for decades. While the deadliest hurricane in US history, it would be far from the costliest. That unfortunate title would go to Hurricane Katrina, which devastated New Orleans in 2005, killing nearly 2,000 people and causing more than $160 billion in estimated damages.

For most of human history, hurricanes/typhoons have been unpredictable phenomena that arrived with little warning. In records prior to the twentieth century, more than a million people in India, China, and Southeast Asia are known to have been killed directly or indirectly by tropical cyclones. Globally, we still haven't figured out how to weather these storms, however: even in our modern satellite weather-forecasting era, comparable numbers of people around the world have been killed by hurricanes because of poor communications, inadequate advance warning, insufficient shelter, and/or poor post-storm response and support.

SEE ALSO San Francisco Earthquake (1906), Big Burn Wildfire (1910), Tri-State Tornado (1925), Dust Bowl (1935), Weather Satellites (1960)

Survivors seeking valuables amid the wreckage of the city of Galveston, Texas, in the wake of the catastrophic hurricane of September 8–9, 1900.

Controlling the Nile

Every summer, melting snow in the mountains of Ethiopia causes the Nile River to swell above its banks and spill over into the surrounding desert. This annual, relatively predictable (even thousands of years ago) inundation brought fertile new sediments to the Nile River Valley, and enabled the development of sustained agriculture. However, the annual flooding of the Nile also prevented the development of cities and infrastructure along much of the river itself, making it harder to exploit the river's resources.

Ideas for taming the river and controlling the flooding go back to at least the eleventh century, focused on trying to dam the river at a narrow point near the city of Aswan, Egypt. By the end of the nineteenth century, after the invasion and occupation of Egypt by the British, the technology had advanced far enough to actually implement the idea. An earthen/masonry dam was designed; construction began in 1898 and was completed in 1902. The Aswan Dam became the largest masonry dam in the world, and was one of the largest modern civil engineering projects ever attempted.

Floodwaters gathered behind the Aswan Dam into a man-made reservoir named Lake Nasser (named after the Egyptian president who initiated the project that created it), used as a source of municipal water and irrigation. Gates released water through the dam to try to keep the downstream flow relatively constant throughout the year. However, the lake supply couldn't keep up with the growing population and agricultural demand in the lower Nile, or with the peak flooding rates from snowmelt either, and so the dam was raised twice in subsequent decades. By the mid-1940s, it had reached its height limit, however, and a new, even larger dam (another earthen embankment) had to be constructed about 4 miles (6 kilometers) upstream. The first dam became known as the Aswan Low Dam, and the second, completed in 1970, the Aswan High Dam.

The benefits of controlling the Nile also came with associated downsides. For example, an estimated 100,000 people were displaced by the new lake, sediment deposition in Upper Egypt has decreased dramatically, water quality is lower because of increased algae levels, and the Nile Delta is eroding faster than ever before. Overall, the human and environmental impact of controlling the Nile is still being critically assessed.

SEE ALSO Invention of Agriculture (c. 10,000 BCE), Civil Engineering (c. 1500), Hydroelectric Power (1994)

A 1912 photo of the original Aswan Dam in Egypt, from the west bank of the Nile River.

San Francisco Earthquake

The boundaries between the several dozen major tectonic plates are the sites of the highest levels of geologic activity on Earth. In some places like the Himalayas, head-on continental collisions are actively building the tallest mountains on the planet. In other places like the Andes or the US Pacific Northwest, oceanic plates slamming into and melting underneath continents are creating chains of earthquakes and highly active volcanoes. In still other places, like farther south along the western coastline of North America, plates sliding parallel to each other sometimes jostle and jolt the surrounding landscape, producing earthquakes but not volcanoes.

Just such a dramatic jostling occurred in the early morning of April 18, 1906, when residents of San Francisco and hundreds of square miles of the surrounding area were awoken by a tremendous earthquake. For nearly 45 seconds, the ground shook violently, uprooting buildings from their foundations and causing especially catastrophic damage to structures built upon sediments surrounding San Francisco Bay, due to a process now known as *liquefaction* (shaking that causes sediments to behave like a liquid and deform/warp dramatically). Hundreds of people were killed instantly by collapsing structures, and perhaps as many as 3,000 people eventually died from the ensuing fires that erupted.

Geologists measure the intensity of earthquakes based on their energy, on a logarithmic scale proposed in 1935 by American seismologist Charles Richter. Small earthquakes, often unnoticeable, that represent small-scale motions within the crust and mantle produce thousands of small earthquakes every day with magnitudes of 1 to 4 on Richter's scale. Magnitude 4 to 6 earthquakes, a few of which occur every day around the world, produce noticeable shaking and moderate damage. Strong-magnitude 6 to 8 earthquakes are rarer, occurring every few days to few weeks, on average, and cause violent shaking that does significant damage to structures in populated areas. The rarest earthquakes with magnitudes above 8.0, which occur on average every one to ten years, can cause massive catastrophic damage and enormous loss of life.

While nowhere close to the deadliest earthquake in history, the San Francisco disaster of 1906 helped to spur significant advances in future building construction and city planning that have no doubt saved countless lives as cities grow and the population expands on our continuously shifting crust.

SEE ALSO Plate Tectonics (c. 4–3 Billion BCE?), The Himalayas (c. 70 Million BCE), Cascade Volcanoes (c. 30–10 Million BCE), The Andes (c. 10 Million BCE), Galveston Hurricane (1900), Valdivia Earthquake (1960), Sumatran Earthquake and Tsunami (2004)

Photograph looking down Sacramento Street at part of the wreckage and ensuing fire from the April 18, 1906 San Francisco earthquake.

Hunting for Meteorites

Daniel Barringer (1860–1929)

The critically important role of impact cratering in altering planetary landscapes has only been understood since the mid-to-late twentieth century, partly because the formation of new craters is such a rare geologic event on the timescale of human history. In general, we are much more familiar with, and appreciative of, the major changes in the Earth's surface that are caused by more frequent volcanic, tectonic, or erosional processes and events.

Nonetheless, impact events *have* shaped the geologic and biologic history of our planet, and coming to that realization represents a major advance in modern geology, paleontology, and planetary sciences. Among the most important natural "classrooms" where this lesson has been learned is a 3,900-foot (1,200-meter) wide, 550-foot (170-meter) deep circular hole in the ground near Winslow, Arizona, that was originally called Coon Mountain. The origin of the circular depression was widely debated among academics, but leading US Geological Survey geologist G. K. Gilbert had come down strongly in the 1880s in support of the hypothesis that it was a volcanic feature, caused by a steam explosion from subsurface magma interacting with groundwater.

Not everyone believed that this was the right answer. Among the detractors was American geologist and mining entrepreneur Daniel Barringer, who believed that the feature was a crater caused by the high-speed impact of a metallic asteroid. Barringer was convinced that he could prove his impact hypothesis, first published in 1906, by searching for and then excavating the remains of the impacting asteroid, which he surmised was buried deep underneath the sediments that had since accumulated in the bottom of the crater. He purchased the land containing the crater and set about surveying and drilling at the site. The search continued unsuccessfully until 1929, yielding no more than a collection of scattered fragments of iron-rich rocks.

Barringer's hypothesis was correct, but what would only be discovered decades later by future geologists such as Eugene Shoemaker is that the tremendous energy of high-speed impact events almost fully vaporizes the impactors (even those made of iron!), as well as a large percentage of the impacted target. Barringer's mother lode of iron had actually mostly vanished into thin air around 50,000 years ago.

SEE ALSO Dinosaur-Killing Impact (c. 65 Million BCE), Arizona Impact (c. 50,000 BCE), The US Geological Survey (1879), The Tunguska Explosion (1908), Understanding Impact Craters (1960), Meteorites and Life (1970), Extinction Impact Hypothesis (1980)

The largest discovered fragment of the small metallic asteroid that created Meteor Crater, on display at the crater's visitor center in Winslow, Arizona.

COLLISION &
IMPACT

Meteor Crater is the Best Preserved and First Proven meteor crater on Earth.

The Holsinger Meteorite is the largest discovered fragment of the 150-foot (45-meter) meteor that created Meteor Crater.

The Tunguska Explosion

Leonid Kulik (1883–1942)

Many residents of remote central Siberia near the Tunguska River were startled awake on June 30, 1908, by a spectacular event. Around 7:15 a.m., the skies erupted in a blinding flash of light, followed by thunderous explosions. The ground shook with the force of a magnitude 5.0 earthquake. Farther away, a fierce hot wind and a rain of fire stripped and felled 80 million trees over more than 811 square miles (2,100 square kilometers)—an area half the size of Rhode Island. Seismic shocks from the event were recorded across Asia and Europe, and night skies around the world glowed with an eerie light for days afterward.

Scientists suspected that the region had experienced a meteoroid impact. The first scientific group to study the remote, uninhabited region where the event actually occurred didn't arrive until 1927, however, when the Russian mineralogist Leonid Kulik searched in vain for the resulting impact crater. Apparently, the event was an airburst explosion, with surface damage caused by shock waves, heat, and fire—but with no associated crater formed like the one at Meteor Crater in Arizona.

Planetary scientists have debated the nature of the Tunguska impactor for more than a century. Was it an icy comet fragment that disintegrated catastrophically from atmospheric entry, or a small, rocky asteroid, perhaps a rubble-pile object that was too weak to survive all the way to the surface? Whatever its origin, the best hypothesis is that an object only about 33 feet (10 meters) across, traveling at around 6 miles (10 kilometers) per second, exploded around 6 miles (10 kilometers) above the surface with an energy of about 10 megatons of TNT—similar to that at Meteor Crater and more than 500 times the yield of a World War II atomic bomb.

A similar, though thankfully less energetic, fireball and airburst explosion occurred over the western Russian city of Chelyabinsk in 2013. Amazingly, no one was killed by the Tunguska or Chelyabinsk explosions (though many were injured by shattered glass in Chelyabinsk). Tunguska and Chelyabinsk are wake-up calls for understanding impact events, especially the catastrophic shock-wave effects that even small objects traveling at extreme speeds can have on our environment when they occasionally slam into our planet.

SEE ALSO Dinosaur-Killing Impact (c. 65 Million BCE), Arizona Impact (c. 50,000 BCE), The US Geological Survey (1879), Hunting for Meteorites (1906), Understanding Impact Craters (1960), Meteorites and Life (1970), Extinction Impact Hypothesis (1980)

Main image: *Artist and planetary scientist William K. Hartmann's impression of the Tunguska forest one minute after the airburst explosion. The painting was made at Mount St. Helens, where the 1980 blast from that volcano produced a Tunguska-like scene.* **Inset:** *1927 photo from the Kulik expedition.*

Reaching the North Pole

Robert Peary (1856–1920)

The desire to push the frontiers of the exploration of our planet over the course of human history has been historically driven by many factors, including nationalism or national expansion, economic exploitation, and scientific advancement. In some instances, however, the drive appears to have been more personal, sometimes even devolving to the level of personal or professional bragging rights. A case in point can be found in the first expeditions to reach and explore our planet's north and south poles.

American Robert Peary was a civil engineer and naval officer with expertise in surveying. Partly self-funded and partly supported by wealthy patrons, he made several trips to the Arctic in the 1880s and 1890s, helping, for example, to prove that Greenland was an island. Between 1898 and 1909, Peary set out on several new expeditions to attempt to be the first to reach the geographic North Pole. In 1909, with a team that included his personal assistant Matthew Henson and four Inuit guides, Peary claimed to have become the first to finally reach the pole, at 90° North latitude.

The claim was not without controversy, as at least one other team of explorers had claimed to have reached the pole a year earlier, and subsequent academic study of Peary's records revealed what some believe to be inconsistencies and/or inaccuracies in his navigational records. Investigative work at the time, however, by organizations such as the National Geographic Society and the newly formed Explorers Club, led to the wide acknowledgement of Peary and his team as first to the pole. Peary was promoted by the Navy to the rank of Rear Admiral and would go on to become president of the Explorers Club.

Because of inherent limits in the accuracy and reliability of navigational equipment around the turn of the twentieth century, as well as the obvious temptations that exaggeration or even blatant falsification of records would have presented to explorers of the time, Peary's claim of being first to reach the North Pole is still mildly debated even today. Indeed, some think that it wasn't until the expedition of a dirigible crew in 1926 (led by Norwegian polar explorer Roald Amundsen) that scientifically verified evidence of truly reaching the North Pole was secured.

SEE ALSO Reaching the South Pole (1911), Exploration by Aviation (1926), Ascending Everest (1953), International Geophysical Year (IGY) (1957–1958)

Main image: *April 1909 photograph by Robert Peary of his exploration crew, at what they believed was the geographic North Pole.* **Inset:** *1909 photograph of Adm. Robert Peary, in full arctic fur garb.*

Big Burn Wildfire

Ever since the first land plants developed on Earth some 470 million years ago, increasing plant abundance combined with increasing atmospheric oxygen levels combined to make wildfires an important part of terrestrial ecology. Lightning strikes, volcanic eruptions or lava flows, and even spontaneous combustion under hot and dry conditions have all been potential natural causes of wildfires throughout history. Of course, they can also be caused by human carelessness, arson, or intentional agricultural clearing.

Once begun, wildfires can quickly consume enormous areas of forest. An extreme example was the so-called "Big Burn," which swept across parts of the western US states of Washington, Idaho, and Montana during a massive firestorm on August 20–21, 1910. During that short time, what had been many smaller fires were merged by strong dry winds into a single enormous inferno that burned an area of almost 3 million acres—roughly the size of the state of Connecticut. The Great Fire of 1910 is the largest wildfire in US history; it killed 87 people, mostly firefighters trying to put out the blaze. A half dozen cities suffered extensive damage, and thousands of people lost their homes, farms, and businesses.

The Big Burn was a defining moment for the newly formed (1905) US Forest Service, which was charged, among other duties, with controlling the damage from forest fires. Uncertain and inconsistent methods for fighting the Big Burn hampered efforts to control it, however. Ultimately, the Great Fire of 1910 helped to establish consistent equipment, methods, and training for firefighters battling wildland fires, as well as the Forest Service's policy to suppress every fire as soon as possible in order to preserve the forest. Ironically, since the Big Burn, some have claimed that the Forest Service's fire-suppression policy may have actually unintentionally promoted the occurrence of fires in some places, by increasing the stores of tinder (dried dead branches and leaves) on the forest floor.

Despite their potential for causing substantial loss of life and property, wildfires are a critical part of Earth's natural plant cycle for many species. The cones of the lodgepole pine, for example, common across the western US, cannot open and release their seeds unless they are heated to the high temperatures typical of a wildfire.

SEE ALSO First Land Plants (c. 470 Million BCE), Oldest Living Trees (c. 3000 BCE), Deforestation (1855–1870), San Francisco Earthquake (1906), Boreal Forests (1992), Temperate Deciduous Forests (2011)

Photograph of residents of Wallace, Idaho, searching through the wreckage of the town after the Great Fire of 1910.

Reaching the South Pole

Roald Amundsen (1872–1928), **Robert Scott** (1868–1912),
Ernest Shackleton (1874–1922)

In the same way that American explorer Robert Peary and others had embarked on a race, of sorts, to be the first people to reach the Earth's North Pole in 1909, there was a similar heated battle that went on among teams from around the world to be the first people to reach the South Pole around the same time. The task was known ahead of time to be quite a different challenge, in that the South Pole, unlike the North, was on an ice sheet above a continental landmass. This meant that an 800-mile (1,300-kilometer) overland trek in harsh wintery conditions would be required.

British polar explorer Ernest Shackleton and a team of three companions on a geographic and scientific expedition aboard the sailing ship *Nimrod* trekked to within about a hundred miles of the South Pole in 1908–1909, but were forced to turn back. Two teams vied for the prize again in the Antarctic summer of 1911. One expedition, funded by the British Royal Society, was led by Royal Navy officer Robert Scott, who sailed to Antarctica on the *Terra Nova* and then proceeded overland via convoy toward the pole. Scott and four team members reached the South Pole on January 17, 1912, only to find a small tent flying the Norwegian flag, placed there five weeks earlier by his rival Roald Amundsen and his team of four men from the expedition of the ship *Fram*. Amundsen was a self- and patron-funded polar explorer who turned his attention to the Antarctic after Peary had reached the North Pole.

Amundsen and his well-prepared and well-stocked crew returned to civilization in March 1912 and became heroes for winning the race. Meanwhile, Scott and his four colleagues, after realizing that Amundsen had won (he wrote in his journal, "The worst has happened . . ."), set out to return to the *Terra Nova* but perished amid poor weather and dwindling supplies along the way. When word of his team's perilous adventure and fate finally reached England in early 1913, he, too, was honored as a hero.

Today, Antarctica is a continent devoted to scientific research, with stewardship shared by many countries, and a US scientific research base located at the pole named, appropriately, the Amundsen–Scott South Pole Station.

SEE ALSO Reaching the North Pole (1909), Exploration by Aviation (1926), Ascending Everest (1953), International Geophysical Year (IGY) (1957–1958)

Photograph of Roald Amundsen (right) and three other members of his team planting the Norwegian flag at the South Pole on December 17, 1911.

Machu Picchu

Hiram Bingham (1875–1956)

The fifteenth- and sixteenth-century Inca civilization stretched through the Andes Mountains along the western coast of South America between what are today the countries of Ecuador and Chile. At its peak, tens of millions of people lived under the rule of the Incan Empire, the largest empire in the pre-Columbian Americas. The Inca mastered masonry techniques to build monumental architectural structures and roads, produced distinct finely woven textiles, and developed innovative agricultural methods specifically adapted to their primarily rugged, mountainous terrains. Most of the major Inca cities and much of their historical, religious, and societal heritage were plundered or destroyed after the arrival of the Spanish conquistadors in the 1520s and 1530s.

One Incan citadel that survived (undiscovered) the Spanish arrival was the small estate of Machu Picchu, located in the mountains high above the former Incan capital of Cusco (now in Peru), at nearly 8,000 feet (2,450 meters) elevation. Machu Picchu was "rediscovered" by Westerners in the late nineteenth and early twentieth centuries. Among the most famous of the early explorers who helped to reconstruct and popularize the history of this now most famous of Peruvian tourist attractions was the American academic and explorer Hiram Bingham. Bingham was not formally trained as an archaeologist, but he studied and taught Latin American history at Yale University and traveled extensively in Peru and surrounding countries, funded by the National Geographic Society. During one particular trip in 1911, Bingham was taken by locals to the site of Machu Picchu, which had become overrun by the jungle in the centuries since its abandonment. Bingham oversaw the clearing and restoration of the site, and his book, *The Lost City of the Incas*, became a best seller.

Archaeologists have since determined that Machu Picchu was essentially a vacation resort for the Incan royal family, housing a support staff of about 750 people responsible for the maintenance, upkeep, farming, and animal husbandry of the estate, and for the import and transportation of additional supplies needed to sustain the community. Researchers continue to study the origin and meaning of many of the stone structures found at Machu Picchu, and many of the artifacts (ceramics, statues, jewelry, and human remains) removed by Bingham and others for research and museum display elsewhere had been returned to Peru by 2012.

SEE ALSO The Andes (c. 10 Million BCE), Stonehenge (c. 3000 BCE), Charting North America (1804), Birth of Environmentalism (1845)

A 2012 photo of the ruins of the fifteenth-century Inca city Machu Picchu, now an archaeological site, in Peru.

Continental Drift

Alfred Wegener (1880–1930)

Even to a child, a map of the Earth's continents can seem like a jigsaw puzzle. If we just move South America next to southern Africa, they line up! Similarly, North America and Greenland line up with northwestern Africa and Europe. *Voilà!* Despite the apparent intuitive nature of the alignments, the reality of the situation for geologists is that there is no obvious way for continents to have moved or "drifted" through the oceanic crust from those positions to their current ones.

And yet, some scholars, such as the German geologist Alfred Wegener, couldn't shake the idea that at some point in the past, the continents *had* indeed been together as one larger landmass. Wegener looked carefully, for example, at rock types and especially plant and animal fossil types on opposite sides of the Atlantic Ocean and found amazing similarities. He also found that some fossil species in tropical India had once flourished in much more temperate latitudes, and surmised that Antarctica, Australia, India, and Madagascar had also once been connected to the eastern side of Africa. In a seminal 1912 research paper, Wegener postulated that all the continents had once been part of a single landmass (an "*Urkontinent*" in German), and that they have been drifting apart from one another since then.

Wegener's continental-drift hypothesis was met with stiff resistance from the established geologic community, many of whom viewed Wegener as an outsider. Geologists had major concerns about the mechanism by which continents could move through rigid oceanic crust (Wegener hadn't solved this), and the rate that Wegener thought the continents were drifting turned out, in later measurements, to be more than 100 times the actual value of relative continental motion. Still, others couldn't shake the idea, either, and research continued.

Wegener's hypothesis was ultimately vindicated, in a way, by the mid-to-late twentieth-century discoveries of island arcs, seafloor spreading, and the eventual realization that Earth's crust is divided into a few dozen large tectonic plates that essentially "float" on the upper mantle and move relative to one another over time. Wegener's *Urkontinent* had existed—geologists now call it Pangea. Sometimes, the solution to a puzzle really is as obvious as it seems, but it takes an outsider's perspective to realize it.

SEE ALSO Plate Tectonics (c. 4–3 Billion BCE?), Pangea (c. 300 Million BCE), The Atlantic Ocean (c. 140 Million BCE), Island Arcs (1949), Mapping the Seafloor (1957), Seafloor Spreading (1973)

Main image: *A 1930 photograph of German meteorologist and geologist Alfred Wegener.*
Inset: *Illustration of an early computer-assisted "fit" of many of the current continents, reconstructing Pangea.*

The Ozone Layer

Henri Buisson (1873–1944), **Gordon M. B. Dobson** (1889–1975), **Charles Fabry** (1867–1945)

As free oxygen began to build up in the Earth's atmosphere, first due to the respiration of microorganisms called cyanobacteria during and since the Archean, and then more recently by the rapid proliferation of oxygen-producing land plants, so too was the abundance of secondary molecules formed because of the presence of that increasing oxygen. One of the most important of those secondary molecules is ozone (O_3).

Ozone is produced high in the atmosphere when high-energy ultraviolet (UV) radiation from the Sun is absorbed by free oxygen (O_2), splitting it into individual oxygen atoms (O), one of which combines with another O_2 molecule to form O_3. Even though the maximum abundance of O_3 is only about ten parts per million, that relatively tiny amount is critical to life on Earth. Specifically, if solar UV light made it unimpeded all the way to the surface, it would quickly break down carbon–hydrogen and other molecular bonds that are critical parts of the organic molecules of life. Indeed, the small fraction of solar UV that makes it to the surface can be harmful (e.g., extreme sunburns) or even destructive to organic molecules, especially at high altitudes. Without ozone, we wouldn't be here.

Ozone was discovered in 1913 by the French physicists Charles Fabry and Henri Buisson, who noted a substantial amount of "missing" ultraviolet radiation in the spectrum of the Sun and who deduced that the culprit was O_3. Building on this discovery, British physicist Gordon M. B. Dobson perfected a special spectroscopy instrument to measure and track atmospheric O_3 from the ground, and established a worldwide network that has been monitoring O_3 since 1958 (in honor of his achievements, meteorologists now measure ozone abundance in "Dobson units"). Since the 1960s, O_3 has been actively monitored from space as well.

In the 1970s, the ozone layer started to become dramatically depleted over the Earth's poles ("ozone holes") due to the interfering effects of chlorofluorocarbons (CFCs), human-produced chemicals used primarily as refrigerants which break down O_3. In a triumphant achievement of environmental stewardship and corporate responsibility, CFCs have been phased out in favor of other chemicals that do not break down O_3, and the ozone layer has slowly recovered.

SEE ALSO The Archean (c. 4–2.5 Billion BCE), Photosynthesis (c. 3.4 Billion BCE), The Great Oxidation (c. 2.5 Billion BCE), First Land Plants (c. 470 Million BCE), Structure of the Atmosphere (1896)

September 2000 image showing the largest recorded "ozone hole" above Antarctica, created from data acquired by the Total Ozone Mapping Spectrometer (TOMS) instrument onboard the NASA Earth Probe satellite.

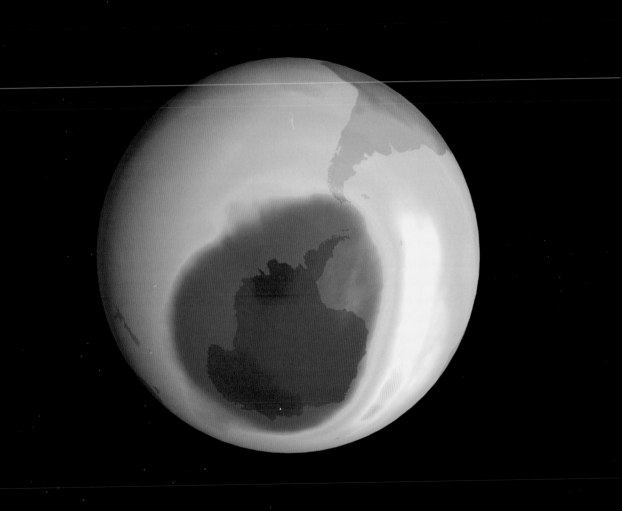

The Panama Canal

One of the reasons that fifteenth- and sixteenth-century European explorers and entrepreneurs were not able to find a route westward to the Spice Islands was that the combined stretch of North, Central, and South America made for a formidable obstacle and a lengthy detour. By the seventeenth and eighteenth centuries, charting and mapping of the eastern and western coastlines of the Americas would reveal that the obstacle was actually extremely narrow in the regions of Central America that are now Panama and Nicaragua. Ideas for a canal began to form.

Specific plans for a canal across Nicaragua go back to the early twentieth century, but have been scuttled by environmental concerns over the potential damage to freshwater Lake Nicaragua, which would be an important part of the waterway. Attention turned instead to the narrow isthmus of Panama, where plans for a potential canal had been discussed since the 1500s. In the late nineteenth century, the US helped to build a railway across the isthmus, to speed the transport of people and cargo from the East Coast to the Gold Rush bonanza of California. Between 1881 and 1884, the French attempted to build a ship canal across Panama, but more than 20,000 men died from malaria and other tropical maladies, and the project went bankrupt.

In what is often cited by historians as a classic "gunboat diplomacy" tactic, the US supported a rebellion in what would be the new nation of Panama (formerly part of Colombia) in exchange for rights to build and indefinitely administer a canal across the country. A plan for a series of locks and manmade lakes that would span the rugged topography between the Atlantic and Pacific Oceans was devised (building on some of the previous French excavation work), and construction of the canal began again in 1906. Construction of the 50-mile (80-kilometer) canal was completed in 1914, and was then the largest American engineering project conducted to date, at a cost of more than $9 billion in today's currency. Even though substantial measures were taken to improve sanitation and worker health, more than 5,000 people, many of them migrant laborers, still lost their lives in the construction of the canal.

The canal cut thousands of miles and weeks of travel time off trips for ships passing through. Today, more than 14,000 ships per year pass through one of the world's busiest shipping lanes.

SEE ALSO The Spice Trade (c. 3000 BCE), Civil Engineering (c. 1500), Circumnavigating the Globe (1519), Controlling the Nile (1902)

1914 photograph of the USS Santa Clara *passing through the Miraflores locks of the Panama Canal.*

Exploring Katmai

Robert F. Griggs (1881–1962)

Frequent volcanic eruptions are a hallmark of the famous "Ring of Fire" that girds the Pacific Ocean rim, including the long, curving chain of the Aleutian Islands extending westward from mainland Alaska. There, the Pacific plate is subducting under the North American plate, forming a deep trench (the Aleutian trench, up to 25,000 feet [7620 meters] deep) just south of the chain of active volcanic islands. Two of the most famous and recently active of these volcanoes are the adjacent summits of Katmai and Novarupta, on the far eastern end of the island arc.

On June 6, 1912, residents of southern and southeast Alaska witnessed an enormous eruption coming from the Katmai region. Over the course of about sixty hours, the eruption ejected more than thirty times as much ash into the stratosphere as the 1980 eruption of Mount St. Helens, and 50 percent more than in the 1991 eruption of Mount Pinatubo. While not quite as prolific as the earlier eruptions of Mount Tambora and Krakatoa, Katmai would turn out to be the largest volcanic eruption of the twentieth century. Ash up to a foot (30 centimeters) deep blanketed nearby Kodiak Island, and ashfall was reported as far south as Seattle.

Eager to investigate the effects of the eruption on the local flora and fauna, American botanist Robert Fiske Griggs embarked in 1915 on a series of expeditions, funded by the National Geographic Society, to explore and document the devastation and ensuing regrowth of the Katmai region. On Griggs's second expedition, in 1916, the group explored (and named) the Valley of Ten Thousand Smokes, a valley in Katmai that was filled with ash from the 1912 eruption and which still contained huge numbers of *fumaroles* (fractures) venting steam through the ash. The 1916 Griggs expedition also discovered a new volcanic mountain adjacent to Katmai that they named Novarupta; geologists in the 1950s discovered that Novarupta, not Mount Katmai, was actually the source of the massive 1912 eruption.

Griggs's widely read *National Geographic* articles and subsequent expeditions generated enormous interest in preserving the environment of Katmai and Novarupta, leading to the creation of Katmai National Monument in 1918 and, eventually, Katmai National Park in 1980.

SEE ALSO Plate Tectonics (c. 4–3 Billion BCE?), Mount Tambora Eruption (1815), National Parks (1872), Krakatoa Eruption (1883), Island Arcs (1949), Mount St. Helens Eruption (1980), Volcanic Explosivity Index (1982), Mount Pinatubo Eruption (1991), Yellowstone Supervolcano (~100,000)

Botanist and explorer Robert F. Griggs and colleagues L. G. Folsom and B. B. Fulton setting up camp in Katmai Village in 1915.

Russian Famine

Human history is replete with a long and sad history of famines (periods of food scarcity) that have claimed hundreds of millions of lives. The causes of the most deadly famines have varied widely, but usually include the individual or combined effects of drought, flooding, crop diseases, rapid population growth, poor economic policies, and/or wars. In addition to starvation, large numbers of people have died in famines from the subsequent effects of malnutrition and the spread of epidemic diseases.

Among the deadliest famines of the modern era was the Russian famine of 1921–1922, which led to the premature deaths of an estimated 5 million people, and which significantly impacted the lives of tens of millions more. While the famine itself began with the devastation of crops during intense flooding of the Volga River in the spring of 1921, followed by an intense drought in the region in the summer of 1921, the stage had really been set for disaster by nearly seven combined years of World War I and the Russian civil wars of 1919–1921. Those conflicts created dramatic economic and political changes in Russia, as well as unrest, violence, and a confrontational environment between Russia's central government and the farmer/peasant communities responsible for growing and transporting food across the newly communist nation. The duration and effects of the famine were also exacerbated by the initial refusal of the Lenin government in Moscow to accept foreign aid.

While there were other, even more deadly, famines in the nineteenth and twentieth centuries (the worst occurring in China and India, killing tens of millions), the timing and circumstances of the Russian famine of 1921–1922 were pivotal in helping to establish a broader network of internationally accepted governmental aid agencies like the American Relief Administration (formed in 1919 and headed by future president Herbert Hoover) and eventually globally administered efforts like the United Nations' World Food Programme, as well as helping to expand the mission and scope of more privately funded aid agencies such as the American Red Cross. The mission of global aid organizations remains critical, as, even in our twenty-first-century world, famines driven by political unrest and climate/weather disasters still cut millions of lives short.

SEE ALSO Invention of Agriculture (c. 10,000 BCE), Population Growth (1798)

Pre-1921 photo of Quaker aid workers from England and the US distributing clothing and food to hungry children in the village of Novosemejkino in the Volga River basin near the city of Samara.

Tri-State Tornado

Tornadoes (also called cyclones, whirlwinds, or twisters) are rapidly rotating columnar storms that extend from clouds to the ground. Tornadoes vary considerably in size and speed, but average about 250 feet (80 meters) across, spin with average wind speeds below 110 miles (180 kilometers) per hour, and move across the landscape over an average distance of a few miles (several kilometers). However, the largest and strongest individual tornadoes can be more than 2 miles (3 kilometers) across, can travel more than 60 miles (100 kilometers), and spin with wind speeds greater than 300 miles (480 kilometers) per hour. At the base of a strong tornado, a cloud of rapidly rotating dust, sand, and natural and human-made debris can cause substantial damage and loss of life.

Indeed, the most deadly tornado in US history was one of these extreme storms. On March 18, 1925, a relatively small funnel cloud–shaped tornado was first spotted in Moore Township in southeastern Missouri. It quickly grew large and began moving northeast at 60–70 miles (96–113 kilometers) per hour, devastating a number of towns and cities along a more than 200-mile (320-kilometer) path that extended into southern Illinois and southwestern Indiana. Cars were thrown into the air, railroad tracks were ripped from the ground, and some buildings were completely knocked off their foundations. By the time the storm had dissipated, only about three and a half hours after it formed, almost 700 people had been killed, more than 2,000 more injured, 15,000 homes were destroyed, and dozens of communities were severely damaged (four completely obliterated), with an estimated cost of the damage approaching $2 billion in today's currency.

Meteorologists would later classify the "Tri-State Tornado" of 1925 as a Class F5 storm (the most damaging, with winds equaling or exceeding 300 miles (480 kilometers) per hour on the scale now used to classify these storms. It covered more ground than any other tornado in recorded history, and it was the deadliest tornado in the world until the Bangladesh tornado of 1989, which killed more than 1,300 people. Scientists continue to debate whether the 1925 tornado was a single storm or a family of tornadoes striking in series; regardless, the massive scale of the devastation and loss of life helped fuel calls for better tornado monitoring, which would finally come decades later with the advent of weather radar and satellite weather forecasting.

SEE ALSO Galveston Hurricane (1900), San Francisco Earthquake (1906), Big Burn Wildfire (1910), Russian Famine (1921), Dust Bowl (1935), Weather Radar (1947), Weather Satellites (1960), Vargas Landslide (1999)

Photo of the ruins of the town of West Frankfort, Illinois, along the path of the March 18, 1925 Tri-State Tornado.

Liquid-Fueled Rockets

1926

Konstantin Tsiolkovsky (1857–1935), **Robert Goddard** (1882–1945), **Wernher von Braun** (1912–1977)

Propelled by the burning of gunpowder, rockets have been around for more than a thousand years. The Chinese are credited with first using rockets in battle as well as for entertainment (fireworks). But in 1903 the Russian mathematician Konstantin Tsiolkovsky wrote the first scholarly work that envisioned them as more than weapons; he also saw them as a potential means of space travel. He worked out much of the theory of rocketry and was among the first to propose using liquid fuels instead of gunpowder to maximize the combustion efficiency as well as the rocket's thrust-to-weight ratio. Tsiolkovsky is widely regarded as the father of modern rocketry in Russia and the Soviet Union.

However, it was the American rocket scientist and Clark University physics professor Robert Goddard who was first able to test Tsiolkovsky's—and his own—theories and show that liquid-fueled rockets were feasible and could provide the thrust needed to lift significant mass to high altitudes. He developed and patented key designs for rockets powered by gasoline and liquid nitrous oxide, as well as designs for the concept of multi-stage rockets, which he claimed could eventually be used to reach "extreme altitudes." Even though the flights of his own rockets were modest by today's standards, Goddard's methods were sound, and others—including a group of postwar space-race engineers led by German-American rocketry pioneer Wernher von Braun—were able to expand on his designs to enable longer, higher, and eventually orbital (and beyond) flights.

Like many inventors, Goddard was a visionary, often working alone and seeing possibilities that others had overlooked. He was an early advocate of rocketry for atmospheric science experiments and, like Tsiolkovsky, for eventual travel into space. It is perhaps ironic that World War II was the impetus for the eventual development of the rockets that would posthumously achieve Goddard's dream of space travel.

SEE ALSO Gravity (1687), Sputnik (1957), Humans in Space (1961), Leaving Earth's Gravity (1968), Settlements on Mars? (~2050)

Robert Goddard poses with his first liquid-fueled rocket, which was launched from Auburn, Massachusetts, on March 16, 1926. Unlike conventional rockets today, this model's combustion chamber and nozzle were at the top, and the fuel tank below. It flew for 2.5 seconds and rose 41 feet (12.5 meters).

Exploration by Aviation

Richard E. Byrd (1888–1957), Roald Amundsen (1872–1928)

During the first few decades of the twentieth century, the pioneers of powered, controlled airplane flight in the US and Europe continued to develop more reliable, longer-range aircraft. These planes would enable a number of important aviation "firsts" in the 1920s (such as the first solo crossing of the Atlantic by Charles Lindbergh in 1927) and would extend the reach of a number of explorer-aviators of that era. Among the most successful was the American naval officer and explorer Richard Byrd, who on May 9, 1926, claimed to be the first (along with fellow Navy pilot Floyd Bennett) to fly over the North Pole.

Byrd and Bennett took off from an airfield on the Norwegian island of Svalbard (78.5°N latitude) and flew about 650 miles (1,050 kilometers) to the North Pole, circling the pole to make sextant measurements before returning to Svalbard. Just a few days later, Norwegian explorer Roald Amundsen (who had been the first to reach the South Pole back in 1911) and crew also flew over the North Pole, on a dirigible expedition from Svalbard to Alaska. Back in America, Byrd became a hero and was promoted to the rank of commander and awarded the Medal of Honor. Some historians debate, however, whether it was Byrd or Amundsen who was first to fly precisely over the pole. Regardless, the era of exploration by aviation had begun.

Byrd competed for but lost the race to Lindbergh to fly first across the Atlantic alone in 1927, but he went on to lead a number of US Navy ship and airplane exploration and mapping campaigns in Antarctica between 1928 and 1956. These expeditions included the first flight over the South Pole (1929), and the most extensive photographic, meteorologic, and geologic surveys yet conducted of the Antarctic continent (including, for example, the documentation of ten previously undiscovered mountain ranges). As part of the preparations for the International Geophysical Year (IGY) of 1957–1958, Byrd commanded a US Navy operation that established permanent Antarctic bases at McMurdo Sound, the Bay of Whales, and the South Pole, marking the beginning of the permanent scientific presence in Antarctica.

SEE ALSO Antarctica (c. 35 Million BCE), Reaching the North Pole (1909), Reaching the South Pole (1911), Exploring Katmai (1915), International Geophysical Year (IGY) (1957–1958).

Left: *Photograph of Richard Byrd in his flight jacket in the 1920s.*
Right: *1929 photo of a Fokker "Super Universal" aircraft used during the Byrd Antarctic Expedition.*

Angel Falls

Jimmie Angel (1899–1956)

As military and commercial aviation became more widely established in the 1920s and 1930s, more and more scientific and resource exploration expeditions began to rely on airplanes for their transportation, photography, and surveying/remote sensing needs. Mining companies, for example, began to use aerial surveys to identify potential new ore deposits, and aerial photos began to become more widely used by geologists and land surveyors prior to going out into the field. A prime example of the use of airplanes in such expeditions is the modern "discovery" of the tallest waterfall in the world, located deep in the jungle of Venezuela.

American James ("Jimmie") Angel learned to fly at a young age, and worked as a flight instructor, test pilot, and stunt pilot/barnstormer in the years following World War I. Eventually his piloting skills took him south of the border, where he worked in Mexico and Central and South America as an aerial surveyor for scientific and government expeditions as well as natural resource companies. During the 1930s, Angel played an important role in the scientific, archaeologic, and geologic exploration of the *Gran Sabana* (Great Savanna) region of southeastern Venezuela, including the many tall table-top mesa or *tepui* scattered across that part of the world. While searching for a particular ore bed on one particular flight on November 18, 1933, Angel flew over an enormous waterfall cascading down from the mesa known as Auyán-tepui. While the indigenous Pemón people likely had known about the falls for thousands of years, Angel had become the first modern Westerner to observe and document what would turn out to be the tallest waterfall in the world. It was named Angel Falls (Salto del Ángel) *in* his honor.

News and photographs of Angel Falls and of Jimmy Angel's subsequent adventures in and around Auyán-tepui and the remote wild jungles and savannas of Venezuela circulated widely, and led to increased scientific and tourist interest in the region. Ultimately, the desire to preserve the spectacular natural landscape and ecologic diversity of Angel Falls, Auyán-tepui, and the surrounding tropical rainforest and savanna led to the Venezuelan government's creation of Canaima National Park in 1962. Canaima is the sixth-largest national park in the world, and visits there and to Angel Falls have become one of the leading tourist attractions in Venezuela.

SEE ALSO First Mines (c. 40,000 BCE), Birth of Environmentalism (1845), Airborne Remote Sensing (1858), Exploration by Aviation (1926), Savanna (2013)

The 3,212-foot-tall (979 meters) cataract of Angel Falls, in the state of Bolivar, Venezuela.

Geology of Corals

Dorothy Hill (1907–1997)

Corals, which are small marine invertebrate organisms that live in underwater colonies and often build shallow-water reefs, appeared rather suddenly in the geologic record (like many other species) around the time of the so-called Cambrian Explosion, some 550 million years ago. Tabulate and rugose (wrinkled) coral species flourished during the Paleozoic era (from about 550 to 250 million years ago), but then went entirely extinct (like 96 percent of all other species) during the "Great Dying" at the Permian-Triassic boundary. The stony corals, which had been only a minor species of corals previously, somehow survived the mass extinction and continue to dwell in shallow ocean-water reef environments today.

Corals are highly sensitive to local environmental conditions (water temperature, salinity, acidity, sunlight levels), and so studies of fossil and living corals can provide important new insights about past climate conditions (like those leading to the Great Dying) as well as current climate variations today. Among the world's leading scientific authorities on corals was Australian paleontologist Dorothy Hill, whose early work at Cambridge University, including a key research paper published in 1934, helped to elaborate the ways that individual coral organisms ("polyps") attached to rocks or previous generations of polyps and grew into elaborate reefs and other structures. She was particularly skilled at figuring out how to infer the role and function of the coral's soft tissues—not preserved by fossilization—in the crystallization of new hard-tissue parts. Her subsequent work in Australia gave her access to both an extensive fossil record of corals preserved on an ancient continent, as well as active, living corals in the Great Barrier Reef and elsewhere.

Hill organized field campaigns, analyzed and interpreted fossil samples, secured research grants, and administered a growing research group—all of which remain among the critical skills of modern paleontology researchers. Just as importantly, she did it while climbing uphill, academically: she was the first female professor hired (in 1946) at an Australian university, and the first female president of the Australian Academy of Science (in 1970). Hill trained dozens of students, postdoctoral researchers, and staff members, leaving a living legacy of her impact on this important subfield of paleontology.

SEE ALSO Cambrian Explosion (c. 550 Million BCE), The Great Dying (c. 252 Million BCE), Reading the Fossil Record (1811), US Geological Survey (1879), Women in Earth Science (1896), The Inner Core (1936), Mapping the Seafloor (1957), Seafloor Spreading (1973), Great Barrier Reef (1981), The Oscillating Magnetosphere (1984)

Main image: *Dorothy Hill (center) supervising a group of field geology students.*
Inset: *Drawings of some of the many kinds of coral structures documented and studied by Dorothy Hill.*

1a Lopholasma
2a Stereolasma
1b
2b
3a
3b Stewartophyllum
4 Amplexiphyllum
5a
5b
5c Saleelasma

Dust Bowl

Water is a key component in most soils on the Earth, partly because it helps bind soil grains together and gives them cohesion, and partly because it helps support the growth of plants that can further promote soil stability, especially on slopes. Remove the water—as during drought conditions—and soils become much less cohesive, much less stable on slopes as plants die off, and much more susceptible to being eroded by wind and rain. Indeed, just such effects were experienced, with devastating consequences, during the so-called "Dust Bowl" period of the mid-1930s in the American Midwest.

In particular, extreme droughts across the Great Plains in Texas, Oklahoma, Kansas, Colorado, and New Mexico during 1934–1936, and 1939–1940 turned much of the previously rich soil and farmland in those areas to dust. Enormous, billowing clouds of dust were lifted and transported by the wind across hundreds of miles, blotting out the Sun, damaging homes and mechanical equipment, and causing countless respiratory problems. Events like the "Black Sunday" dust storms (stretching from Canada to Texas) of April 14, 1935, were widely reported, spawning a new name for the region and for the circumstances: the Dust Bowl. Tens of thousands of families, already poverty-stricken by the economic effects of the Great Depression, were forced to abandon their farms and migrate to other states. In some places, more than 75 percent of the arable topsoil was blown away by the wind.

Drought was only partially responsible for the Dust Bowl of the 1930s, however. Another important factor was the adoption of plowing and farming practices that removed the network of plant roots that holds the soil together in the grasslands of the Great Plains. Using newly mechanized farm equipment to substantially increase their crop sizes, but without understanding in detail the ecological importance of plants in stabilizing the soil, farmers helped to create the circumstances that would lead to rapid soil breakdown during subsequent periods of drought.

As a result of the enormous human and economic damage of the Dust Bowl, the US government became an active participant in land management and soil conservation efforts in the Great Plains and elsewhere. Today, farmers in the Great Plains are more savvy and educated, and they work closely with academic and government colleagues to adopt planting and harvesting practices that are more robust to (inevitable) events like occasional droughts.

SEE ALSO Soil Science (1870), Galveston Hurricane (1900), San Francisco Earthquake (1906), Big Burn Wildfire (1910), Russian Famine (1921), Tri-State Tornado (1925), Weather Radar (1947), Weather Satellites (1960), Grasslands and Chaparral (2004)

April 18, 1935, photograph of a towering dust cloud approaching the town of Stratford, Texas.

The Inner Core

Inge Lehmann (1888–1993)

The structure of the interior of the Earth has been revealed since the beginning of the twentieth century via the way earthquake seismic waves travel through the planet. Starting around then, seismologists—geologists who study earthquakes and the Earth's interior—have been placing sensitive earthquake monitoring systems around the world, building up a network that can detect not only strong local earthquakes, but also the weak waves from distant temblors that have occurred across the globe. By monitoring how long it takes waves to travel through the Earth, and by triangulating signals from the same earthquake observed at many widely spaced stations, seismologists have been able to learn that the Earth is divided into onionskin-like layers of core, mantle, and crust.

The initial model of the Earth's interior was relatively simplistic, and a number of seismologists were puzzled by conflicting signals in the ever-more-sensitive network of seismometers being deployed. A key discovery was made in 1936 by the Danish seismologist Inge Lehmann, who analyzed earthquake seismic data and hypothesized that the Earth's core was actually divided into two zones: a liquid (molten) outer core comprising about 60 percent of the core's volume, surrounding a solid inner core all the way to the center. Previously, partly because of the strength of the Earth's magnetic field, the core was thought to be a single molten object. Lehmann's calculations and analyses were heavily scrutinized but widely accepted by geologists within only a few years. She was widely honored in the scientific and geological community during her career, mostly for her seismological studies, but also for overcoming the significant challenges of what was still a mostly male-dominated field.

Today's global seismometer network continues to expand in coverage and sensitivity, and even small earthquakes send waves through the Earth that are monitored and studied to refine our model of the Earth's internal structure. The composition of the inner and outer core is now known to be mostly iron (with some trace heavy metals), and the mantle is now known to be divided into an inner and outer zone as well. The depths of the crust, inner and outer mantle, and inner and outer core boundaries are now known to high accuracy.

SEE ALSO Earth's Core Forms (c. 4.54 Billion BCE), Earth's Mantle and Magma Ocean (c. 4.5 Billion BCE), Women in Earth Science (1896), San Francisco Earthquake (1906), Earth Science Satellites (1972), Earth's Core Solidifies (~2–3 Billion)

Main image: *Illustrative cross-section of earthquake seismic waves traveling through the Earth's interior, including the different ways that the waves are refracted (bent) at the boundaries between the mantle, outer core, and inner core.* Inset: *A 1932 photo of Danish geophysicist Inge Lehmann.*

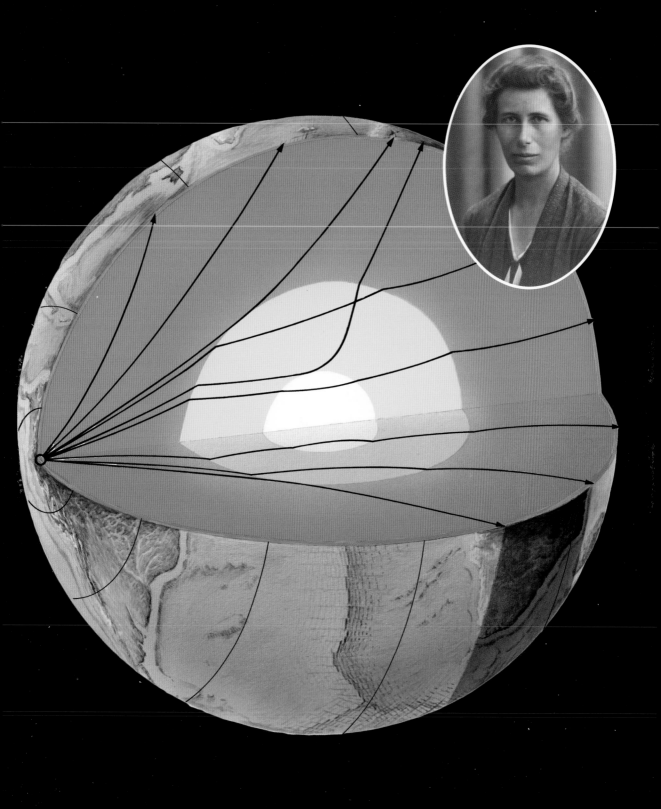

Landfills

As settlements and populations grew with the advent of organized agriculture through the Industrial Revolution, towns and cities inevitably had to start worrying about what to do with all their trash. Archaeologists have discovered evidence of a variety of different kinds of *middens*—dump sites for domestic trash—from prehistoric through pre-industrial societies. These refuse sites, which often contain food and plant waste, animal bones, excrement, pottery or stone tool fragments, or other artifacts, provide important information about the daily lives of people who lived there. Some middens are just holes in the ground quickly dug by nomadic societies; others consist of enormous mounds of shells, piles of animal manure, or trash heaps built up over generations on the (then) outskirts of towns and villages.

Accumulating trash or human waste within or near towns and villages began to become a more significant human health issue, however, as those places grew into cities. Open trash piles, manure pits, and sewage gutters led to the spread of deadly diseases in many cities, and the practice of burning trash also led to significant health problems, especially for people living nearby. Organized, governmental-scale responses were clearly needed to handle the world's growing trash problems.

One solution, building on the ancient concept of a midden, was for cities and other municipalities to organize a dedicated local *landfill*—a site for the efficient and sanitary burial of waste materials. Among the first such official government-funded landfills was an innovative facility built by the city of Fresno, California, in 1937. Unlike previous municipal landfills, which were really no more than holes in the ground where trash was dumped and then (when the hole was full) buried by topsoil, trash deposited into a series of trenches in the Fresno Sanitary Landfill was mechanically compacted and covered by a layer of soil every day. This significantly reduced the typical problems with rodents/birds, loose debris, and offensive smells that previous traditional landfills experienced, and likely significantly sped up the breakdown of much of the organic and compostable waste being buried there. The Fresno landfill was closed after reaching capacity in 1987, but in recognition of its significant role as a prototype for modern landfills, it is now listed as a US National Historic Landmark as well as being in the National Register of Historic Places.

SEE ALSO Invention of Agriculture (c. 10,000 BCE), Civil Engineering (c. 1500), Population Growth (1798), Industrial Revolution (c. 1830), Birth of Environmentalism (1845), Deforestation (c. 1855–1870), The Anthropocene (c. 1870), Soil Science (1870)

A 1995 photo of trash being buried at Fresh Kills landfill, on Staten Island, New York. Opened in 1948 and closed in 2001, Fresh Kills was once the largest landfill in the world.

Exploring the Oceans

Jacques-Yves Cousteau (1910–1997), **Émile Gagnan** (1900–1979)

While countless explorers throughout history had made discoveries by epic treks across the land, climbs up mountains, expeditions across vast glaciers and sheets of ice, and in ships on the surface of the ocean, until the twentieth century very little exploration was being done *within* the oceans themselves. That would change dramatically with the development and advancement of new technologies that enabled individuals and small teams of people to stay and work underwater for extended periods of time, including visits all the way to the deep ocean floor.

As far back as the fifteenth century, people had been using various forms of snorkels and diving bells to be able to spend more time underwater, and the technology was useful in helping to repair boats and moorings or to salvage valuables from shipwrecks. The first military submarines were developed in the early nineteenth century, and the first "closed-circuit" Self-Contained Underwater Breathing Apparatus (SCUBA) using compressed oxygen was developed in 1876. A major innovation occurred in 1943, however, when French naval lieutenant Jacques Cousteau and engineering colleague Émile Gagnan invented (and eventually patented) an easy-to-use SCUBA set known as an Aqua-Lung. Cousteau used the Aqua-Lung for underwater filming and mine clearance, and the device quickly became popular and commercially successful around the world. After he left the Navy, Cousteau himself set out to use the device to explore the oceans.

Starting in 1950, Cousteau and his crew of the ship *Calypso* began conducting diving and filming expeditions, using the ship as a mobile laboratory for field research and exploration. He became an international celebrity and global spokesperson for ocean conservation when millions of people around the world tuned in to watch the Frenchman in the red beret host the popular ocean exploration documentary TV shows *The Undersea World of Jacques Cousteau* (1966–1976) and *The Cousteau Odyssey* (1977–1982). The shows, and Cousteau himself, received numerous awards and helped to create enormous interest in protecting and preserving the resources and habitats of the Earth's oceans. Indeed, the Cousteau Society, founded in 1973, today boasts more than 50,000 members worldwide dedicated to understanding and appreciating the fragility of life on our ocean world.

SEE ALSO Birth of Environmentalism (1845), Mapping the Seafloor (1957), Mariana Trench (1960), Extremophiles (1967), Earth Day (1970), Deep-Sea Hydrothermal Vents (1977), Great Barrier Reef (1981), Underwater Archaeology (1985)

A 1955 photo of French oceanographer and explorer Jacques Cousteau, using his Nautilus underwater propulsion device.

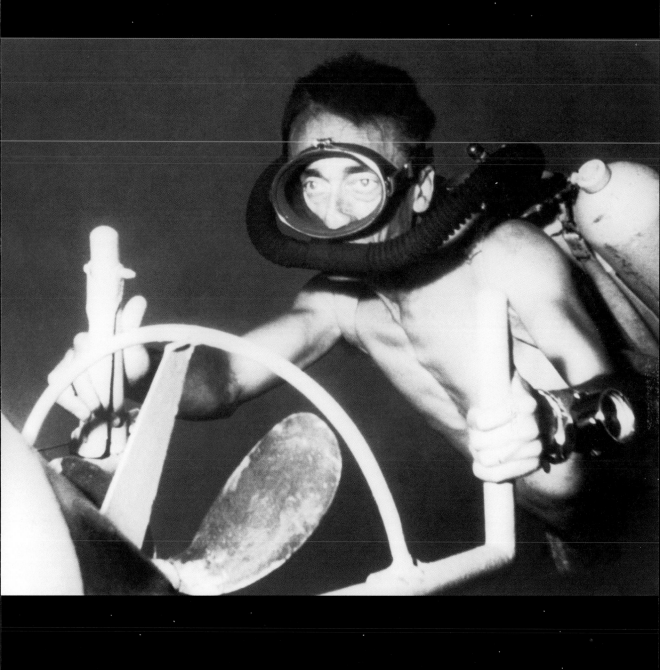

Sky Islands

Both explorers and evolutionary biologists have discovered that geographic isolation, for example on islands far from a continent, often leads to significant changes in plant and animal species. An early case study was Charles Darwin's analysis of variations in the beaks of finches from the islands in the Galápagos, which partly led to the concept of natural selection as a major force of evolution. Isolation need not be restricted to islands in the middle of the ocean, however.

Geologists have long studied the nearly sixty relatively isolated mountains and small mountain ranges that dot the desert landscape near the southern borders of the US states Arizona and New Mexico and the northern borders of the Mexican states Chihuahua and Sonora. These peaks are part of the Sierra Madre (Madrean) range, and rise from 5,000 to above 10,000 feet (1,525 to 3,050 meters) in elevation. As Alexander von Humboldt and others first noted back in the 1840s, the ecology of the local flora and fauna varies dramatically with elevation in tall mountain peaks. Indeed, in terms of their biology and biodiversity, tall, isolated mountain peaks are in many ways like isolated oceanic islands. Starting around 1943, travel guides began referring to the Madrean peaks as *sky islands*, and the term stuck.

The environments of the Madrean sky islands are dramatically different from those of the lowlands that surround them. Ascending from the dry and hot Sonoran desert, the ecology transitions first to grasslands, then to woodlands of scattered oak and pine, then to pine forest, and finally to forests of spruce, firs, and aspens. Average temperatures drop by many tens of degrees at higher elevations. Some species have migrated upward since the last glacial period as the lowlands turned into deserts, and many are known as *relict species* to biogeographers.

While the Madrean sky islands of the North American desert are the most famous and well studied of the genre, hundreds of other similarly isolated and similarly biologically divergent continental sky islands exist around the world, spanning a wide range of climate zones or biomes. Sky islands serve as living laboratories for the study of climatic and evolutionary trends related to unique (and often endangered) species.

SEE ALSO The Rockies (c. 80 Million BCE), Sahara Desert (c. 7 Million BCE), Galápagos Islands (c. 5 Million BCE), Foundations of Geology (1669), Birth of Environmentalism (1845), Natural Selection (1858–1859), Exploring the Grand Canyon (1869), Basin and Range (1982)

2009 photo of Mount Graham, the tallest of the "sky islands" in southern Arizona, near Tucson.

Geosynchronous Satellites

Hermann Oberth (1894–1989), **Herman Potočnik** (1892–1929),
Arcthur C. Clarke (1917–2008)

Newton's Laws of Gravity and Motion and Kepler's Laws of Planetary Motion in particular apply to artificial satellites just as well as to planets in orbit around a star or moons in orbit around a planet. Rocket technology and astronautics—the study of navigating through space—advanced quickly after the first liquid-fueled rockets capable of reaching high altitudes were developed in the 1920s by Robert Goddard.

Several of Goddard's contemporaries were already beginning to think about the mechanics and dynamics of orbital (and beyond) rocket flights. Two of those contemporaries were the Hungarian-German physicist Hermann Oberth and the Austro-Hungarian rocket engineer Herman Potočnik, who expanded on and worked out the details of concepts first described by the Russian mathematician Konstantin Tsiolkovsky. One of those concepts was the idea of a geostationary, or geosynchronous, orbit.

A satellite in a geosynchronous orbit will complete one orbital revolution in the same amount of time that it takes the body the satellite is orbiting (in this case, Earth) to rotate once on its axis. From the vantage point of an observer on the Earth's surface, such a satellite would appear to be parked high in the sky, never moving. Given the Earth's mass and rotation rate, Newton's second law can be used to derive the orbital altitude of a geosynchronous satellite—it turns out to be about 22,000 miles (36,000 kilometers) above the surface.

The British science fiction author and futurist Arthur C. Clarke was one of the first to grasp what would ultimately be one of the most practical applications of such satellite orbits: global telecommunications, described in a 1945 magazine article called "Extra-Terrestrial Relays—Can Rocket Stations Give Worldwide Radio Coverage?" Clarke's popularization of the idea helped it to gain wide attention and support. Starting in 1964, the actual use of geosynchronous satellites has now gone far beyond just radio relays. Today they also relay TV, Internet, and global positioning system (GPS) signals and help us to monitor Earth's weather and climate.

SEE ALSO Laws of Planetary Motion (1619), Gravity (1687), Liquid-Fueled Rockets (1926), Weather Satellites (1960), Earth Science Satellites (1972)

Main image: *A snapshot of satellites currently being tracked by the NASA Orbital Debris Program Office. Earth's ring of geosynchronous satellites can be clearly seen.*
Inset: *Space shuttle* Discovery *deploying the AUSSAT-1 communications satellite in 1985.*

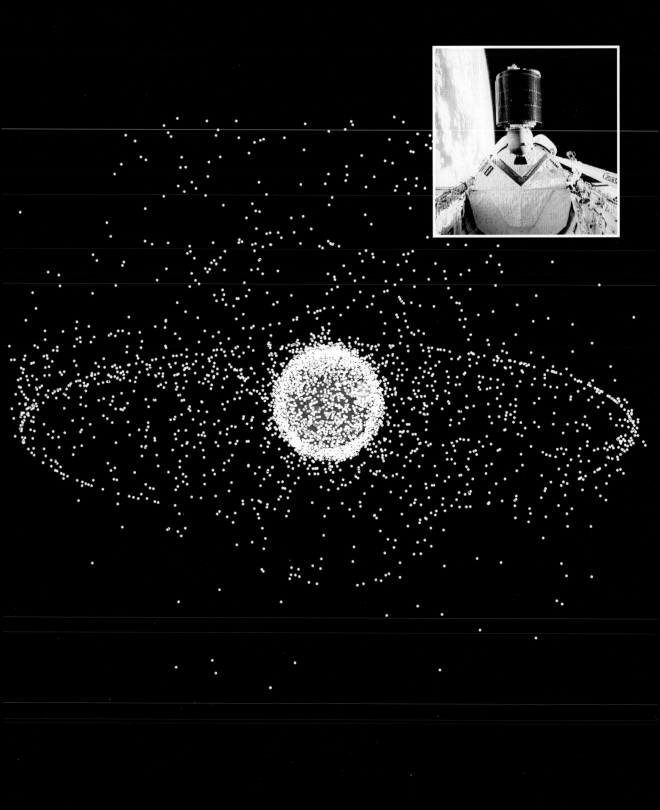

Cloud Seeding

Vincent Schaefer (1906–1993), Bernard Vonnegut (1914–1997)

Ideas about controlling the weather date back to antiquity, but most often it was relegated to gods or demi-gods, not mere mortals, hurling thunderbolts or changing the course of the winds. The idea that humans could actually influence the weather is relatively new. Most notably, theoretical ideas for ways to cause clouds to produce rain were first developed in the 1890s to 1930s, and the first experimental confirmations of the theory happened in the 1940s.

Specifically, in 1946 American chemist and meteorologist Vincent Schaefer, working at General Electric Laboratories, was trying to develop a solution to the problem of aircraft wings getting coated with ice when flying through cold clouds or snow/sleet. He and colleagues built a "cloud chamber" in the laboratory to try to understand the physics of condensation and evaporation of water under controlled conditions. Schaefer accidentally noticed that while absolutely pure water could exist as a vapor well below its freezing temperature (a state known as *supercooled*), introducing even a minor amount of fine-grained dry ice would spontaneously cause the supercooled water to condense out into rain droplets. In a field experiment that same year, Schaefer was able to create the first "man-made snow" by dumping dry ice out of an airplane and into the clouds above the Berkshire Mountains in western Massachusetts.

Shaefer's GE colleague and fellow chemist Bernard Vonnegut (older brother of novelist Kurt Vonnegut Jr.) quickly discovered that supercooled clouds could be "seeded" to form snow and rain by other fine-grained particles as well (he used silver iodide grains), and that such grains serve as "nucleation sites" for the initial growth of water droplets. It was subsequently discovered that even simple table salt could be used to initiate the growth of raindrops in supercooled clouds. In nature, fine particles of dust or ice often serve the same function.

Shaefer and Vonnegut's methods quickly gained commercial and government/military interest, and "cloud seeding" to initiate snowfall or rain has since become a viable way to actually control the weather in many countries around the world, but especially in arid regions or during times of drought. In the United Arab Emirates, for example, cloud seeding is used by the government to create artificial rain storms in the Dubai and Abu Dhabi deserts.

SEE ALSO Weather Radar (1947), Weather Satellites (1960)

A 1946 photo of cloud-seeding pioneer Vincent Schaefer working in his laboratory at General Electric Research in Schenectady, New York.

Weather Radar

Catastrophic natural storms like the hurricane that destroyed Galveston, Texas, in 1900 or the devastating tri-state tornado of 1925 were partly as disastrous as they were because of the lack of significant advance warning or of active tracking/monitoring of those kinds of meteorological events. With the advent of airborne remote sensing in the late nineteenth and early twentieth centuries, it became possible to monitor some kinds of storms from balloons, airships, or airplanes. However, such deployments were sparse and expensive, and often couldn't yield reliable information on where it was raining or snowing because the observers were "clouded out." Meteorologists needed a way to see through the clouds to truly track the storms.

That way would come with the advent of radar, an acronym that stands for "**RA**dio **D**etection **A**nd **R**anging." Active radar systems broadcast radio waves in a certain direction using an antenna and then measure how much of that radio energy gets reflected back to a receiver that uses the same antenna. The technique was pioneered in the 1930s and then significantly enhanced for use in World War II as a way to detect and track enemy airplanes and ships. Radar technology was found to have many other important military and civilian applications, one of which is weather forecasting.

Thick clouds are opaque to visible light, but the tiny aerosol particles that they are made of are transparent to radio waves. However, much larger raindrops, sleet, and snow particles that form within clouds are not transparent to radio waves. Thus, when a radar system sends a pulse into a cloud, much of that signal passes right through, except for a fraction that bounces off rain, snow, or sleet and gets reflected back to the radar antenna. Radar can thus "see" weather that human eyes cannot.

Once the technology was declassified, the first dedicated commercial weather radar stations began operation in 1947 under the control of the US Weather Bureau (now the National Weather Service). The first station was commissioned in Washington, DC, and was quickly followed by several dozen more stations around the country, deployed to help provide warnings and monitoring of extreme weather events. The systems were enhanced over time and expanded in coverage, and today a global network of advanced weather radar stations provides critical data for forecasting, aviation, shipping, and weather-related academic research.

SEE ALSO Airborne Remote Sensing (1858), Galveston Hurricane (1900), Tri-State Tornado (1925), Geosynchronous Satellites (1945), Cloud Seeding (1946), Weather Satellites (1960), Earth Science Satellites (1972).

Weather radar map of rain bands in Hurricane Katrina as it approached the southeastern coast of Florida on August 25, 2005.

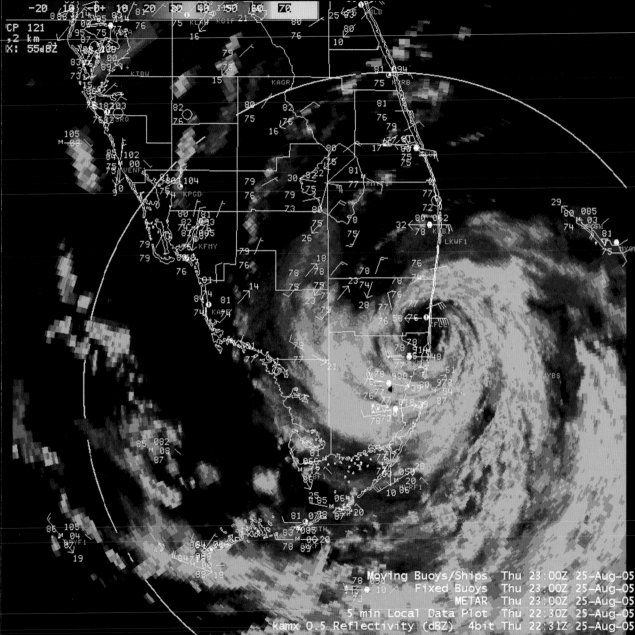

Tracing Human Origins

Mary Leakey (1913–1996), **Louis Leakey** (1903–1972)

One of the major successes of modern anthropology (the study of human origins and past human behavior) is the realization that our species, *Homo sapiens*, likely originated in Africa some 200,000 years ago, and has spread across almost the entire planet since then. Coming to this conclusion has required centuries of dedicated geologic, paleontologic, and genetic forensic work to connect the dots between societies and tribes across the globe today to back-out their origins, lifestyles, and patterns of migration.

Among the leading early figures in the modern study of human origins was the husband and wife paleoanthropologic team of Mary and Louis Leakey. Louis was originally from Kenya and studied archaeology at Cambridge University; Mary was British and had studied archaeology and other related subjects at University College London, eventually taking a job as an illustrator for articles and books about anthropology. After they married in 1936, the couple traveled extensively in eastern Africa, searching for and excavating Stone Age and Neolithic fossils and other clues about early humans who dwelled in what are now Kenya and Tanzania.

Mary Leakey made a major career-defining discovery in 1948 near Lake Victoria in Kenya, finding and identifying fossil remains (including the skull) of an ape called *Proconsul africanus* that lived in the Miocene geologic epoch, between about 23 and 14 million years ago. *Proconsul* was a quadruped primate with many similarities to monkeys, chimpanzees, and bonobos, and Mary and Louis—and much of the archaeological community—thought it would end up being an important link between early primates and eventual humans. The couple continued making important hominid fossil and tool discoveries in the Serengeti plains and elsewhere throughout the 1940s and 1950s.

Decades of subsequent work, some of which was conducted by researchers such as Dian Fossey and Jane Goodall who had been mentored by the Leakeys, revealed a much more detailed hominid family tree than the Leakeys or others of the mid-twentieth century had imagined. While *Proconsul* was indeed an important find, it was but one of what are known now to be many links in the chain that led to modern *Homo sapiens*.

SEE ALSO Primates (c. 60 Million BCE), Stone Age (c. 3.4 Million BCE TO 3300 BCE), Lake Victoria (c. 400,000 BCE), *Homo sapiens* Emerges (c. 200,000 BCE), Beringia Land Bridge (c. 9000 BCE), Reading the Fossil Record (1811), Women in Earth Science (1896), *Gorillas in the Mist* (1983), Chimpanzees (1988)

A photo from the 1960s of Mary Leakey (right) showing Louis Leakey (left) one of the places where she discovered fossils of early hominid species in Olduvai Gorge, Tanzania.

Island Arcs

Kiyoo Wadati (1902–1995), **Hugo Benioff** (1899–1968)

Prior to the development of our modern understanding of plate tectonics, early- to mid-twentieth-century geologists and geophysicists were able to use networks of seismometers to determine that the places where most earthquakes occur followed specific patterns. For example, in the late 1920s, Japanese geophysicist Kiyoo Wadati showed conclusively that some earthquakes occur very deeply below the surface, and that many occur in inclined zones extending underneath deep ocean-floor valleys like the Marianas Trench. In 1949, building upon Wadati's work, American seismologist Hugo Benioff showed that these inclined zones of earthquakes occur all over the Pacific Rim along curving planes that dip down at angles of up to 45 degrees.

The surface expression of a curving plane plunging into a sphere is a curved arc—which is exactly the pattern observed among volcanic islands in the Aleutians, Japan, the Caribbean, and elsewhere around the world. These curving chains of volcanoes are now called island arcs, and they are the telltale sign of slabs of oceanic crust *subducting* (diving down) under other nearby crust, melting surrounding rocks under the intense heat and pressure, and then erupting as volcanoes on the plate in front of the subducting slab. The zones along the subducting slab where the earthquakes occur are now referred to as Wadati-Benioff zones.

By the 1950s, the observations of curving volcanic arcs above ocean-continent boundaries, deep seafloor trenches at those boundaries, and deep earthquakes along the subducting oceanic slabs all provided compelling evidence in support of Alfred Wegener's once-ridiculed Continental Drift hypothesis. Clearly there was significant relative crustal motion at these boundaries, but questions still remained, especially about what was driving the drift. It would take the complete mapping of the seafloor and the discovery of mid-ocean volcanic ridges, and then the discovery of magnetic signatures that revealed that new oceanic crust was being created at those seafloor-spreading ridges, for all the puzzle pieces of the modern theory of plate tectonics to fall into place.

SEE ALSO Plate Tectonics (c. 4–3 Billion BCE?), Cascade Volcanoes (c. 30–10 Million BCE), Hawaiian Islands (c. 28 Million BCE), The Andes (c. 10 Million BCE), Mount Tambora Eruption (1815), Krakatoa Eruption (1883), Continental Drift (1912), Mapping the Seafloor (1957), Mariana Trench (1960), Seafloor Spreading (1973)

Illustrative cross-section showing a higher-density plate (right) diving or subducting under a lower-density plate (left), producing a zone of earthquakes and volcanoes along the collisional zone. On a spherical planet, these volcanoes trace out an arc-shaped curve across the surface.

Ascending Everest

Edmund Hillary (1919–2008), **Tenzing Norgay** (1914–1986)

In 1923, reporters asked British mountain climber George Mallory why he was trying to ascend Mount Everest, the tallest peak on Earth. His response has become the rallying cry for mountaineers and explorers worldwide: "Because it's there." Mallory would never reach the summit, nor would any climbers of the 1930s or 1940s. Everest was proving to be a singularly difficult climb, as explorers had to overcome ice crevasses, sheer cliffs, and a dizzying lack of oxygen at an elevation greater than 29,000 feet (8,800 meters) approaching the summit.

British and Swiss climbing expeditions in the 1920s through the early 1950s had failed to reach the summit. In 1953, however, a new British expedition was organized, sending pairs of climbers on treks to incrementally advance trailblazing progress to the summit. The second team to attempt the ascent, consisting of New Zealander Edmund Hillary and Nepalese Sherpa guide Tenzing Norgay, reached the summit on May 29, 1953.

It was a monumental achievement, which both Hillary and Norgay readily acknowledged was a team effort. Not only had significant trailblazing been involved by their precursor team (and numerous climbers that had attempted the ascent through Nepal from the south), but they relied on significant assistance from technology that was not available to earlier climbers, including sophisticated oxygen-supply systems. In addition, Norgay had been part of the Swiss team's unsuccessful attempt to reach the summit in 1952, giving him valuable experience that he could exploit in the 1953 climb.

Hillary and Norgay became instant global celebrities. Their team had adopted a careful, incremental strategy of approaching the summit, establishing five camps between 19,400 and 26,000 feet (5,900 to 7,925 meters) before the final ascent. Hillary was knighted by the queen in England, and Norgay received high honors from the British and Indian governments. But most of the accolades from international scientific and exploration societies were rightly awarded to the entire team.

By the end of 2017, more than 4,800 different people had climbed to the summit of Mount Everest, and nearly 300 have died (including guides) in the attempt. Pressures are mounting to regulate ascent attempts, as they may be having an environmental toll on the summit region, and are certainly having a human toll on the local guides who attempt to make a living guiding climbers to the top of the world.

SEE ALSO Birth of Environmentalism (1845), Reaching the North Pole (1909), Reaching the South Pole (1911), Machu Picchu (1911), Exploration by Aviation (1926), Exploring the Oceans (1943)

Sir Edmund Hillary (left) and Tenzing Norgay (right) make their way up the slopes of Mount Everest, 1953.

Nuclear Power

The world's demand for mass-produced energy has risen dramatically since the Industrial Revolution in the nineteenth century. Initially, enough energy capacity could be generated from the action of rivers and streams turning turbine generators (hydroelectric power), or heating water with wood or coal fire to drive steam turbines (steam power). As demand increased, the internal-combustion engine, powered by fossil fuels (gasoline, oil, and natural gas) to drive turbines, was added as a way to create electricity. Petroleum fuels, natural gas, and coal are all *nonrenewable* resources used to produce energy, as they or their starting products were formed long ago via geologic processes. To address the need for sustainable long-term energy sources, many governments and industries are seeking ways to increase the use of alternate sources of energy, such as solar, wind, and, most recently, nuclear power.

Physicists in the early 1930s were finally able to firm up the basic structure of the atom, at which time they realized that it was possible to "breed" enriched radioactive elements like plutonium in nuclear-fission reactions. The technique was used to create the first nuclear weapons during World War II, and after the war some of that technology was declassified and adopted by the civilian and private sectors as the basis for a new alternate source of energy. Nuclear power uses the heat generated by the decay of radioactive elements inside a reactor to boil water that drives steam turbines. The first electricity-generating (~100 kilowatt) research-grade nuclear reactor came online in the US in 1951, and in 1954 the world's first nuclear power plant to generate large-scale electricity (~5 megawatts) for a city power grid came online in the USSR city of Obninsk. Today, nuclear power plants generate about 10 percent of the world's electricity.

While it seems like it should be relatively simple to just boil water using radioactive heat, substantial engineering innovations had to be developed and tested so as to not allow the nuclear materials to overheat and "meltdown" the reactor. Despite such efforts, a few notable meltdowns have occurred, including at the Three Mile Island plant in the US in 1979 and the Chernobyl plant in Russia in 1986. The human and environmental impacts of high-profile accidents like these and the longstanding general problem of how to dispose of spent nuclear-reactor fuel in an environmentally safe way have kept nuclear power from becoming even more widespread.

SEE ALSO Industrial Revolution (c. 1830), Radioactivity (1896), Controlling the Nile (1902), Wind Power (1978), Solar Power (1982), Chernobyl Disaster (1986), Hydroelectric Power (1994), End of Fossil Fuels? (~2100)

A 1979 photo of the Three Mile Island nuclear generating station near Harrisburg, Pennsylvania. The large structures are cooling towers, and the smaller cylindrical structures with rounded tops are the reactors.

Mapping the Seafloor

Marie Tharp (1920–2006), **Bruce Heezen** (1924–1977)

For most of human history, the nature of the seafloor—representing more than 70 percent of the surface of our planet—was a complete mystery. With the advent of oceangoing research expeditions in the twentieth century, however, the detailed and ultimately surprising topographic and geologic diversity of the seafloor was finally revealed.

Initial seafloor mapping started during World War I using early sonar (an acronym for **SO**und **N**avigation **A**nd **R**anging) technology to try to identify mines, shipwrecks, and submarines. Sonar sends sound waves from a ship to the ocean floor, and the timing of the reflected sound echoes can be used to determine the topography. Advanced versions of the technology were developed and used in World War II, and then afterward for civilian oceanographic research.

Among the most talented and prolific of those early researchers was the American oceanographer Marie Tharp, who worked at Columbia University starting in the late 1940s with American geologist Bruce Heezen. Heezen and a team of colleagues conducted numerous research expeditions to collect *bathymetry* (ocean depth) data using advanced sonar equipment. Tharp worked to assimilate and interpret the data being sent back from these cruises (women were not allowed on research ships at the time). By 1952, she had mapped enough data to recognize the mid-Atlantic ridge, and interpreted it as potential evidence of the continental drift hypothesis. Heezen and others were skeptical at first, but in 1957 Tharp and Heezen published a comprehensive map outlining the physiography (physical features, such as mountains, valleys, canyons, and so forth) of the North Atlantic seafloor, and she convinced many of the validity of her ideas.

Tharp's maps of the seafloor were revolutionary, and they provided a key part of the story for the then-emerging idea of global-scale plate tectonics. The Atlantic mid-ocean ridge, for example, is now recognized as the largest continuous mountain range on the planet. Major advances in sonar technology since then, along with dedicated bathymetry expeditions designed to map the entire seafloor around the world, are enabling detailed geologic studies of Earth's dynamic seafloor, including underwater mountains, trenches, volcanoes, landslides, and earthquakes.

SEE ALSO Plate Tectonics (c. 4–3 Billion BCE?), Reading the Fossil Record (1811), Women in Earth Science (1896), Continental Drift (1912), Reversing Magnetic Polarity (1963), Seafloor Spreading (1973)

Top: Marie Tharp in 2001. **Bottom:** *Section of a modern map of the topography of the seafloor of the southern Atlantic Ocean.*

Sputnik

Sergei Korolev (1906–1966)

Americans tend to vividly recall where they were and what they were doing during key events that have come to define certain generations or eras. Examples include the bombing of Pearl Harbor, the assassination of John F. Kennedy, the explosion of the space shuttle *Challenger*, and of course the traumatic terrorist events of 9/11. For one generation of Americans, the defining event came in the fall of 1957.

On October 4 of that year, the Soviet Union became the first nation to successfully launch an artificial satellite into space. Leading Soviet rocket engineer Sergei Korolev headed the team that had created the USSR's first intercontinental ballistic missile (ICBM), and he lobbied the government to allow him and his team to modify the R-7 rocket so that it could launch a small scientific payload (a set of instruments or experiments) into Earth's orbit. The Soviet government approved Korolev's plan in the hopes that they could beat the Americans into space. The payload was called *Sputnik*, Russian for "satellite." The Space Age had officially begun.

Sputnik 1 circled Earth every 96 minutes for three months (until it burned up in the atmosphere when its orbit decayed), emitting a telltale beep-beep-beep on its single-watt radio, which could easily be picked up by ham radio operators around the world. The satellite created a sort of mild hysteria in the United States, where the public was acutely aware of the Soviet Union's ability to launch ICBMs—outfitted with nuclear warheads—to any target on the planet. The US government stepped up its own space efforts, and America's first satellite, *Explorer 1*, was successfully launched about two weeks after Sputnik burned up.

Sputnik also launched an unprecedented mini-revolution in science and technology funding and education in the United States, the effects of which are still experienced today. The Americans most influenced by Sputnik—often referred to as the Apollo generation—went on to see the US win the space race, with twelve people walking on the Moon between 1969 and 1972, followed by decades of other stunning achievements in space.

SEE ALSO Liquid-Fueled Rockets (1926), Geosynchronous Satellites (1945), Earth's Radiation Belts (1958), Humans in Space (1961)

A replica of Sputnik 1, *the world's first artificial space satellite, housed in the National Air and Space Museum of the Smithsonian Institution in Washington, DC. The metallic sphere is about 23 inches (58 centimeters) in diameter, and the antennae (only partially shown here) extend out 112 inches (9 feet; 285 centimeters).*

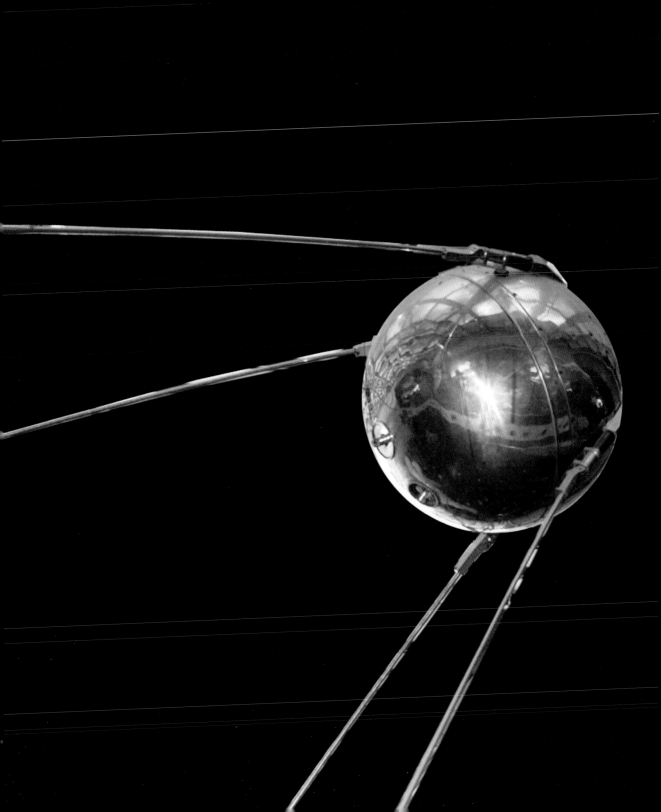

International Geophysical Year (IGY)

After World War II and especially after the Korean War in the early 1950s, relationships among the world's leading technological countries chilled dramatically. Indeed, the tensions of this geopolitical Cold War extended beyond just governments, even influencing the level of communication and collaboration that could occur among scientists worldwide. Some leading scientists decided to take matters into their own hands, organizing a global effort to bring scientists together to advance earth science despite the political turmoil of the times.

Scientific societies of the late nineteenth and early twentieth centuries had organized several "International Polar Years" to promote collaboration on polar research and exploration. Following that model, the International Council of Scientific Unions proposed that the world's scientists come together again to study not only the polar regions, but also many other aspects of Earth's surface, oceans, and atmosphere. They dubbed it the "International Geophysical Year" or IGY, and ultimately 67 different nations participated in a variety of collaborative projects. The "year" actually ran for 18 months, from July 1, 1957, to December 31, 1958, and the earth science projects spanned the fields of geology, seismology, geomagnetism, meteorology, ionospheric physics, oceanography, and heliophysics (studies of the Sun).

Among the scientific successes and legacies of the IGY were the launches of the first successful earth science satellites, *Sputnik 1* and *Explorer 1*, the latter of which discovered the Van Allen radiation belts, and the establishment of a substantial and lasting scientific presence and collaboration on the continent of Antarctica. Specifically, the British, French, Belgians, Japanese, and Americans all set up collaborative research stations on the continent shortly before, during, and/or shortly after the IGY, including the Amundsen-Scott South Pole Station, which is administered by the US. The IGY also led directly to the Antarctic Treaty of 1959, which permanently established Antarctica as an environmentally protected continent devoted to peaceful purposes and cooperative scientific research.

IGY proved that scientists (and governments) could work effectively together to make discoveries and solve important problems in the earth sciences, even during politically difficult times. Much of the work started during the IGY continues today, and many scientists and scientific societies still work hard to promote peaceful scientific collaboration among all nations.

SEE ALSO Antarctica (c. 35 Million BCE), Reaching the North Pole (1909), Reaching the South Pole (1911), Exploration by Aviation (1926), Exploring the Oceans (1943), Sputnik (1957), Earth's Radiation Belts (1958)

Photographer Galen Rowell is reflected in the mirrored surface of the ceremonial South Pole outside the Amundsen-Scott South Pole Research Station.

Earth's Radiation Belts

James Van Allen (1914–2006)

After the Soviet Union surprised the world with the successful launch and operation of *Sputnik 1* in the fall of 1957, the US government scrambled to catch up. The goal of launching and operating a small satellite with a minimal, simple science payload was passed along to a joint team composed of the Army Ballistic Missile Agency, responsible for the launch of the satellite on a modified Jupiter-Redstone intermediate-range ballistic missile, and an Army-Caltech (California Institute of Technology) facility near Pasadena called the Jet Propulsion Laboratory (JPL), which was responsible for the satellite— *Explorer 1*—and its scientific experiments.

Explorer 1's scientific payload—the brainchild of the American space scientist James Van Allen—consisted of a cosmic ray counter, a micrometeoroid impact detector, and some temperature sensors. The experiments were more complex than Sputnik's simple radio transmitters, but low enough in mass, power, and volume to be deployed into orbit by the Jupiter missile.

America's first artificial satellite (and the world's third, after *Sputnik 2* in November 1957) was successfully launched from the Cape Canaveral Missile Annex in Florida on January 31, 1958. *Explorer 1* settled into a 115-minute elliptical orbit and operated its science instruments for more than three and a half months before its batteries died. During the mission, the science instruments radioed back streams of data in real time to the JPL science team.

The data from *Explorer 1* were at first puzzling, because they appeared to be caused by dramatic increases in the number of energetic particles at certain altitudes and in certain locations around the Earth. Van Allen and his team interpreted the data to reveal the presence of a zone or belt of high-energy particles or plasma being confined by Earth's magnetic field. The results were confirmed a few months later, by *Explorer 3*. This was the first major space science discovery made by a satellite, and in honor of the science team's leader, the region of enhanced energetic particles in near-Earth space is now called the Van Allen Radiation Belt. *Explorer 1* was the first of what are now nearly 100 successful missions to date in the *Explorer* small-spacecraft series.

SEE ALSO Solar Flares and Space Weather (1859), Sputnik (1957), International Geophysical Year (IGY) (1957–1958), The Oscillating Magnetosphere (1984)

The aurora borealis, or northern lights, gleaming brightly over Bear Lake, Alaska, in January 2005. Auroral displays like this are the result of high-energy solar-wind particles interacting with Earth's magnetic field and with particles trapped in the Van Allen radiation belts.

Weather Satellites

1960

When they were first introduced in the late 1940s, ground-based weather radar systems almost immediately had a profoundly strong impact on the ability to track and forecast storms and to save lives during the most extreme weather events. Still, even though they could be linked to larger networks, they mostly provided only a relatively limited local perspective on the weather. Balloons, aircraft, and sub-orbital sounding rockets could help provide a broader perspective, but they could be deployed only infrequently, and still might only provide a relatively local view of the weather. What was needed was a truly global perspective—like a view of the Earth from space.

With the dawn of the space age and the successful launch of the world's first scientific satellites in the late 1950s, it became possible to deploy cameras and other instruments into Earth orbit to monitor the weather and other characteristics of our planet's surface and atmosphere. As a result, in 1960 a joint team from the Radio Corporation of America (RCA) and the US Army's Signal Research and Development Laboratory ("Army Signal Corps"), under the direction of a newly created federal agency called the National Aeronautics and Space Administration (NASA), launched and operated the first successful weather satellite, called *TIROS-1 (for Television InfraRed Observation Satellite)*. Even though its mission was short (just 78 days), it proved so successful and so useful for weather forecasters that it spawned a 20-year series of seven new weather satellites called *Nimbus 1 to Nimbus 7*.

These early weather satellites were in short-period low-Earth orbits. Starting in the late 1960s and early 1970s, however, another relatively new federal agency, the National Oceanic and Atmospheric Administration (NOAA) began to operate satellites in geosynchronous orbits, effectively "parking" them above different parts of North America so that they could continuously monitor the same large region all the time. Among the most successful workhorses of the US weather-satellite world are the NOAA Geostationary Operational Environmental Satellite (GOES) series, the first of which was launched in October 1975, and the seventeenth of which was launched in March 2018. The Russian, Japanese, Chinese, Indian, and European space agencies have also launched and operated many geosynchronous weather satellites to provide critical forecasting data for their parts of the planet as well.

SEE ALSO Airborne Remote Sensing (1858), Galveston Hurricane (1900), Tri-State Tornado (1925), Geosynchronous Satellites (1945), Cloud Seeding (1946), Weather Radar (1947), Sputnik (1957), Earth's Radiation Belts (1958), Earth Science Satellites (1972)

A 1960 photo of engineers testing the Television Infrared Observation Satellite (TIROS), the world's first successful weather satellite, at RCA in Princeton, New Jersey.

Understanding Impact Craters

Eugene Shoemaker (1928–1997)

There is plenty of evidence around us that the surface of the Earth is constantly being changed by erosional, volcanic, and tectonic forces and processes. What is not so obvious, however, is how much our planet has been influenced by the other major surface geologic process at work in our solar system: impact cratering. One need not look any farther than the Moon and its crater-covered surface to realize that the Earth must also have been similarly bombarded.

It took geologists a very long time to realize the importance of impact cratering as a geologic process on the Earth, as well as on other planets and moons. Leading USGS geologist G. K. Gilbert, for example, was convinced that what was then named Coon Mountain in Arizona was an explosive volcanic crater, even though he acknowledged that many of the circular depressions on the Moon were likely to be impact craters, and even though others like the mining entrepreneur Daniel Barringer were convinced that the feature had been caused by the impact of a small iron-rich asteroid. The definitive proof of the impact origin of what is now called Meteor Crater, and the key pieces of evidence that would eventually help identify nearly 200 other impact structures across the world, would come from another USGS geologist, Eugene ("Gene") Shoemaker.

Shoemaker was a consummate field geologist and a keen observer and interpreter of the stories the rocks were telling him. In order to better understand the way impact craters could modify a planetary surface, he spent time studying small craters made from nuclear-explosion tests in Nevada. He learned to recognize the telltale signs of such high-energy explosive processes, including the creation of special minerals formed at high shock pressures. In his PhD dissertation in 1960, Shoemaker used his experience to convince the remaining skeptics that Meteor Crater is indeed the result of an impact of a small metallic asteroid some 50,000 years ago. Shoemaker devoted the rest of his career (much of it in collaboration with his wife, the planetary astronomer Carolyn Shoemaker) to searching for and discovering impact structures on the Earth, and to searching for and studying the nature of the large population of near-Earth asteroids that could eventually pose potential future impact hazards to life on our planet.

SEE ALSO Dinosaur-Killing Impact (c. 65 Million BCE), Arizona Impact (c. 50,000 BCE), The US Geological Survey (1879), Hunting for Meteorites (1906), The Tunguska Explosion (1908), Meteorites and Life (1970), Extinction Impact Hypothesis (1980), Torino Impact Hazard Scale (1999)

Geologist Eugene Shoemaker (center, holding hammer) lectures on the geology of impact craters to a group of astronauts at Meteor Crater, Arizona, in 1967.

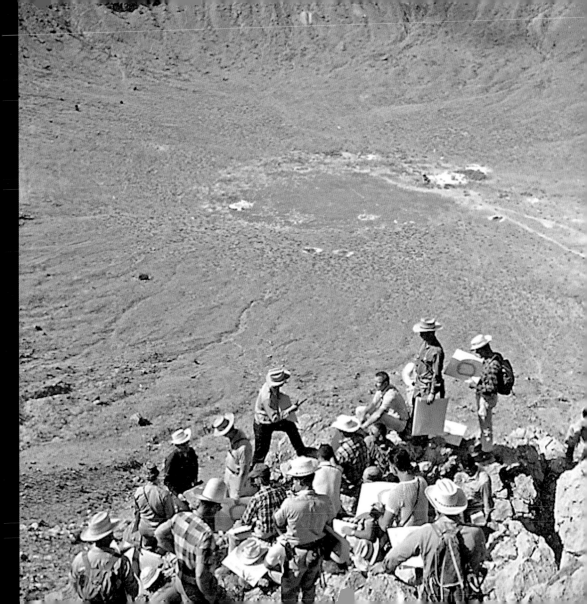

Mariana Trench

Donald Walsh (b. 1931), **Jacques Piccard** (1922–2008)

It is perhaps obvious why collisions among tectonic plates on the Earth's surface produce the tallest mountains in the world, but perhaps not so obvious that such collisions also produce the deepest valleys. This is because collisions between plates of vastly different densities (like between denser oceanic crust and lighter continental crust) cause the denser plate to sink or *subduct* under the other, dragging the sinking crust at the collision boundaries downward to form deep, long trenches.

The deepest of these, and the lowest known elevation below sea level on the Earth, occurs in a valley called the Challenger Deep within the Mariana Trench (also called the Marianas Trench).

The Mariana Trench is a 1,580-mile (2,550-kilometer), 43-mile-wide (69 kilometers) crescent-shaped scar in the Earth's crust on the western Pacific Ocean floor. The trench marks the collisional boundary where the large Pacific plate is subducting under the smaller Mariana (or Marianas) plate to the west. Even though the collision is between two relatively dense oceanic plates, the Pacific is subducting under the Mariana because that part of the ocean floor is much older, cooler, and thus denser than the younger crust to the east.

The average depth of the Pacific Ocean floor is about 13,740 feet (4,188 meters) below sea level, but the Mariana Trench extends down to about 36,200 feet (11,000 meters)—deeper under the ocean than Mount Everest is above sea level. The first modern bathymetric soundings of the depths of the trench were made using sonar in the early 1950s by the crew of the British Royal Navy vessel *Challenger II* (hence, the deepest point being named the Challenger Deep). In January 1960, American oceanographer Don Walsh and Swiss oceanographer and engineer Jacques Piccard took the first-ever voyage to the bottom of the Challenger Deep, in a specially designed US Navy self-propelled deep sea submersible called the *Trieste*.

Walsh and Piccard only spent about twenty minutes on the ocean floor, but they noted a surprising number of fish living at those depths (at pressures more than 1,000 times atmospheric pressure at sea level). Two subsequent robotic descents, and a solo descent by filmmaker James Cameron in 2012, have also revealed unexpected biologic diversity in one of our planet's most extreme environments.

SEE ALSO Plate Tectonics (c. 4–3 Billion BCE?), Continental Drift (1912), Exploring the Oceans (1943), Island Arcs (1949), Mapping the Seafloor (1957), Extremophiles (1967), Seafloor Spreading (1973)

The US National Oceanic and Atmospheric Administration's remotely operated robotic submersible Deep Discoverer, *being used in 2016 to explore the geology of layered rocks in the Mariana Trench, at depths of around 20,000 feet (6100 meters).*

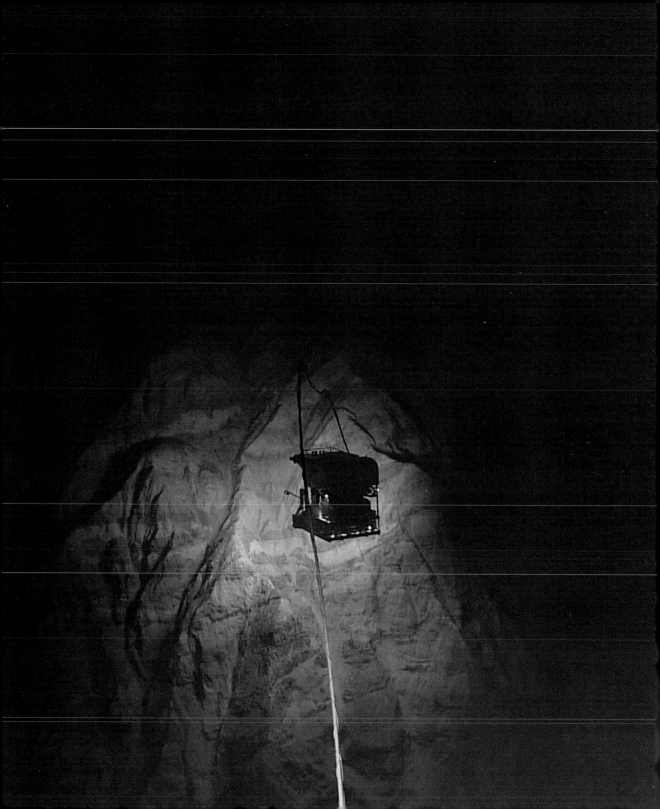

Valdivia Earthquake

Seismologists began trying to classify the strengths or magnitudes of earthquakes in the 1930s, relying on the growing network of seismometers being deployed around the world. Among the first ways to measure earthquake strength was the local magnitude scale, developed in 1935 by American seismologists Charles Richter and Beno Gutenberg, and referred to as the *Richter Scale*. It was limited to classification of earthquakes very close to seismometer stations, however, and was not well able to distinguish among the intensities of the very largest events. Thus, the scale has been modified over time, and since the 1970s the *Moment Magnitude Scale* (**M**) has been used to estimate the strengths of all earthquakes (even though it is still often called the Richter scale by the media). **M** is a logarithmic energy scale: one jump in **M** corresponds to an increase in energy by a factor of 32; two jumps corresponds to a factor of 1,000.

The largest **M** earthquake in recorded history occurred on May 21, 1960, and was centered near the city of Valdivia, Chile. With an **M** estimated between 9.4 and 9.6, the Valdivia earthquake released more than fifty times as much energy in a violent ten-minute period of shaking than the famous 1906 earthquake that devastated San Francisco. Valdivia and many other nearby towns were reduced to rubble. The quake generated an offshore tsunami that battered the coast of Chile with waves more than 80 feet (25 meters) high. Tsunami waves traveled across the entire Pacific Ocean, and waves more than 35 feet (11 meters) high devastated the city of Hilo on the Big Island of Hawaii. Somewhere between 1,000 and 7,000 people are estimated to have been killed from the effects of the earthquake and tsunami, with many billions of dollars of associated damages.

The Valdivia earthquake was caused by the ongoing subduction of the Pacific plate under the South American plate along the western coast of South America, and the resulting tsunamis formed as a result of the upward thrusting of significant amounts of crust into the ocean waters. Partly as a result of devastating earthquakes like the 1960 Valdivia event, substantial enhancements in global earthquake response and tsunami prediction systems in the decades since have helped to save many lives.

SEE ALSO Plate Tectonics (c. 4–3 Billion BCE?), The Andes (c. 10 Million BCE), San Francisco Earthquake (1960), Sumatran Earthquake and Tsunami (2004)

Part of the ruins of the city of Valdivia, Chile, after the colossal earthquake of May 21, 1960.

Humans in Space

Yuri Gagarin (1934–1968), **Alan Shepard** (1923–1998)

Ever since the earliest science fiction and then the early history of rocketry, visionaries have dreamed about humanity's becoming a spacefaring species. It would take the technological push of World War II and the political push of the Cold War, however, to make the dreams of spaceflight pioneers like Robert Goddard and Konstantin Tsiolkovsky come true, with the development of rockets powerful enough to be able to carry not just satellites, but people, to escape velocity and into Earth orbit.

The US and the USSR engaged in a heated battle to become the country (and political system) that would be the first to launch a person into space. Both nations employed leading scientists and engineers and expended significant financial resources in the race. In the spring of 1961, the Soviets won the race, but just barely. Cosmonaut Yuri Gagarin became the first person in space on April 12, 1961, and Astronaut Alan Shepard became the second only about three weeks later.

Gagarin was launched on an Intercontinental Ballistic Missile (ICBM) that was modified to accommodate his capsule, *Vostok 1*, which spent about 108 minutes in space, completing a little more than an orbit of the Earth. Shepard was also launched on a converted Redstone ICBM, and his capsule Freedom 7 completed a 15-minute sub-orbital flight. Both missions were successful, and both nations followed up with more ambitious orbital missions designed to demonstrate continuing advances in rocketry and spaceflight navigation and control technology. Indeed, within only a few weeks of Shepard's flight, US President John F. Kennedy was calling on Congress to approve an ambitious plan "before this decade is out, of landing a man on the Moon and returning him safely to the Earth." The Russians, too, would set their sights on the Moon, but would ultimately end up losing that particular race in July 1969.

Both Gagarin and Shepard became celebrities who shared their spaceflight experiences widely with the media and the public. Gagarin would not fly in space again, but Shepard would go on to become one of just twelve people to walk on the Moon, commanding the Apollo 14 mission in 1974. Since 2001, an international celebration called "Yuri's Night" has been held in many nations on April 12 to commemorate Gagarin's and others' achievements in space exploration.

SEE ALSO (Geo) Science Fiction (1864), Liquid-Fueled Rockets (1926), Sputnik (1957), Leaving Earth's Gravity (1968), Geology on the Moon (1972), Settlements on Mars? (~2050)

Soviet cosmonaut Yuri Gagarin, the first human in space, prepares for his April 12, 1961 space flight.

Terraforming

Carl Sagan (1934–1996)

During the pre–twentieth-century exploration of our solar system, it was not so far-fetched to believe that nearby planets like Venus or Mars might be quite earthlike, sporting breathable atmospheres, oceans, and habitable (and perhaps inhabited) surfaces. The reality has proven far different, however, with Venus having since been discovered to be a hellishly hot world where CO_2 greenhouse heating has evaporated any possible past ocean, and Mars to have an atmosphere too cold and thin to allow liquid water to be stable on the surface today.

The idea that planets like these could be *turned* into earthlike worlds, however, started receiving serious scholarly study in the 1960s, via the concept of planetary-scale engineering activities known as terraforming ("Earth-shaping"). The goal of terraforming would be to modify the atmosphere and surface environment (temperature, pressure, humidity) of a planet to make it like that of the Earth, and thus potentially suitable for life as we know it.

The American astronomer and planetary scientist Carl Sagan was among the first to hypothesize serious ideas for the terraforming of terrestrial planets. In a landmark paper in 1961, Sagan postulated that the atmosphere of Venus could be "seeded" with algae that would slowly remove CO_2 from the atmosphere and sequester it on the surface as graphite, reducing the greenhouse effect and making the surface habitable. Subsequent discoveries about the atmosphere of Venus (including the presence of sulfuric-acid clouds) have made Sagan's ideas untenable, however. He also proposed ideas about terraforming Mars in 1973, spurring wide scientific-community interest and eventually a series of serious academic workshops and conferences on terraforming in the 1970s and 1980s.

Numerous other scientists, philosophers, and visionaries have helped to carry Sagan's terraforming torch since then, and it has become clear that dramatically changing the characteristics of a planetary environment would be an expensive endeavor that would take centuries or millennia to implement, and that would rely on many kinds of technologies that can be imagined but have not yet been invented. Regardless, there are dreamers out there who still imagine our planetary neighbors being engineered into more earthlike worlds in the future.

SEE ALSO Many Earths? (1600), (Geo) Science Fiction (1864), The Greenhouse Effect (1896), Cloud Seeding (1946), Extremophiles (1967), Earthlike Exoplanets (1995), Settlements on Mars? (~2050)

Artist's impression of a potentially terraformed planet Mars (many thousands of years in the future), with oceans and a thick atmosphere capable of supporting life.

Reversing Magnetic Polarity

Motonori Matuyama (1884–1958), **Lawrence W. Morley** (1920–2013),
Frederick J. Vine (b. 1939), **Drummond H. Matthews** (1931–1997)

Deep inside the Earth, electric currents created in the outer core—a spinning shell of molten iron—create powerful magnetic fields that form a strong magnetosphere around our planet. Earth's magnetic field is the force that makes compass needles point to the north or south. The subfield of geophysics called *paleomagnetism* focuses on the study of changes in the Earth's magnetic field over time, and on the ways that those changes can be used to better understand Earth's surface and interior processes.

Volcanic rocks contain a small fraction of magnetic minerals, like magnetite, that, when melted, line up like little compass needles parallel to the Earth's magnetic field. When the rock cools, it preserves a record of the Earth's magnetic field at the time of solidification. Geologists have been studying magnetized rocks since the early twentieth century, noting reversals in the polarity (north or south) of some rocks relative to others. In 1929, Japanese geophysicist Motonori Matuyama discovered that the most recent reversal in the Earth's magnetic field occurred about 780,000 years ago. Subsequent researchers found that Earth's field has flipped hundreds of times since the Jurassic, though the exact cause is still unknown.

An even more important discovery was made in 1963, when two studies (one by Canadian geophysicist Lawrence Morley, the other by the British geophysicists Frederick Vine and Drummond Matthews) independently interpreted magnetic field reversals that had been mapped since the 1950s in stunning magnetic "stripes" on the seafloor. The fact that the polarity of the stripes was symmetric on opposite sides of mid-ocean ridges led them to conclude that the ridges must be the source of new molten rock being erupted onto the seafloor. The rock solidifies and preserves the magnetic-field polarity at the time, but then those rocks get carried away from the ridges and eventually new lava that erupts later can preserve the opposite magnetic polarity. Magnetic-field reversals were thus one of the keys to discovering that mid-ocean ridges are also centers of *seafloor spreading*.

SEE ALSO Earth's Core Forms (c. 4.54 Billion BCE), Continental Crust (c. 4 Billion BCE), Plate Tectonics (c. 4–3 Billion BCE?), Magnetite (c. 2000 BCE), Solar Flares and Space Weather (1859), The Inner Core (1936), Mapping the Seafloor (1957), Seafloor Spreading (1973), The Oscillating Magnetosphere (1984)

As new seafloor is created by volcanoes at mid-ocean ridges, the polarity of Earth's magnetic field at the time is recorded in magnetic minerals in the volcanic rocks. As the seafloor spreads away from the ridges, it acts like a geologic "tape recorder" to reveal changing patterns of magnetic polarity over time.

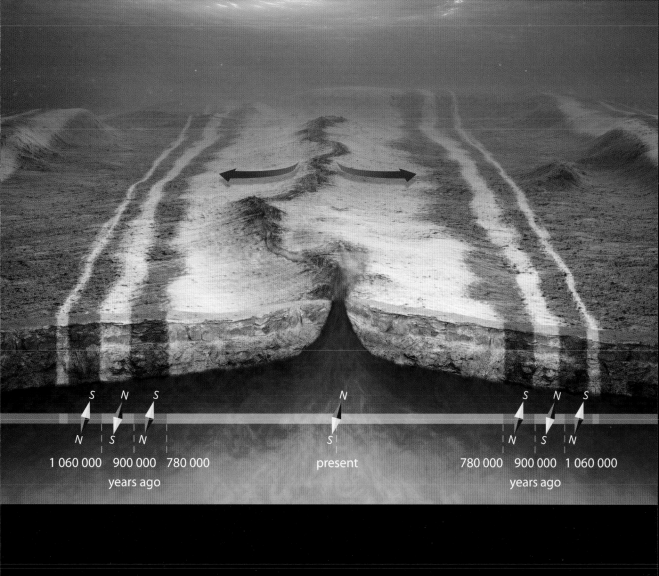

S	S	N	S		N		S	N	S
N		S	N		S		N	S	N

1 060 000   900 000   780 000   present   780 000   900 000   1 060 000

years ago   years ago

Endosymbiosis

Konstantin Mereschkowski (1855–1921), **Lynn Margulis** (1938–2011)

In biology, *symbiosis* is the close and long-term relationship and interaction between two or more dissimilar organisms. Examples include the symbiotic interactions that clownfish have with sea anemones, or that we have with a variety of microorganisms in our digestive tract. Evolutionary biologists have taken the concept to heart in trying to understand how the first eukaryotic cells (those with complex internal cell structures) and the first complex multicellular organisms came to be. A leading hypothesis for the origin of cellular complexity is called *endosymbiosis* (or sometimes *symbiogenesis*), which postulates that eukaryotic organisms evolved from symbiotic relationships with simpler, single-celled prokaryotic precursor organisms.

The basic outline of endosymbiosis was pioneered by Russian biologist Konstantin Mereschkowski in the early twentieth century. Mereschkowski was strongly influenced by the symbiotic relationship between fungi and algae in the functioning of lichen. Partly because of the lack of adequate lab technology, and partly because of his rejection of natural selection as an explanation for this symbiotic relationship, his ideas were ridiculed or ignored by many at the time.

Using new field and laboratory techniques, however, and a deeper understanding of the role of natural selection at the cellular level, the endosymbiotic hypothesis was resurrected by American evolutionary biologist Lynn Margulis, who wrote a key research paper in 1966 (and a popular science book in 1970) that thrust the debate back onto the international evolutionary biology stage. Like Mereschkowski, Margulis's ideas about the origin of eukaryotic mitochondria from the symbiotic union of prokaryotic bacteria, or the origin of eukaryotic photosynthetic chloroplasts from the symbiotic union of prokaryotic cyanobacteria, faced resistance from the scientific community. By the 1980s, however, and driven partly by Margulis's stamina and persistence in advocating her ideas broadly, that hard data would come from many experimental evolutionary biologists in the form of detailed genetic analyses of the DNA of mitochondria and chloroplasts—which turned out to be different from the DNA of their symbiotic hosts. Endosymbiosis went from a hypothesis to a widely accepted theory that encompasses all of organogenesis (the phase of embryonic development when specialized organs are formed).

SEE ALSO Photosynthesis (c. 3.4 Billion BCE), Eukaryotes (c. 2 Billion BCE), The Origin of Sex (c. 1.2 Billion BCE), Complex Multicellular Organisms (c. 1 Billion BCE), Natural Selection (1858–1859)

Evolutionary biologist Lynn Margulis, one of the founders of the theory of cellular endosymbiosis.

Earth Selfies

"Selfies," in which people turn their cameras around to take pictures of themselves (and possibly friends and family), provide a sense of perspective and inward-looking context. Turns out the same has been true in space photography almost from the beginning.

Some of the earliest cameras carried into space were mounted on captured German V-2 rockets and launched by the US Army into sub-orbital flights above the White Sands Missile Range in New Mexico in the late 1940s. The photos were grainy and crude by today's standards, but from a hundred miles up, they were the first to show the graceful curvature of the Earth from those altitudes.

Space photography became more common with the advent of the first weather satellites in the 1960s. However, it wasn't until 1966 that the first real "deep space" selfie of the Earth was taken by a space mission. *Lunar Orbiter I* was the first of a series of five robotic spacecraft sent to orbit the Moon between 1966 and 1967 to take photos of eventual landing sites for the Apollo astronauts. Since there were no digital cameras back then, they had the equivalent of a darkroom, scanner, and fax onboard so that they could process their own black-and-white film negatives, convert them to digital data files, and downlink them to the Earth.

Seizing the unprecedented opportunity to gain the perspective and inward-looking context of viewing the Earth from deep space, the *Lunar Orbiter I* team lobbied NASA to turn the spacecraft away from looking down at the Moon for just a short time, to look back at the Earth. It was risky—what if the spacecraft didn't turn back to the Moon afterward? But it was ultimately recognized as a unique and visionary opportunity, and so on August 23, 1966, the team was given approval to turn their camera around, back toward themselves and all of us, and take a selfie. The resulting first photo of the Earth rising above the lunar surface was an immediate public and media sensation. It, along with subsequent Earth selfies, helped spur an awakening environmental movement.

SEE ALSO Weather Satellites (1960), Leaving Earth's Gravity (1968), Earth Day (1970), Earth Science Satellites (1972), Geology on the Moon (1972)

The first "selfie" of the Earth from deep space, taken on August 23, 1966 by the Lunar Orbiter I *spacecraft and reprocessed in 2008.*

Extremophiles

Thomas Brock (b. 1926)

Astrobiology is the study of the origin, evolution, and distribution of life and habitable environments in the universe. It is perhaps a unique discipline in that it has only one data point with which to conclusively justify its existence. That is to say: so far, we know of only one example of life in the universe—that is, life on Earth—all of which is fundamentally similar, based on similar RNA, DNA, and other carbon-based organic molecules.

The search for life elsewhere is more than just the search for complex life forms like us, however. It is a search for other planetary environments that could be suitable for the most dominant form of life on our own planet—bacteria and other "simple" life forms. The best place to start a search for those conditions is right here on our own planet, where significant advances in our understanding of habitability have been made in the past fifty-plus years.

For example, in 1967, the American microbiologist Thomas Brock wrote a landmark paper describing heat-tolerant bacteria (*hyperthermophiles*) that flourished within hot springs at Yellowstone National Park. He challenged the prevailing wisdom that the chemistry of life requires moderate temperatures to operate. Brock's work helped to spur the study of *extremophiles*—life forms that survive and even thrive in harsh environments.

Hyperthermophilic bacteria have since been identified in very hot water near deep-sea hydrothermal vents as well; on the opposite extreme, *psychrophiles* have been found that live and thrive in near- or below-freezing temperatures. Life forms have also been found that exist over extremes of salinity (*halophiles*), acidity (*acidophiles* and *alkaliphiles*), high pressure (*piezophiles*), low humidity (*xerophiles*), and even high levels of UV or nuclear radiation (*radiophiles*).

The message for astrobiologists from the history of life on our planet is clear: life can thrive in an enormous range of environments. Thus, searching for evidence of past or present extremophiles or their habitable environments in extreme places such as Mars, the deep oceans of Jupiter's moons Europa and Ganymede, the sub-surface waters of Saturn's moon Enceladus, or the frigid, organic-rich surface of its moon Titan, is not as crazy an idea as it used to be.

SEE ALSO Life on Earth (c. 3.8 Billion BCE?), Complex Multicellular Organisms (c. 1 Billion BCE), Cambrian Explosion (c. 550 Million BCE), Meteorites and Life (1970)

Morning Glory Pool, a hot spring in Yellowstone National Park, Wyoming. The colors along the spring's outer edges are from a variety of hyperthermophilic bacteria that can survive and thrive even at the high temperatures of the spring (above 176°F [80°C]).

Leaving Earth's Gravity

William Anders (b. 1933), **Frank Borman** (b. 1928), **James Lovell** (b. 1928)

An estimated 100 billion or so humans have lived on planet Earth since our species first appeared, and until the late 1960s exactly zero of them had left the gravitational influence of their home planet. The first people to do that, in late 1968, were the crew of the Apollo 8 mission, one of the dress-rehearsal missions designed to prove that the critical spacecraft, navigation, and life-support capabilities needed to send people to the Moon (and return them safely to the Earth) would work. Frank Borman, Jim Lovell, and Bill Anders, all former US military pilots and officers, became the first humans to reach Earth escape velocity on December 21, 1968, as the first crew to ride atop NASA's massive *Saturn V* rocket. They were also the first humans to see the Earth rise over the horizon of another world, and their famous color "Earthrise" photo eclipsed the popularity of the previous *Lunar Orbiter I* black-and-white version from 1966.

Twenty-one additional astronauts over the course of eight more Apollo missions between 1969 and 1972 would follow the crew of Apollo 8 to escape Earth's gravity and travel to the Moon and back. Since 1972, about 520 more people have traveled into space, but none have traveled farther than low Earth orbit (a realm within 1,200 miles [2,000 kilometers] of the surface), and thus none have left the Earth's gravity. Part of the reason for the end of human deep-space travel was the retirement (and lack of replacement) of the *Saturn V* rocket in 1973, and the USSR's cancellation of their N-1 heavy-lift rocket program in 1976. NASA subsequently focused on the Space Shuttle—which could deliver people, cargo, and spacecraft only to low Earth orbit— and the International Space Station, which orbits only about 250 miles (400 kilometers) above the surface.

Renewed interest is brewing, however, in once again launching people out of Earth's "gravity well," both back to the Moon and on to Mars and other destinations. NASA's replacement for the *Saturn V*, called the *Space Launch System*, began assembly in 2014 and is expected to make its first test flight in 2019.

SEE ALSO Airborne Remote Sensing (1858), (Geo) Science Fiction (1864), Liquid-Fueled Rockets (1926), Exploration by Aviation (1926), Humans in Space (1961), Earth Selfies (1966), Earth Day (1970), Geology on the Moon (1972)

Computer-generated model of the Apollo 8 Command Module, which became the first crewed spacecraft to leave Earth's gravity and orbit the Moon in December 1968. The scene depicts the approximate relative orientation of the spacecraft, Earth, and Moon when the crew took the famous color "Earthrise" photo of our home world.

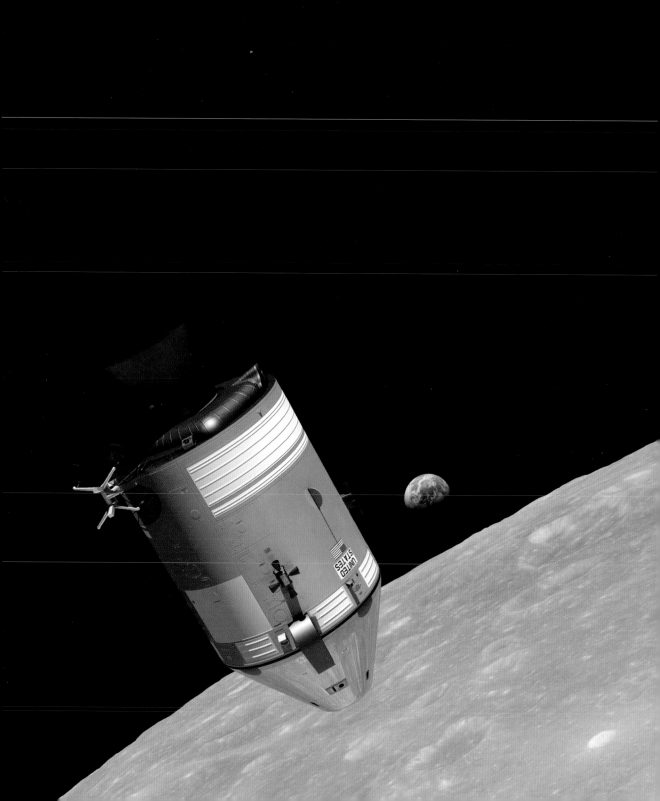

Meteorites and Life

One of the motivators of space exploration is the search for life beyond our home planet. But how do we conduct such a search? One way is to search for the chemical elements that occur in life on our own planet—elements such as carbon, hydrogen, nitrogen, oxygen, phosphorus, and sulfur. But these elements occur throughout the cosmos, in many places and environments (such as the insides of stars) that are unlikely to be conducive to life. A more effective strategy might be to search not for specific elements but for certain arrangements of them—molecules—that could reveal evidence for life's basic chemistry.

Life on Earth is based on organic molecules. Some organic molecules are simple, like methane (CH_4), methanol (CH_3OH), or formaldehyde (H_2CO); and others are much more complex, like proteins, amino acids, ribonucleic acid (RNA), and deoxyribonucleic acid (DNA). Over the last half century, astronomers have identified many simple organic molecules in dense interstellar clouds, comet tails, icy moons and rings of the outer solar system, and in the atmospheres of Titan and the giant planets and even in the active plumes of Saturn's small moon Enceladus.

Sometimes, organic molecules are delivered to us from space. For example, on September 28, 1969, a meteor and fireball streaked across the daytime sky and crashed to the ground near the town of Murchison in Victoria, Australia. More than 220 pounds (100 kilograms) of meteorite samples were found in the area. After detailed analysis of the samples, scientists announced in 1970 that the meteorite contained some common amino acids, important organic molecules in the chemistry of life. Later studies found that the Murchison meteorite contained more than seventy kinds of simple amino acids, plus many other simple and complex pre-biotic organic molecules.

Life on Earth as we know it requires liquid water, sources of energy like heat or sunlight, and abundant, complex organic molecules. The discovery of amino acids in Murchison and other meteorites supports the idea that molecules that are critical to life can form nonbiologically in environments such as in a solar nebular disk, on a comet, or on a newly forming planetesimal. Life may or may not be abundant in the cosmos, but at least the stuff of life appears to be everywhere.

SEE ALSO Life on Earth (c. 3.8 Billion BCE?), Extremophiles (1967)

X-ray image of magnesium (red), calcium (green), and aluminum (blue) in the Murchison meteorite, a more than 4.55-billion-year-old carbonaceous chondrite. This ancient rock contains primitive minerals condensed from the solar nebula, water, and complex organic molecules, including more than 70 kinds of amino acids.

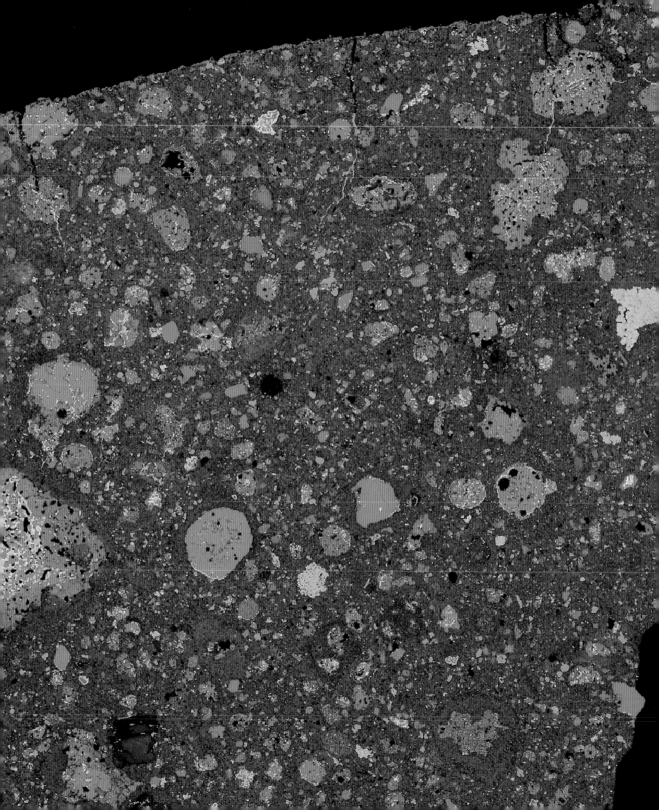

Earth Day

Rachel Carson (1907–1964), Morton Hilbert (1917–1998)

Ecology is the study of the interactions of organisms with their environment. Humans have always had to practice ecology, whether we knew it or not, because an awareness of our environment and of the threats and opportunities that it poses is key to survival. As the world's population began to soar in the nineteenth and then twentieth centuries, and as the Industrial Revolution and rapid technological advances began to make it easier for people (especially in cities) to detach from their historical connections to the environment, the health of our planet's biosphere began to decline. Deforestation, pollution, and irresponsible resource extraction started to threaten the health of individual societies and people.

The situation began to stir significant public reaction and engagement in the 1960s with the spread of rapid global mass communication and the growth of environmental activism—people and organizations who wanted to make a stand against damaging and unsustainable practices that harm our environment. A landmark event was the 1962 publication of *Silent Spring*, an influential book by American conservationist Rachel Carson, who pointed out many of the negative effects that people and industries were having on the natural world. She, and other ecologists, conservationists, and scientists helped to create what would eventually become a global movement focused on environmental awareness and stewardship.

An important and lasting manifestation of that growing environmental movement was the establishment, by American professor of public health Morton Hilbert and others, of Earth Day, a specially designated international day of recognition of humanity's role in changing our environment, and of our responsibility to ensure that we pass on a safe and sustainable world to our children. Hilbert, his students, and various US government officials and organizations started the first Earth Day in April 1970. Starting in 1990, Earth Day became an annual event, now celebrated every April 22, and expanded to include educational symposiums, performances, community clean-up events, and other activities worldwide designed to increase people's environmental awareness, and to hold accountable those individuals or organizations that would threaten our fragile ecosystem. Today's Earth Day events are organized by more than 5,000 environmental groups worldwide, engaging millions of people in environmental activism.

SEE ALSO *Homo sapiens* Emerges (c. 200,000 BCE), Population Growth (1798), Industrial Revolution (c. 1830), Birth of Environmentalism (1845), Deforestation (c. 1855–1870), The Sierra Club (1892), Earth Selfies (1966), Leaving Earth's Gravity (1968)

Crowd in New York City at the first Earth Day celebration, April 22, 1970.

Earth Science Satellites

The application of space-based imaging technology to better monitor and forecast the weather was an obvious and important first step in the use of satellites for the remote sensing of the Earth. However, the technology expanded significantly in the 1960s and early 1970s to the point where not only simple black-and-white or visible-wavelength color cameras could be deployed, but cameras that could also image the planet in diagnostic ultraviolet or infrared wavelengths, as well as other instruments such as spectrometers that could identify atomic, molecular, or mineral components of the atmosphere and surface.

A major advance in the use of space for remote sensing studies of the Earth was the 1972 launch of the first US government *Earth Resources Technology Satellite*—later renamed *Landsat 1*. *Landsat 1* carried two advanced (for the time) imaging cameras and was launched into a polar orbit so that it could map the entire planet as the surface spun below the satellite. *Landsat 1* was operational for five years and compiled an enormous dataset that was used to assess agricultural crop health, fisheries and shoreline health, deforestation rates and forest health, changes in water resources (rivers, lakes, and seas), recovery rates from natural disasters like fires and floods, and global-scale climate changes (mountain snowpacks, glaciers, and ice caps).

Because of its impact on our understanding of the planet, follow-on satellites from *Landsat 2* through *Landsat 8* (still operational today) have been launched by NASA and operated by the US National Oceanic and Atmospheric Administration (NOAA), with sensors of increasing capability included on each newer satellite. The US has also launched a number of additional earth science satellites specifically designed to study the oceans, atmosphere, and gravitational and magnetic fields, to monitor characteristics of our planet such as ocean wave heights, wind speeds, precipitation and clouds, and surface moisture.

Other nations and space agencies have also been deeply involved in space-based earth science. Since 1972, earth science and weather monitoring satellites have been launched and operated by Argentina, Brazil, China, the European Space Agency, France, India, Japan, Morocco, Nigeria, Pakistan, the Philippines, Russia, South Korea, Sweden, Thailand, Turkey, and Venezuela. More nations are getting involved all the time; and, most recently, many private companies have also become involved in the collection, processing, and distribution of earth science satellite data.

SEE ALSO Airborne Remote Sensing (1858), Structure of the Atmosphere (1896), The Ozone Layer (1913), Geosynchronous Satellites (1945), Weather Satellites (1960), Earth Selfies (1966), Oceanography from Space (1993)

False color composite image from the Landsat-1 mission taken on July 25, 1972, of the Dallas/Fort Worth metropolitan area in Texas. The colors are intended to indicate levels of vegetation cover (shades of red) and plant health.

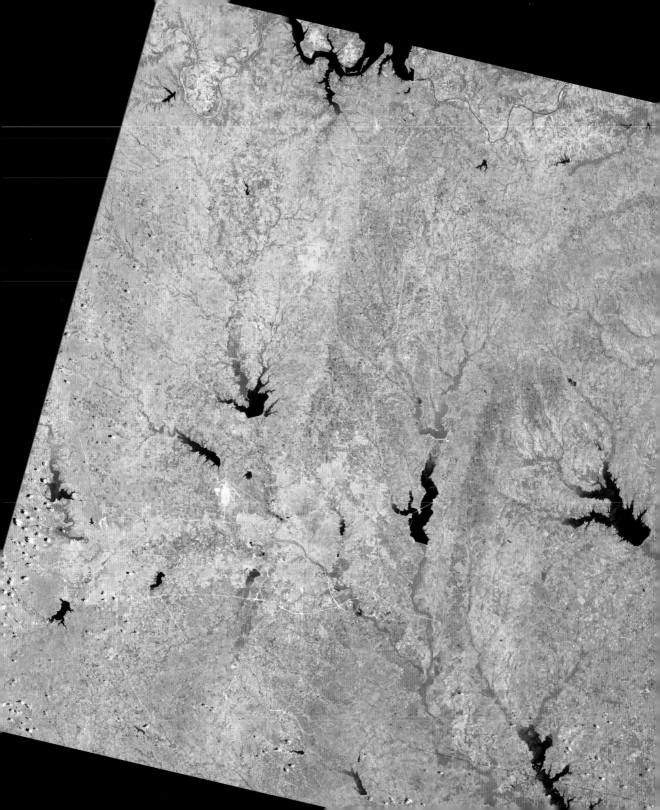

Geology on the Moon

Harrison H. Schmitt (b. 1935)

The foundational principles of geology had been worked out in the seventeenth century and significantly refined and expanded since then. But could those principles apply to the study of the geology of other worlds? Would Earth prove to be our laboratory for understanding other planets, and would the study of other worlds provide new and useful information about the Earth?

Humanity's first chance to answer such questions directly came from the Apollo missions to the Moon between 1969 and 1972. During six separate missions, twelve NASA astronauts explored a diverse set of landing sites and their environs on the near side of the Moon (the side always facing Earth). The Apollo astronauts brought back more than 800 pounds (363 kilograms) of rocks and soils from the Moon, much of which has ended up playing an important role in the favored hypothesis for the origin of the Moon, which is thought to have formed from the debris of a giant impact between the young Earth and a Mars-sized protoplanet around 4.5 billion years ago.

While all the Apollo astronauts were train ed in basic geological science and sampling methods, they were mostly trained as test pilots and in other areas needed for military aviation. During the final Apollo mission to the Moon in late 1972, however, one NASA astronaut became the first (and, so far, only) scientist to explore the Moon. Harrison H. "Jack" Schmitt was trained in geology at Caltech, Harvard, and the United States Geological Survey (USGS) before joining the astronaut corps in 1965. As a scientist-astronaut, Schmitt worked with fellow USGS geologist Gene Shoemaker to train the rest of the Moon-bound astronauts in geology and field methods.

Schmitt's deep geologic background paid off during the mission—for example, when he realized, based on his earth-bound geology experience and expertise, that layers of colored glass beads had come from ancient fire-fountaining volcanic eruptions early in the Moon's history. Schmitt and other planetary scientists have since learned that fundamental terrestrial geologic processes like volcanism, tectonism, erosion, and even impact cratering modify planetary surfaces across the solar system. The in-depth exploration of the final frontier requires and benefits significantly from a deep expertise and understanding of the geology of our home planet.

SEE ALSO Birth of the Moon (c. 4.5 Billion BCE), Foundations of Geology (1669), Exploring the Oceans (1943), Understanding Impact Craters (1960), Humans in Space (1961), Leaving Earth's Gravity (1968), Earth Science Satellites (1972), Oceanography from Space (1993)

The first scientist on the Moon, geologist Jack Schmitt, studies and takes samples from soils and rocky outcrops during the Apollo 17 mission in December 1972. Photo by Apollo 17 mission commander Gene Cernan.

Seafloor Spreading

Tanya Atwater (b. 1942)

As more information, maps and topographic data, and fossil evidence became available over the course of the twentieth century, support continued to grow for Alfred Wegener's once-ridiculed idea of continental drift. The problem had been that there was no obvious explanation or physical process that could explain how a continental mass could plow through an oceanic plate; but this became a moot point in the 1960s with the discovery of seafloor magnetic stripes of reversing polarities and the realization that the mid-ocean ridges were spreading centers where new oceanic crust was being formed. The uppermost crust of the Earth appeared to be operating like a kind of global-scale conveyor belt, with new crust appearing at the ridges and old crust being subducted and destroyed at the trenches and collisional plate boundaries.

But how was it all connected? And did the math all work out: was new crust being created at the same rate that old crust was being destroyed? And what is driving it all? Among the leading geologists working to find those answers is American oceanographer Tanya Atwater, who in a key research paper and several book chapters published in 1973 set out to stitch the world together, calculating and synthesizing information that showed that the seafloor was spreading apart along the world's mid-ocean ridges (at a typical rate from 5 to 9 centimeters per year), that the seafloor was being subducted at plate boundaries, and that plates were sliding past each other (rather than colliding) along faults such as the famous San Andreas in California.

Atwater applied her and her colleagues' observations of the topography and geology of the seafloor to make discoveries about how Earth's dozen or so tectonic plates interact with one another, how continents like North America were built up over time by collisions of ancient, long-gone tectonic plates, and how the active geology and volcanology of much of the ocean floor could provide interesting environments, such as hydrothermal vents, for life to potentially thrive. Indeed, she found, Earth's plates are all connected in one way or another, and a rich history of those past connections and interactions is still preserved in the geologic record today.

SEE ALSO Plate Tectonics (c. 4–3 Billion BCE?), Continental Drift (1912), Island Arcs (1949), Mapping the Seafloor (1957), Reversing Magnetic Polarity (1963), Global Positioning System (1973), Oceanography from Space (1993)

Global maps of the ages of the seafloor crust. Red is the youngest (including new crust being created today); blue is the oldest oceanic crust remaining on the planet, formed between about 150 and 180 million years ago.

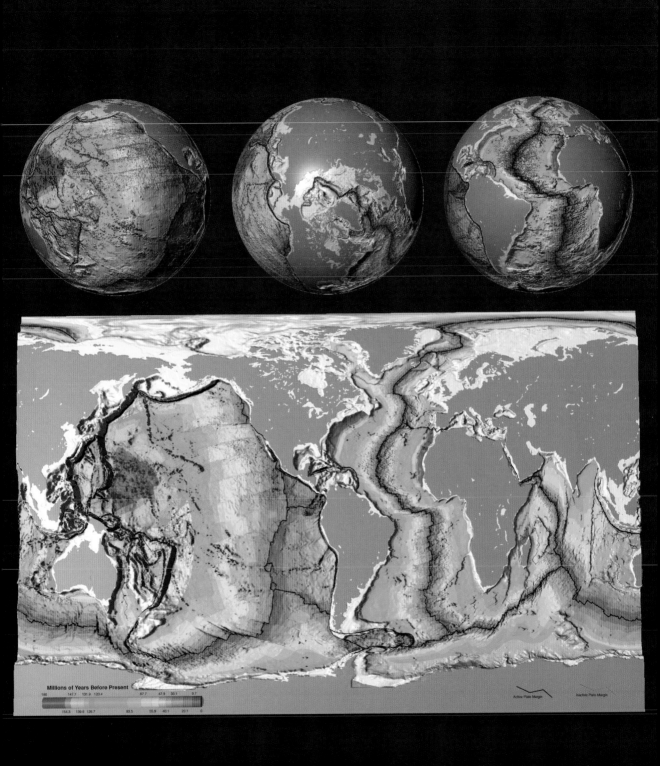

Millions of Years Before Present

190   147.7   131.9   120.4   87.7   47.9   33.1   9.7

154.3   139.6   126.7   93.5   55.9   40.1   20.1   0

Active Plate Margin   Inactive Plate Margin

Tropical Rain/Cloud Forests

The Earth's ten or so major kinds of ecological community types or *biomes* are not randomly distributed across the surface. Rather, they strongly correlate with other factors, including elevation, proximity to the ocean, proximity to major mountain ranges, and, perhaps most importantly, latitude. At low latitudes near the equator, sunlight is most intense and precipitation is heaviest; at high latitudes near the poles, it is coldest and driest.

Tropical rainforests occur exclusively in those equatorial regions of the greatest heating and rainfall, and as a result exhibit the highest density of plant growth and the highest diversity of species of any of the biomes on Earth. By definition, a tropical rainforest is a place where it rains at a roughly constant (and high) rate all year round—there is no "dry season" like in many other biomes. A closely related class called cloud forest is a biome that extends to subtropical latitudes where temperatures are hot but slightly cooler and rainfall is still abundant and continuous. Cooler temperatures and often slightly higher elevations lead to nearly continuous ground fog, which is where cloud forests get their name.

Tropical rainforests and cloud forests are constantly at risk for unsustainable human exploitation because of their enormous resource potential (wood, minerals, game) and the difficulty of monitoring and managing such huge expanses of wild terrain like the Amazon. An important milestone in the realization of the importance of these places was the establishment, in 1973, of the Monteverde Cloud Forest Reserve, in the western Cordillera de Tilarán mountains of Costa Rica, about 10° north of the equator. Monteverde is a privately funded 26,000-acre nature preserve that is estimated to contain over 2,500 species of plants, 100 species of mammals, 400 bird species, 120 species of reptiles and amphibians, and thousands of species of insects, all in an area only about the size of Walt Disney World in Florida.

The success of Monteverde (more than 70,000 people visit each year) attests to widespread interest in the preservation of unique places like rainforests and cloud forests. Dozens of public and private preserves and national parks now aim to protect as many of these hot, wet, and wild places as possible around the world.

SEE ALSO Sahara Desert (c. 7 Million BCE), Amazon River (1541), Deforestation (c. 1855–1870), Angel Falls (1933), Temperate Rainforests (1976), Tundra (1992), Boreal Forests (1992), Grasslands and Chaparral (2004), Temperate Deciduous Forests (2011), Savanna (2013)

A 2013 photo taken of lush jungle flora deep within the Monteverde Tropical Rainforest Preserve in Costa Rica.

Global Positioning System

Since the late 1950s, satellites have been important tools for studying the Earth, having been used to discover the Earth's radiation belts, to forecast the weather, and to enable new discoveries in geology, geophysics, and oceanography. At the same time, an enormous global industry has sprung up around the use of satellites for communications and entertainment. And at the same time as that, early military satellite networks were being established to transmit super-precise atomic clock signals to the ground, enabling accurate positional determination and navigation capabilities with properly equipped ground receivers. An important hybrid triple-use of these kinds of satellite applications was started in 1973 with the development of the concepts that became the Global Positioning System, or GPS, network.

GPS was initially called "Navigation System Using Timing and Ranging," or Navstar, and it consisted of ten satellites launched by a joint team from the US Department of Defense that incorporated technologies and concepts developed by individual precursor satellites and programs from the US Army, Navy, and Air Force. The first Navstar network (called "Block I") became fully operational in 1985, providing the US military with the ability to locate or track specific assets to within about 30 to 60 feet (about 10–20 meters) anywhere on the Earth.

After the tragic shooting-down by the Russian military of a civilian airliner that had strayed off course in 1983 (Korean Air Lines Flight 007), the US government decided to declassify signals from the GPS network and make them freely available, globally, for civilian use. Initially, civilian signals were degraded; but starting in 2000, access was granted to the full potential positional capabilities to all users. Improvements to new Blocks over time allows modern GPS receivers using signals from today's dozen or so Block IIIA satellites to typically locate a user's position to within about 10 feet (about 3 meters). Planned advances into the early 2020s should bring that accuracy down to within about 1 foot (30 centimeters).

Dozens of other nations now share or operate their own GPS networks. In addition to navigation for cars, boats, planes, and people, GPS signals provide time information for cell phones and computers, meteorological data on the atmosphere and ionosphere, and geologic data on tectonic plate motions and earthquake displacements.

SEE ALSO Airborne Remote Sensing (1858), Geosynchronous Satellites (1945), International Geophysical Year (IGY) (1957–1958), Weather Satellites (1960), Earth Science Satellites (1972), Oceanography from Space (1993)

Illustration (not to scale) showing the more than two dozen GPS satellites, which orbit around 12,500 miles (20,000 kilometers) above the surface at an orbital tilt of about 55°. The system is designed to ensure that at least four satellites are visible at least 15° above the horizon at any given time anywhere in the world.

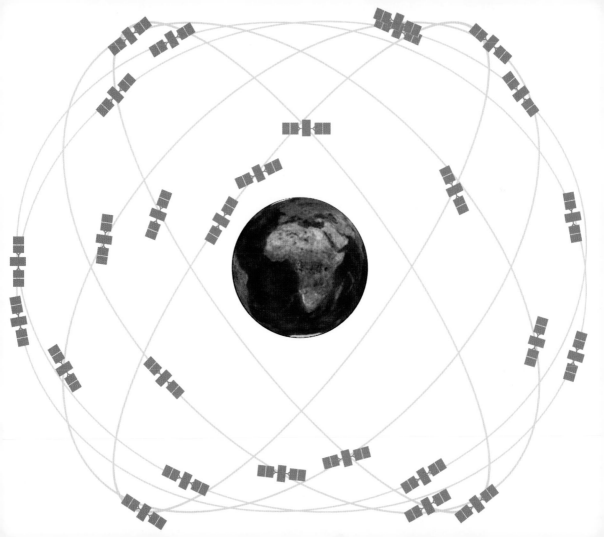

Insect Migration

Fred Urquhart (1911–2002)

Many species (including early humans) migrate—move from one environment to another—in a persistent and relatively repeatable way in response to seasonal changes in weather, food supply, predation conditions, or other factors. Even relatively small animals like insects can exhibit impressive and surprisingly long migratory patterns.

Among the most famous and mysterious of the insect migrations is the story of the Monarch butterfly, subspecies of which can be found around the world. For example, each year in September or October, Monarchs from the eastern US and southern Canada migrate 2,000 miles (3,200 kilometers) or more to wintering sites presumably in Mexico. Canadian zoologist Fred Urquhart, a pioneer in the study of Monarch migration, had been searching since the 1930s for the Monarch overwintering sites, and finally hit the jackpot in 1975 when several residents of central Mexico responded to his newspaper advertisement seeking help finding the butterflies. The locals took Fred to a mountaintop in Michoacán state, where huge numbers (hundreds of millions) of Monarchs were found in the first widely known wintering refuge. Other nearby sites were found, compelling the Mexican government to protect the region as a unique ecological preserve, now known as the Monarch Butterfly Biosphere Reserve.

The annual journey of the Monarchs is a multi-generational migration. The parents die on the voyage south, yet somehow their offspring born along the way know how to reach the traditional wintering sites. None of *them* will survive the voyage back north starting in March, but their children (or grandchildren) will complete the return trip.

Many species of moths, dragonflies, locusts, and beetles also migrate seasonally, and some cover similarly large distances as the Monarchs. Many of these migrations are also multi-generational, spanning six generations to make the round trip, in some cases. How exactly the (instinctual?) knowledge of precisely where to go is passed along from generation to generation is unknown. Perhaps it is simply an inherited instinct, or perhaps they follow the Sun, or the Earth's magnetic field lines, or specific geographic features along the way. Figuring out one of the most fascinating migratory behaviors on Earth is the subject of intense research and debate among entomologists and others.

SEE ALSO Magnetic Navigation (1975), Boreal Forests (1992), Large Animal Migrations (1997)

Spectacular clusters of Monarch butterflies share the same small spaces while "overwintering" among trees in the Monarch Butterfly Biosphere Reserve in Angangueo, Mexico.

Magnetic Navigation

Salvatore Bellini (1925–2011), **Richard Blakemore** (b. 1942)

Animals have evolved the relative power of their senses (sight, hearing, smell, touch, taste) via natural selection processes over time that made it advantageous to, for example, be able to see better at night, or to be able to smell or hear things better than the predator hunting them. It makes sense that a species would thrive by taking advantage of specific environmental conditions in its prime habitat. It is even the case that some animals have developed specialized senses that other species lack and that give them unique capabilities—such as echolocation in bats and dolphins.

An even more surprising example of a specialized new sense was the unpublished 1963 observation, by an Italian medical doctor named Salvatore Bellini, that some bacteria can react and move in response to a magnetic field—a navigational process now known as *magnetotaxis*. A fuller understanding of how these bacteria sense the magnetic field would come in 1975, in a peer-reviewed paper by American microbiologist Richard Blakemore, who discovered that these bacteria contain tiny chains of crystalline magnetic minerals such as magnetite, a strongly magnetic iron oxide, surrounded by membranes and proteins. Blakemore dubbed the magnetite chains *magnetosomes*, speculating that they represent a totally new type of sensing organ that was, at least then, unique to what he called *magnetotactic* bacteria.

Since then, the ability to sense the strength and direction of the Earth's magnetic field and move or make other decisions based on that sense (which is known as *magnetoreception*) has also been discovered in a number of more complex species, including arthropods, mollusks, and certain vertebrates. Homing pigeons, for example, use their orientation relative to the Earth's magnetic field as part of their navigation system. Tiny grains of magnetite have been found within part of their beaks, but their role and function in the birds' navigation is unclear. Even some mammals (such as certain mouse and bat species, but not humans) appear to have magnetoreception capabilities. To date, however, there has been no obvious "organ" (like a nose or a tongue) identified in magnetoreceptive animals that appears to be an obvious sensory receptor. How these animals relay their "feel" for their relationship to Earth's magnetic field is a mystery.

SEE ALSO Magnetite (c. 2000 BCE), Reversing Magnetic Polarity (1963), Magnetic Navigation (1975), The Oscillating Magnetosphere (1984)

High-resolution photo of a single-celled bacterium known as Magnetospirillum magnetotacticum, *or MS-1, containing a linear chain of magnetosomes made of the mineral magnetite. The cell is about 0.00008 inch (2 microns) long.*

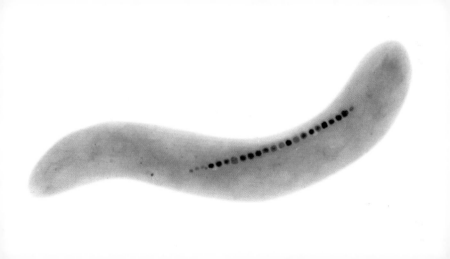

Temperate Rainforests

Tropical rainforests (which generally occur within about 25 degrees of the equator) contain the highest densities of living organisms and the greatest range of species diversity of any biome on the Earth. Not far behind, however, are different kinds of 102between the Tropic of Cancer and the Arctic Circle (23.5°N to 66.5°N) and in the southern hemisphere between the Tropic of Capricorn and the Antarctic Circle (23.5°S to 66.5°S).

Temperate rainforests are densely wooded areas that receive heavy annual rainfall (greater than 55 inches [140 centimeters]) and are populated with both conifer and broadleaf trees and a wide variety of mosses and ferns. The trees form a canopy (top-level habitat which collects most of the rain and sunlight) that can rise 300 feet (100 meters) or more above the forest floor and that can shade as much as 95 percent of the floor during the summer. Little sunlight and cool temperatures keep the forest floor damp year-round, and promote the growth of mosses, ferns, lichens, slugs, and a variety of shade-tolerant shrubs.

Temperate rainforests occur in only a few places around the world, and are generally associated with regions that receive significant moisture from the ocean; among the largest regions are along the coast of the North American Pacific Northwest, along the southwestern coast of South America, and in southern China and northern North Korea. Much of the temperate rainforest in the Pacific Northwest, including the Hoh rainforest (the largest temperate rainforest in the US) became protected land when it was incorporated into a national monument in 1909 and then the Olympic National Park in 1938. Recognizing its importance as one of a rare and dwindling number of special ecosystems in the world, UNESCO designated the Olympic National Park as a Biosphere Preservation District in 1976.

The lush Hoh rainforest of Washington state's Olympic Peninsula, which receives over 127 inches (323 centimeters) of rain each year, is a classic example of "old growth" temperate rainforest. The Hoh's canopy consists of enormous (23-foot/7-meter diameter) spruce and hemlock trees as well as Douglas fir, cedar, maple, and cottonwoods, many of which are between 150 and 300 years old.

SEE ALSO Flowers (c. 130 Million BCE), Oldest Living Trees (c. 3000 BCE), Deforestation (c. 1855–1870), National Parks (1872), Tropical Rain/Cloud Forests (1973), Boreal Forests (1992), Temperate Deciduous Forests (2011)

A 2007 photo of lush temperate vegetation along the Hall of Mosses Trail, in the Hoh Rainforest, Olympic National Park, in the US state of Washington.

Voyager Golden Record

About once every 176 years, the giant outer planets Jupiter, Saturn, Uranus, and Neptune line up along a gentle astronomical curve that enables a single spacecraft to potentially efficiently visit all of them, using the gravity of each planet to slingshot from one to the next. Scientists realized in the 1960s that the next opportunity for such a "Grand Tour" would occur in the 1970s. Are atmospheric processes on the giant planets similar to those on the Earth? And how? Have their moons been habitable or geologically active in the past, or are they in the present? Answering such questions relatively quickly would require a mission to take the Grand Tour.

Just such a mission was devised in the late 1960s and launched in 1977. The NASA *Voyager 1* and *Voyager 2* flyby probes completed the first detailed reconnaissance of the giant outer planets within twelve years of launch, and since 1989 have been drifting slowly outward to explore the farthest reaches of the Sun's magnetic field. Now, because they are going fast enough to escape the Sun's gravity, they are exploring interstellar space.

Each spacecraft carries a "message in a bottle," a time capsule, in the form of a gold-plated copper phonograph record (with stylus) upon which are encoded greetings in 55 languages from children to world leaders, 21 different sounds (like wind, rain, crickets, dogs, tractors, trains, and a kiss), 90 minutes of music in 27 songs from around the world (including Chuck Berry's "Johnny B. Goode"), and 115 images, including mathematical symbols, people, food, buildings, and natural wonders. The Voyager Golden Record was the brainchild of astronomer Carl Sagan and a small team of scientists, artists, and visionaries tasked by NASA with succinctly representing the 1970s-era planet Earth to whoever or whatever might one day find the spacecraft.

Because they are encased in gold-plated aluminum casings, and because neither spacecraft is expected to encounter anything more than an occasional hydrogen molecule or perhaps a speck of cosmic dust on its journey around the Milky Way galaxy, the Golden Records are predicted to last billions of years or more, perhaps even outlasting the Earth and Sun. Might these "messages in a bottle," thrown into the cosmic sea, one day be all that remains from our Pale Blue Dot?

SEE ALSO Laws of Planetary Motion (1619), Sputnik (1957), Earth's Radiation Belts (1958), Earth Selfies (1966), Long-Duration Space Travel (2016).

Main image: *An example of the 12-inch-diameter (30.5 centimeters) gold-plated copper disc known as the Voyager Golden Record, two of which are carrying recorded songs, greetings, and images from Earth on a voyage across the galaxy.* **Inset:** *Artist's concept of the* Voyager *spacecraft.*

Deep-Sea Hydrothermal Vents

One of the ways Earth releases its internal heat is through volcanism. We tend to think of volcanoes as places where molten lava or ash erupt onto the surface, but the reality is that volcanic vents and fissures can also release enormous amounts of water and other volatile gases into the atmosphere and the oceans. Places where superheated water, steam, and other gases are being volcanically released onto the surface are called hot springs or *hydrothermal vents*.

With the discovery of seafloor spreading and the realization that the mid-ocean ridges and hotspots are the sites of the active volcanic eruption of new oceanic crust, many scientists realized that there were likely to be submarine (deep-sea) hydrothermal vents associated with the mid-ocean-ridge spreading centers. In the early-to-middle 1970s, research crews from the Scripps Institute of Oceanography, Woods Hole Oceanographic Institution, and other leading oceanography institutions searched for and eventually discovered the first evidence for submarine hydrothermal vents, on the seafloor near the Galápagos Islands.

A three-person crew of marine geologists in the deep submersible vehicle *Alvin* made the first dive to directly sample the Galápagos rift hydrothermal vents in February 1977. At the vents, they discovered that cold seawater that had percolated down through fractures in the crust and come into contact with subsurface magma was being released back onto the seafloor at temperatures between about 5°C to 15°C above the average surrounding seafloor water temperature of about 2°C. Surprisingly, they also discovered a stunning diversity of *extremophile* organisms living at high pressures near the vents, including simple bacteria and archaea and complex organisms such as tube worms, clams, and shrimp.

Subsequent expeditions led to the discovery at mid-ocean ridges around the world of very high temperature hydrothermal vents and *black smokers*—tall columns of stalagmite-like deposits where superheated sea water (over 700°F, or 370°C) emerges from the vents and cools quickly, precipitating out dark minerals like sulfides. Lower-temperature *white smokers* were also discovered in places where lighter-colored silicate minerals are being precipitated. Marine biologists have since discovered that the base of the food chain for these thriving deep-sea ecosystems are bacteria that use sulfur compounds coming from the superheated seawater and a process called *chemosynthesis*—rather than photosynthesis—as their energy source.

SEE ALSO Plate Tectonics (c. 4–3 Billion BCE?), Photosynthesis (c. 3.4 Billion BCE), Cascade Volcanoes (c. 30–10 Million BCE), Hawaiian Islands (c. 28 Million BCE), Galápagos Islands (c. 5 Million BCE), Mapping the Seafloor (1957), Extremophiles (1967), Seafloor Spreading (1973), Underwater Archaeology (1985)

Black smokers and tube worm communities at "Strawberry Fields" in the Endeavour Hydrothermal Vent Field in the northeast Pacific Ocean, southwest of Vancouver Island, about 7,380 feet (2,250 meters) below sea level.

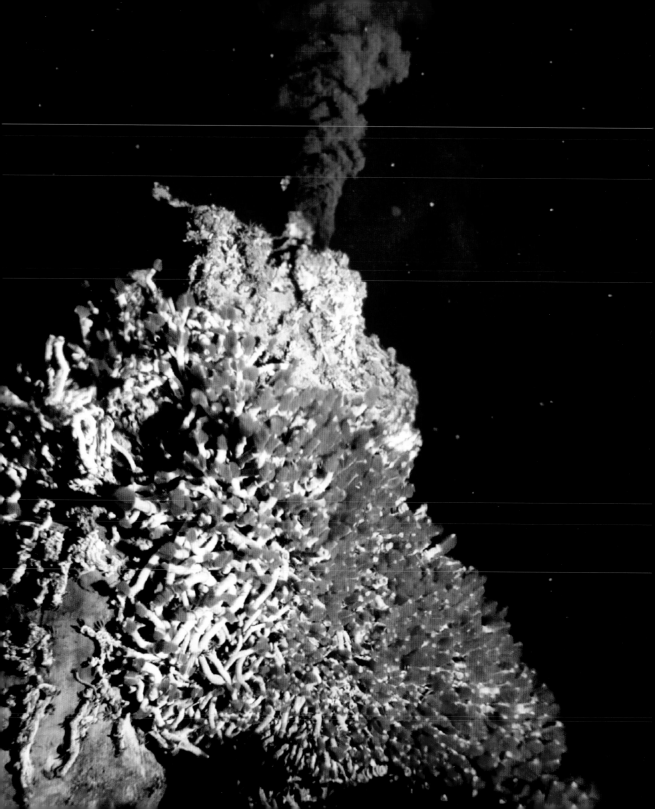

Wind Power

Like sunlight, the wind seems like a potentially never-ending supply of energy, if it could be properly harnessed. Wind has powered sailing vessels since prehistoric times, and historical evidence for the first experimental windmills goes back a few thousand years, to the Roman Empire. Seventh-century Persian engineers developed the first widely used windmills for practical purposes, like pumping water or grinding (milling) grain.

The use of windmills for pumping and milling continued to grow over the centuries, and by the late nineteenth century the first windmills to generate electricity began to appear. At first these wind turbines were generating only small amounts of power (~5 to 25 kilowatts, enough for individual farms or small communities). In certain persistently windy places, however, the incentive of potentially abundant, sustainable, "low carbon" energy compelled governments, companies, and even communities to develop larger, higher-capacity wind turbines. A watershed milestone in the development of wind turbine technology came in 1978, when the first reliable multi-megawatt (2 megawatt capacity) wind turbine went online in the community of Tvind, along the windy North Sea coast of Denmark.

The upgraded Tvind turbine continues to generate power for the region today. Since the 1970s, however, increasing costs of fossil fuels (and predictions about their eventual depletion), concerns about the safety of nuclear power, and the slower-than-expected development of high-capacity solar power generation facilities have incentivized turbine companies to build larger and more efficient systems with even higher power capacity. Today's largest wind turbines, for example, are up to 600 feet (180 meters) tall and can generate up to 8 megawatts of electricity, enough from a single windmill to potentially power thousands of homes.

According to British government studies, wind power, and specifically wind power generated from large multi-megawatt offshore wind farms (some anchored in shallow water, others floating in deep water), is now the lowest cost option for large-scale, low-carbon power. Globally, electricity generation from wind power now represents about 8 percent of the world's total, but that number is expected to increase to 15 percent or more by 2040 as the costs (economic as well as environmental) of fossil fuels continue to increase.

SEE ALSO Industrial Revolution (c. 1830), Nuclear Power (1954), Controlling the Nile (1902), Solar Power (1982), Hydroelectric Power (1994), End of Fossil Fuels? (~2100)

A 2006 photo of part of a wind-turbine farm between the San Jacinto and San Bernardino mountains near Palm Springs, California. More than 4,000 turbines provide enough electricity to power Palm Springs and the nearby Coachella Valley.

A World Wide Web

The invention of the first computers in the 1950s made it possible not only for faster computations for scientific and engineering applications in Earth sciences and other disciplines, but also for wider communications among individual scientists and their research groups. The earliest computers were so-called *mainframe* systems with a bulky central core of processors that was accessed from a *terminal*. At first the terminals were co-located with the mainframe, but then as they started to get more distant (an office on another floor, a building across campus), computer scientists had to devise protocols and standards for multiple terminals to communicate with the central hub, as well as with each other.

In the 1960s, the initial concepts and methods of establishing an expanded interconnected networking system, or *internet*, were developed. These included the idea of sending short packets of data formatted in ways that everyone agreed upon, and assigning special computers called *routers* to deliver these packets to their intended destinations. During the 1970s, small internets were being established between leading computer-science universities as well as among US government organizations such as the Defense Advanced Research Projects Agency (DARPA). And then, in 1979, computer scientists developed a specific standard way to send text-based news stories and messages among specific users on the internet. That same year, a company called CompuServe began to offer internet users of both mainframe and newly emerging personal computers the ability to send electronic mail using this standard. The internet as we know it today, and the Information Age, were born.

Major advances in speed, bandwidth, and routing have followed in the decades since. Another important milestone was the development in 1989 of software tools and protocols that would allow information on the expanding internet to be stored and viewed by others on specific nodes or pages of what was dubbed the World Wide Web. Today, "the web" has gone global, and the internet is accessible everywhere on the planet (though it is censored in some countries) to anyone who can afford a "terminal" such as a cellular telephone, laptop, tablet, or personal computer. Science, education, politics, and our civilization in general have all been profoundly changed by enabling access to the expanding web of all the world's knowledge.

SEE ALSO Population Growth (1798), Industrial Revolution (c. 1830), The Anthropocene (c. 1870), Geosynchronous Satellites (1945), Global Positioning System (1973)

Like a spider's web, the world is now more electronically interconnected than ever before.

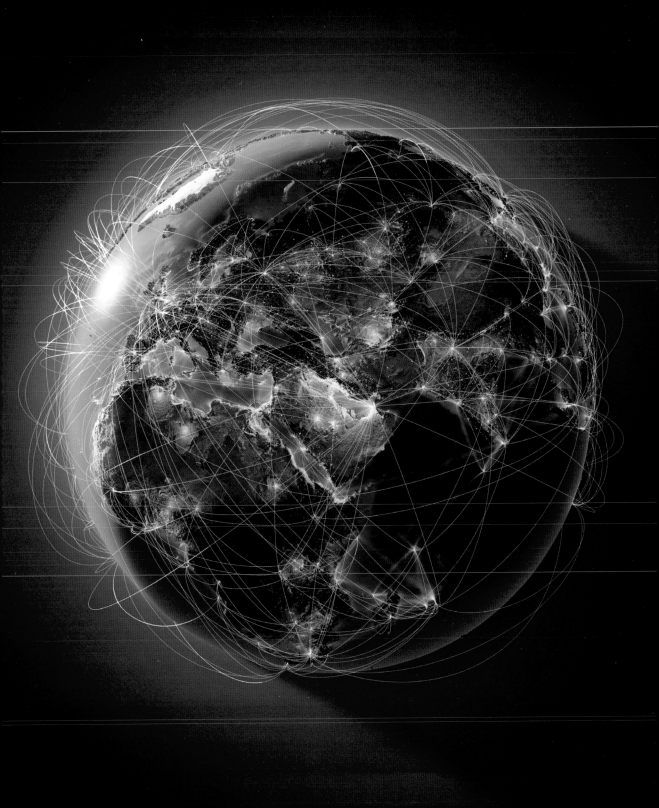

Mount St. Helens Eruption

Volcanic island arcs and tall mountain chains are common features on the Earth at tectonic plate boundaries where one plate is sliding under (subducting) the other. Famous examples in the western hemisphere include the Cascade volcanoes in the northwestern US and southwestern Canada.

Volcanoes like those in the Cascades are fundamentally different from the kinds of volcanoes that form at mid-ocean ridges or above intra-plate hotspots, like in the Hawaiian Islands or Iceland. In ridge or hotspot volcanism, magma from the upper mantle erupts relatively gently directly onto the surface, producing relatively fluid lava flows that slowly build up the heights of the ridges or islands. In subduction-zone volcanism, however, melting of the sinking slab of oceanic crust causes melting of the overlying continental crust and thus a lower-iron, higher-silicon, more viscous (thicker) magma. That thicker magma plugs up subsurface fractures and causes gases to build up, until they result in an explosive eruption at the surface that releases huge volumes of pulverized ash and dust.

One of the most dramatic examples in US history of such an explosion occurred on the morning of May 18, 1980, when Mount St. Helens in southern Washington erupted violently. Within minutes, the top fifteen percent of the mountain was blown off, and parts of a thick, roiling cloud of ash some 40 miles (64 kilometers) wide were ejected 15 miles (24 kilometers) high into the stratosphere. Trees, other vegetation, animals, and more than 50 people in a blast zone extending as far as 19 miles (31 kilometers) from the volcano were wiped out, and ashfall up to 5 inches thick settled over parts of Washington, Oregon, and Idaho. The landscape in and around the huge new crater caused by the blast looked more like the surface of the Moon than the Earth.

Fortunately, focused swarms of strong earthquakes in the weeks before the eruption had allowed geologists to predict that an eruption was imminent, and most people in the area evacuated. The mountain has continued to periodically erupt again, however, in smaller events as recently as 2008. Exactly when Mount St. Helens or one of the other Cascade volcanoes will next erupt is unknown, but it is certain that more eruptions will happen along that active plate margin.

SEE ALSO Plate Tectonics (c. 4–3 Billion BCE?), The Sierra Nevada (c. 155 Million BCE), Cascade Volcanoes (c. 30–10 Million BCE), Hawaiian Islands (c. 28 Million BCE), The Andes (c. 10 Million BCE), Pompeii (79), Huaynaputina Eruption (1600), Krakatoa Eruption (1883), Island Arcs (1949), Volcanic Explosivity Index (1982), Mount Pinatubo Eruption (1991), Eyjafjallajökull Eruption (2010)

The violent May 18, 1980 ash and steam eruption of Mount St. Helens, a stratovolcano in southwestern Washington State, USA.

Extinction Impact Hypothesis

Luis Alvarez (1911–1988), **Walter Alvarez** (b. 1940)

The idea that one or more of the five large mass extinctions discovered in the fossil record since the end of the Ordovician (~ 450 million years ago) might have been caused by the impact of a large asteroid or comet was essentially impossible for geologists to accept before the middle of the twentieth century, because the role of impact cratering in modifying planetary surfaces and their environments was not yet developed.

It was an idea, however, keenly in the mind of American geologist and archaeologist Walter Alvarez, whose studies of patterns of Roman settlement via the analysis of the geochemistry of various artifacts took him to the mountains near Gubbio, Italy, in the 1970s. There, he discovered an extensive and anomalously dark, thin layer of clay with the same age as one of the extinction events at the boundary between the Cretaceous and Paleogene (K-Pg) periods about 65 million years ago, when all of the non-avian dinosaurs went extinct.

Alvarez had an advantage in the analysis of these clay samples: his father, the Nobel Prize-winning American nuclear physicist Luis Alvarez, had access to special laboratories that could make the most sensitive measurements of the trace elements in those samples. The father-and-son team discovered that they were heavily enriched in the platinum-group heavy-metal element iridium relative to "normal" Earth rocks. Since most of Earth's heavy metals sank into the core when the Earth first formed, they and colleagues hypothesized, in a famous 1980 research paper, that the iridium enrichment came from the impact of an extraterrestrial asteroid or comet, and that impact was the proximal cause of the K-Pg extinction of the dinosaurs.

The impact extinction hypothesis was met with significant skepticism because it lacked adequate proof. However, in 1990 a large and heavily eroded impact crater with an age of about 65 million years was discovered near Chicxulub, on the Yucatán peninsula of Mexico. It was the smoking gun that the community needed, and the scientific consensus today is that the Chicxulub impact was at least part of the cause of the K-Pg extinction, perhaps in combination with other major geologic events such as the nearly contemporaneous volcanic eruptions of the Deccan Traps.

SEE ALSO Mass Extinctions (c. 450 Million BCE), The Great Dying (c. 252 Million BCE), Triassic Extinction (c. 200 Million BCE), Deccan Traps (c. 66 Million BCE), Dinosaur-Killing Impact (c. 65 Million BCE), Arizona Impact (c. 50,000 BCE), Platinum Group Metals (1802–1805), The US Geological Survey (1879), Hunting for Meteorites (1906), The Tunguska Explosion (1908), Understanding Impact Craters (1960)

Luis (left) and Walter Alvarez in 1981, sampling layered sedimentary rocks in Gubbio, Italy, at the 65-million-year-old boundary between the Cretaceous (lower) and Paleogene (upper, formerly called Tertiary) geologic time periods.

Great Barrier Reef

Ecologists rely on certain species of *bioindicators* (or indicator species) to reveal the characteristics of different parts of the environment, both in the deep past as well as in the present. For example, corals that build reefs in shallow-water colonies are highly sensitive to the temperature, salinity, and acidity of the waters in which they grow. This environmental sensitivity enables paleobiologists to reconstruct the details of ancient shallow-water oceanic environments going back to the first appearance of corals around the time of the Cambrian Explosion 550 million years ago. Even though the vast majority of coral species went extinct (along with 96 percent of all other species) about 250 million years ago during "The Great Dying" mass-extinction event at the end of the Permian era, a small fraction of coral species survived and continue to survive and serve as oceanic bioindicators today.

The largest coral reef on Earth today occurs in the Great Barrier Reef, in the shallow waters off the northeastern coast of Australia. The reef system is almost 1,400 miles (2,300 kilometers) long and is composed of thousands of individual reefs and many hundreds of small islands covering an area about the size of Germany. It is the largest single structure in the world made by living organisms. In 1975 the Australian government established the Great Barrier Reef Marine Park to recognize the special nature of this region to the nation. Perhaps more importantly, the Great Barrier Reef was elevated to a United Nations World Heritage Site in 1981 in recognition of its global status as a unique and special ecosystem in need of constant protection.

Corals provide a perfect example of endosymbiosis, as they rely on algae and other organisms to provide energy via photosynthesis and to help with the calcification that enables reefs to grow. However, because of rising ocean temperatures, overfishing, ocean acidification, and pollution, corals worldwide are suffering because they are unable to provide adequate amounts of CO_2 and other nutrients for the algae, breaking down the symbiotic relationship. The result is "bleaching" (whitening) of corals worldwide as the algal population decreases, leading eventually to the slow starvation and death of the corals themselves. Corals can recover from bleaching, but perhaps not without our (worldwide) help.

SEE ALSO Earth's Oceans (c. 4 Billion BCE), Cambrian Explosion (c. 550 Million BCE), The Great Dying (c. 252 Million BCE), Geology of Corals (1934), Exploring the Oceans (1943), Endosymbiosis (1966), Oceanography from Space (1993), Ocean Conservation (1998)

A 2014 photo of shallow-water tropical corals and fish at Flynn Reef in the Great Barrier Reef Marine Park, off the coast of Queensland, Australia.

Genetic Engineering of Crops

People have been manipulating the genes of other organisms since prehistoric times, for example via selective breeding methods designed to increase the level of domestication in many animals, or via similar methods designed to enhance the hardiness and yield of agricultural crops. These kinds of *genetic engineering* methods rely on the exploitation and enhancement of naturally occurring variations in animal behavior or plant size/robustness, and thus represent a sort of human-guided method of natural selection.

As the human population has grown since (and partly as a result of) the Industrial Revolution, and especially as that population could reach more than 10 billion by the middle of this century, traditional "natural" methods of genetic engineering have not proven adequate to meet the world's demand for food. This reality, combined with dramatic increases in our understanding of microbiology starting in the early twentieth century, has fueled the advancement of modern *biotechnology*—the use of living organisms to make or modify products or processes.

A globally important application of biotechnology is the artificial genetic engineering of agricultural crops, in order to increase their tolerance to pests, pesticides, diseases, and/or environmental extremes like drought, as well as to increase their shelf life after harvesting and/or their nutritional value as food products. The first genetically modified organism (GMO) with agricultural applications was an antibiotic-resistant tobacco plant, developed in the laboratory in 1982 by researchers at the Monsanto chemical company. Additional laboratory and field testing of other modified versions of tobacco throughout the 1980s produced crops resistant to insects and herbicides. Farmers around the world were quick to adopt the technology as it was applied to a variety of crops, because crop yields were increased on average by more than 20 percent, pesticide use was decreased by more than 35 percent, and thus farm profits were increased dramatically.

Genetic engineering has certainly helped to feed the world's growing population. While the scientific consensus is that GMO foods are safe to eat, much of the public is skeptical. Some nations regulate the cultivation or importation of GMO crops heavily, and some ban or restrict them outright. Even scientific advocates agree that much work remains to be done, for example in proving that GMO crops and foods are environmentally safe in the long run.

SEE ALSO Domestication of Animals (c. 30,000 BCE), Invention of Agriculture (c. 10,000 BCE), Fermentation of Beer and Wine (c. 7000 BCE), Population Growth (1798), Industrial Revolution (c. 1830), Natural Selection (1858–1859), The Anthropocene (c. 1870), Soil Science (1870), Terraforming (1961), Plant Genetics (1983)

Examples of unusually colored and shaped maize from Latin America that is being genetically combined with domestic US corn crops to help increase their genetic diversity.

Basin and Range

John McPhee (b. 1931)

Anyone who's ever flown over, driven through, or seen satellite photos of the American desert southwest can't help but notice the undulating, almost rhythmic nature of the landscape. Across large swaths of desert lowlands in southern New Mexico, Arizona, and California as well as north into Nevada and south into Mexico, the terrain alternates in a series of long, narrow, parallel ridges separated by long, narrow, parallel valleys (the most famous of which is Death Valley). Because of those valleys and hills, geologists call this region of the world the Basin and Range Province.

Despite agreeing on a name, geologists can't agree on the cause of the Basin and Range topography, partly because the region has had a very complex geologic history. The leading hypothesis is that subduction of the former Farallon plate under North America starting around 155 million years ago ushered in a long period of uplift and the creation of compressional mountain ranges like the Sierra Nevada and the Rockies. The remains of that plate are now fully underneath the continent, sinking into the mantle, and now causing the crust to stretch out instead of compress. As extensional forces pull on the crust, parallel faults are formed perpendicular to the direction of stretching. Blocks of crust that drop down between adjacent faults (called *graben*) form long valleys or basins; the ridges between the down-dropped blocks (called *horst*) form long hills or ranges.

The sparse vegetation of the desert southwest accentuates the dramatic geologic story that has unfolded in the Basin and Range. Among the most lauded of those storytellers is American writer and Princeton professor John McPhee. His 1982 book *Basin and Range* was the first of his four-volume Pulitzer-prize winning *Annals of the Former World* series, in which he recounts, in layman's language but with a geologist's eye for the natural world, the geologic history of North America and the many characters who have helped to reveal, and revel in, that history. McPhee's writing has helped to communicate to the general public both consensus positions as well as ongoing debates within the earth sciences community. Science, as a human endeavor, needs not only great observationalists and theorists, but also great storytellers.

SEE ALSO Continental Crust (c. 4 Billion BCE), Plate Tectonics (c. 4–3 Billion BCE?), Greenstone Belts (c. 3.5 Billion BCE), The Appalachians (c. 480 Million BCE), The Sierra Nevada (c. 155 Million BCE), The Rockies (c. 80 Million BCE), Cascade Volcanoes (c. 30–10 Million BCE), The Grand Canyon (c. 6–5 Million BCE), Death Valley (c. 2 Million BCE).

Main image: *A view of Basin and Range National Monument, Nevada.*
Inset: *John McPhee, author of numerous popular science books in earth sciences.*

Solar Power

In the late nineteenth century, physicists discovered that light shining on some materials would cause an electric current to flow through them, and in the early twentieth century Albert Einstein won the Nobel Prize for explaining the physics behind this so-called *photoelectric effect.* By the 1940s and 1950s, the first *solar cells* were being invented to create electricity from sunlight, and since the late 1950s they have been deployed into telecommunications, weather, and earth and planetary science satellites.

Solar cells for generating electrical power for terrestrial applications began to be more widely developed and deployed in the 1970s, fueled by the oil embargo and energy crises that occurred worldwide during that time. A major milestone was the development of the first pilot megawatt-scale solar power plant in 1982, in the Mojave Desert outside Hesperia, California. The plant was built by the Atlantic Richfield Company (ARCO), a major fossil-fuel supplier, on the assumption that the price of oil would skyrocket in the decades ahead. The pilot plant paved the way for an even larger (5.2 megawatt) ARCO solar power plant built nearby, which consisted of more than 100,000 separate arrays of solar cells. The price of oil did not increase dramatically, however, and so these early forays into industrial-scale solar power production were abandoned in the late 1990s.

Since then, however, major technological advances in the efficiency of solar cells and the distribution of electricity, the slowly increasing costs of fossil fuels, and growing environmental concerns over non-renewable energy sources have sparked a resurgence in solar-power generation. Residential and commercial rooftop solar panels have become more common and less expensive, and solar-power "farms" have been built around the world that generate from several hundred up to more than 850 megawatts of electricity each. The worldwide capacity to create electricity from sunlight has quadrupled over the past five years, and today about 1.7% of global electricity production is from sunlight. That is a small fraction compared to other sources like wind, hydroelectric, or nuclear power, but solar-power production is the fastest-growing renewable energy source, and many analysts think that as efficiency increases and solar cell costs continue to drop, solar power could be the world's main energy source by sometime in the second half of this century.

SEE ALSO Industrial Revolution (c. 1830), Controlling the Nile (1902), Nuclear Power (1954), Weather Satellites (1960), Earth Science Satellites (1972), Wind Power (1978), Hydroelectric Power (1994), End of Fossil Fuels? (~2100)

A 2017 photo of a large solar collector power station in Andalusia, Spain.

Volcanic Explosivity Index

Volcanic eruptions have impacted people's lives since prehistoric times. Even in relatively more recent history, famous eruptions such as Mount Vesuvius in 79, Krakatoa in 1883, Mount St. Helens in 1980, Mount Pinatubo in 1991, and Iceland's Eyjafjallajökull in 2010 have had significant local, regional, and global effects, including dramatic short-term changes in climate associated with the larger eruptions. Even relatively "mild" volcanic eruptions can cause significant disruptions to local communities.

Just like geophysicists thought it was important to develop a scale to measure and catalog the energy of our planet's many earthquakes over time, volcanologists came to the conclusion that doing something similar for volcanic explosions could provide a way to understand the nature of individual volcanic events, as well as the potential danger that individual volcanoes might pose to life or property. Thus, in 1982, a standard system of measurement of the power of individual volcanoes was established, based primarily on the volume and style of material erupted from the volcano's vent or caldera. The resulting Volcanic Explosivity Index (VEI) ranges from 0 for relatively gentle and continuous, low-volume, small-plume eruptions like those from Kilauea and Mauna Loa volcanoes in Hawaii, to 8 for extremely rare mega-colossal explosions with plumes that reach heights greater than 12 miles (20 kilometers) and inject huge amounts of ash and dust in the stratosphere, where it significantly influences the global climate.

Eruptions with VEI values of 0 to 3 are constantly occurring somewhere on Earth, with the largest ones in that range exploding about every three months and most only having local effects on the weather and climate. VEI 4–5 eruptions occur about once a year to once a decade and can inject significant amounts of ash and dust into the atmosphere (and onto local cities and towns: Vesuvius, which destroyed the city of Pompeii in 79, was a VEI 5 event) and have significant regional effects on the weather. The largest eruptions, VEI from 6 to 8, occur only from about once or twice per century to once or twice per 100,000 years, and not only cause massive local and regional damage, but can have a major influence on the global climate.

SEE ALSO Pompeii (79), Huaynaputina Eruption (1600), Mount Tambora Eruption (1815), Krakatoa Eruption (1883), Exploring Katmai (1915), Island Arcs (1949), Valdivia Earthquake (1960), Mount St. Helens Eruption (1980), Mount Pinatubo Eruption (1991), Sumatran Earthquake and Tsunami (2004), Eyjafjallajökull Eruption (2010), Yellowstone Supervolcano (~100,000)

A 1970 photo of a 30-foot-high (10 meters) lava fountain eruption from the Mauna Ulu vent of the Kilauea volcano on the Big Island of Hawaii.

Gorillas in the Mist

Dian Fossey (1932–1985)

Research into the origin and evolution of primates relies heavily on archaeological discoveries of fossils, tools, and other artifacts. However, since many primate species are still extant, genetic, zoological, and sociological research of existing populations can also provide significant information on their evolution and interrelationships. A specialized subfield called *primatology* is dedicated to the study of both living and extinct species of primates. Primatologists have expertise in anthropology, biology/genetics, zoology, veterinary medicine, anatomy, psychology, and sociology; they work in museums, zoos, animal research/rescue facilities, and out in the field.

Among the leading twentieth-century authorities on primates, and especially on gorillas, was the American zoologist and primatologist Dian Fossey. Trained as an occupational therapist, Fossey took a transformational trip to central Africa in 1963, where she met archaeologists Mary and Louis Leakey and first encountered wild mountain gorillas. Three years later, at the invitation of the Leakeys and with primary funding from the National Geographic Society, Fossey moved to the Congo to begin what would be a nearly twenty-year study of endangered mountain gorillas.

In her articles for *National Geographic* magazine, and especially in her landmark 1983 book, *Gorillas in the Mist*, Fossey revealed intimate details of the social lives, behaviors, and habits of this most mysterious of the great ape species. Poaching and habitat encroachment had reduced the size of the mountain gorilla population to only around 800 individuals; through her work and her writing, Fossey became an ardent advocate of conservation and preservation efforts aimed at permanently protecting this population. She made enemies among the local poachers and others who made their living off trapping or killing gorillas, however; and sadly, she was murdered in her field camp in Rwanda in 1985, likely in retribution for her vigorous anti-poaching efforts.

Fossey's legacy of protecting mountain gorilla populations lives on in governmental and privately funded conservation efforts in Rwanda, Uganda, and the Democratic Republic of Congo, where national parks have been created to protect them. Despite a slowly growing population and the work of international conservation organizations, mountain gorillas remain critically endangered and in need of continuing conservation efforts to survive.

SEE ALSO Primates (c. 60 Million BCE), First Hominids (c. 10 Million BCE), *Homo sapiens* Emerges (c. 200,000 BCE), Tracing Human Origins (1948), Tropical Rain/Cloud Forests (1973), Chimpanzees (1988)

Zoologist and conservationist Dian Fossey, at home with some mountain gorilla friends in Rwanda, central Africa.

Plant Genetics

Gregor Mendel (1822–1884), Barbara McClintock (1902–1992)

Genes—the fundamental units of inheritance in all forms of life on Earth—were essentially discovered in the late nineteenth century by Austrian botanist Gregor Mendel. While Mendel did not have the technological tools to visualize actual genes, he deduced their presence by a series of clever experiments on pea plants between 1856 and 1863. By carefully following the expression of various traits such as seed and flower color and plant height, he was able to show that inherited traits could be *dominant* or *recessive*, and that a series of *rules of heredity* could be used to predict the traits of offspring from the known traits of parents. While not recognized as such at the time, Mendel is now widely regarded as the founder of modern genetics.

Among the most important early advances in genetics in the twentieth century was the ability to finally visualize individual chromosomes, the carriers of the DNA inside cells. Some of the most important early discoveries in this field were made by American botanist and geneticist Barbara McClintock, who studied the chromosomes of maize (corn) from the late 1920s to the early 1950s. McClintock developed novel microscopy methods to directly observe changes in chromosomes during cell division and to characterize the roles that different parts of the chromosome play in transmitting genetic information to offspring. She was among the first to advocate that parts of the DNA structure could control, or regulate, the expression of certain genes or genetic traits. The idea was not understood or accepted by her contemporaries, however. Many of her concepts were ultimately vindicated in the 1960s and 1970s as even more advanced technology provided the molecular evidence in support of her hypotheses.

Like Mendel (who was an Augustinian friar), McClintock did her painstaking observational research mostly in solitude, and was not widely recognized for her contributions at the time. However, unlike Mendel, she was ultimately recognized in her lifetime, receiving the Nobel Prize in Physiology/Medicine in 1983 for her early genetics work. That same year, a popular biography of McClintock (*A Feeling for the Organism*, by E. F. Keller) was published, helping to further bring her critical contributions to genetics to light.

SEE ALSO The Origin of Sex (c. 1.2 Billion BCE), First Land Plants (c. 470 Million BCE), Flowers (c. 130 Million BCE), Genetic Engineering of Crops (1982)

Main image: *Plant geneticist Barbara McClintock in her laboratory in 1947. McClintock made important discoveries about the ways that genes are responsible for switching the physical traits of an organism on or off.*
Inset: *A 2008 photo of a variety of plant chromosome pairs from the science museum in South Kensington, London.*

The Oscillating Magnetosphere

Margaret Kivelson (b. 1928)

Earth is surrounded by a protective "bubble" of magnetic fields called the *magnetosphere*. Our planet's magnetism is generated deep in its interior, as Earth's partially molten, metallic iron core spins on its axis. Like the invisible field lines from an iron bar magnet, Earth's field lines extend far past the surface. And like the water in front of a fast-moving boat, Earth's magnetic field lines are compressed into a bow wave where the strong "flow" of high-energy particles streaming out of the Sun (the *solar wind*) plows into our magnetosphere.

The basic outline and properties of the magnetosphere have been studied since the seventeenth century, through laboratory experiments with magnets as well as long-term observations of *auroras* in the polar regions, which are one manifestation of our magnetosphere's interactions with the solar wind. However, only with the use of satellites that have been directly probing the magnetosphere since the late 1950s have we really begun to understand this critical part of Earth's system.

Among the leading magnetosphere researchers is American space physicist Margaret ("Margie") Kivelson, who began studying magnetic fields in the late 1950s, and has since been involved in a number of space satellite missions studying the magnetosphere. Kivelson led a key research paper in 1984 that discovered the presence of large ultra-low-frequency waves in the magnetosphere. Apparently, these waves are created by the pulsating solar wind crashing into the outer—most tenuous—layers of the magnetosphere. Like waves breaking on a beach, these huge magnetic waves help accelerate and transport particles from the solar wind and outer magnetosphere onto the "shores" of the inner magnetosphere, where they can be studied in detail.

Kivelson, who has won numerous prestigious honors and awards for her work, also uses her expertise in terrestrial magnetism to study the magnetic fields of the giant planets and their moons. Indeed, she led the team that discovered that the solar system's largest moon, Ganymede, has a magnetic field of its own (and is the only moon with one), and that Jupiter's magnetic fields provide strong evidence for the existence of a subsurface liquid ocean on its moon Europa.

SEE ALSO Earth's Core Forms (c. 4.54 Billion BCE), Earth's Mantle and Magma Ocean (c. 4.5 Billion BCE), Women in Earth Science (1896), The Inner Core (1936), Earth's Radiation Belts (1958), Extremophiles (1967), Earth Science Satellites (1972), Earth's Core Solidifies (~2–3 Billion)

Main image: *Illustrative depiction of the Earth's magnetosphere (blue lines), which "holds off" the onrushing solar wind (orange, at left), protecting life on our planet from much of that high-intensity radiation.* **Inset:** *Margie Kivelson in 2017.*

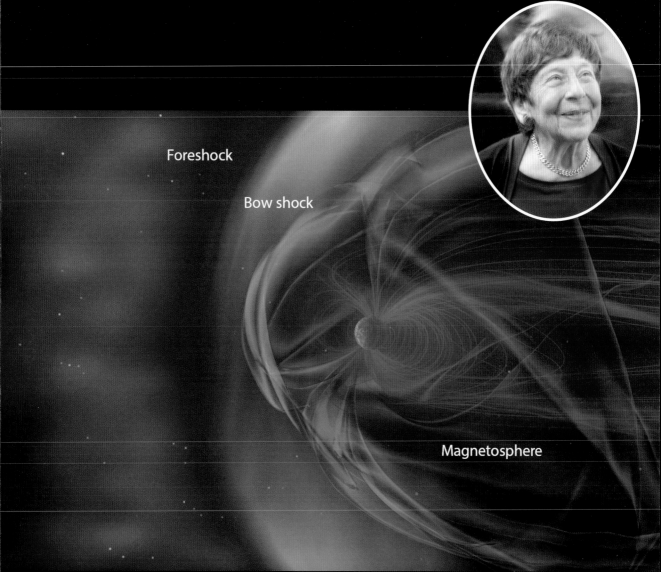

Foreshock

Bow shock

Magnetosphere

Underwater Archaeology

Robert Ballard (b. 1942)

Archaeologists typically excavate historical and prehistoric artifacts from the ancient world that have been covered or buried by jungle, sediments, or volcanic deposits over time. Part of the ancient (and even modern) history of the world, however, is under water.

Marine archaeology has thus emerged as a specialized subfield dedicated to the study of the physical remains (shipwrecks, submerged settlements, and other artifacts) of past interactions between humans and the world's oceans, seas, lakes, and rivers. While having been practiced for centuries, a huge advance was the development of diving bells, helmets, SCUBA equipment, and deep-sea submersible vehicles that enable people to spend much more time under water.

Among the most successful practitioners and popularizers of modern underwater archaeology is the American oceanographer and explorer Robert Ballard. Ballard led the 1977 expedition in the submersible *Alvin* that discovered the first black smoker deep-sea hydrothermal vents, and frequently turned his interests towards searching for and exploring shipwrecks or other artifacts that could help fill important historical holes in the study of the past. His most famous find came in 1985, when he led an expedition using the underwater robotic vehicle *Argo* that successfully found the wreckage of the RMS *Titanic*, which sank in 1912 and was lying 12,500 feet (3,800 meters) deep on the North Atlantic seafloor off Newfoundland. Despite Ballard's plea "that it be left unmolested by treasure seekers," numerous other expeditions since then have photographed and removed artifacts from the famous ship.

Ballard and other marine archaeologists have since explored numerous other shipwrecks and submerged ancient coastal sites, aircraft, and other artifacts around the world, dating from the Bronze Age to ancient Greece, Scandinavia, and Rome to modern industrial times. Particularly important underwater historical finds have been made in the Mediterranean, Black Sea, and even among the maritime Pacific battlefields of World War II. Many of these artifacts are extremely well preserved in deep oxygen-poor water and can thus provide unique "time capsules" for studying the people, places, and events at the time the artifacts were submerged.

SEE ALSO Exploring the Oceans (1943), Mapping the Seafloor (1957), Mariana Trench (1960), Deep-Sea Hydrothermal Vents (1977), Oceanography from Space (1993), Ocean Conservation (1998)

Main image: *View of the bow of the RMS* Titanic, *photographed in June 2004 by the ROV* Hercules *during an expedition returning to the shipwreck site.* **Inset:** *Oceanographer and explorer Robert Ballard.*

Chernobyl Disaster

The advent of nuclear power as an alternative to fossil fuels for the generation of electricity starting in the second half of the twentieth century has resulted in the construction of more than 450 nuclear power stations now in operation around the world, accounting for about ten percent of the world's total electricity generation. While the safety record of this growing industry is generally outstanding, because the systems being used often involve high pressures, temperatures, and (by definition) radioactivity, accidents do occur.

The worst nuclear power accident to date occurred in April 1986 at the Chernobyl nuclear power plant near the town of Pripyat, in the Ukrainian Republic of what was then the Soviet Union. During a safety test, one of the station's reactors overheated and caused an uncontrolled nuclear chain reaction, or *criticality accident*, that led to a catastrophic steam explosion and an intense open-air fire that released large amounts of radiation high into the atmosphere, as well as radioactive fallout directly onto the local environment. More than 50,000 people were evacuated from the regions around the plant, including the entire towns of Chernobyl and Pripyat. More than fifty people, including many employees and first responders, died either immediately or shortly after the accident, and as many as 4,000 additional cancer deaths are predicted to occur from people who had been living in the region near the facility. Shortly after the accident and evacuation, a 1,000-square-mile (2,600-square-kilometer) "exclusion zone" was set up around Chernobyl. The Zone, as it is called, remains a vast, unpopulated, semi-urban landscape to this day, and is a solemn reminder of the potentially devastating effects of a nuclear power plant accident.

In 1990, the International Atomic Energy Agency (IAEA) established the "International Nuclear Event Scale," rating nuclear accidents from 0 (no safety concerns) to 7 (major accident with widespread health and environmental effects). Chernobyl was the only level-7 accident until 2011, when some of the Fukushima Daiichi nuclear power plant's reactors outside Tokyo, Japan, overheated as a result of a large earthquake and tsunami. The worst nuclear accident in the US, the 1979 partial meltdown of a reactor at the Three Mile Island nuclear power plant in Pennsylvania, rates as a 5 on the IAEA scale.

SEE ALSO Industrial Revolution (c. 1830), Radioactivity (1896), Nuclear Power (1954), Wind Power (1978), Solar Power (1982), Hydroelectric Power (1994), End of Fossil Fuels? (~2100)

A 2017 photo of the Chernobyl Victims Memorial, in front of the nuclear power station's reactor #4 building, part of which is now a cement sarcophagus entombing the original reactor chamber.

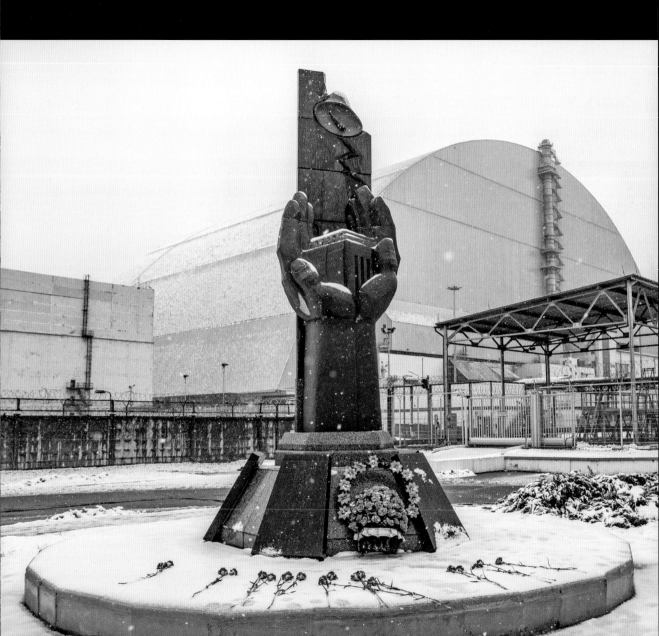

California Condors

Just like the oceans harbor certain diagnostic indicator species like the corals of the Great Barrier Reef, a number of land animals provide important information about the health of continental or island ecosystems, as well as lessons about the potentially catastrophic effects that people can have on those ecosystems. A prime example comes from the plight of the California condor, a bird species that once roamed widely across much of North America prior to the first human settlement of the continent, but which was reduced to extinction in the wild by 1987.

The California condor is in the vulture family and is the largest land bird in North America, with a typical wingspan of around 10 feet (3 meters). Like other vultures, condors are scavengers that mostly eat carrion—the remains of dead and decaying animals. Fossil evidence indicates that the California condor population began to decline around the end of the last glacial maximum (around 12,000 years ago), coincident with the extinction of many of the large megafauna species (mammoths, tapirs, bison, horses, etc.) around the same time. Although controversial, many researchers believe that overhunting by Stone Age peoples may have played a major role in these post-glacial extinctions. Regardless, the already-stressed population of California condors was eventually driven to near-extinction during the European settlement of North America by poaching, lead poisoning, pesticide poisoning, and habitat destruction.

By 1987, only 22 California condors were left, all in captivity. A US government captive breeding program, carried out by organizations including the San Diego Wild Animal Park and the Los Angeles Zoo, was initiated to attempt to educate the public about the plight of this species and to repopulate condors to their former Southwestern habitats. The process has been slow, partly because California condors live long lives (up to 60 years) and only start breeding relatively late compared to other species, partly because they choose a single mate for life, and partly because females typically lay just one egg every other year.

Despite the obstacles, the California Condor Recovery Plan is building momentum. Starting in the 1990s, captive-bred California condors began to be reintroduced to former habitats in California and Arizona, and today the total California condor population is estimated to be around 500, with about 300 living in the wild.

SEE ALSO The First Birds (c. 160 Million BCE), End of the Last "Ice Age" (c. 10,000 BCE), Population Growth (1798), Deforestation (c. 1855–1870), Earth Day (1970), Great Barrier Reef (1981), Savanna (2013)

A 2013 photo of a California condor perched on a rock at the Grand Canyon in Arizona, one of the native habitats for this critically endangered bird. The California condor has one of the widest wingspans of any living bird species, at around 10 feet (3 meters).

Yucca Mountain

While nuclear power has great potential as an alternative source of energy, it also has a number of challenges that have prevented it from providing more than about ten percent of the world's energy needs. One challenge is safety, and major nuclear accidents like those at Chernobyl in 1986 attest to the enormous potential harm of nuclear power stations to people and the environment. Another major challenge might prove just as hard to overcome: what to do with nuclear waste.

Since the first nuclear reactors came online in the 1950s, there have been many ideas about what to do with the spent fuel. Some of it is reprocessed to produce new radioactive materials for nuclear weapons, but this also yields high-level radioactive waste products that must be disposed of safely. Some of the spent nuclear fuel is temporarily stored in deep underground geological caverns or in other shallow underground storage facilities. Much of the accumulating waste fuel, however, is stored in steel or concrete casks on-site at individual nuclear plants. The situation is not sustainable, as it will take more than 10,000 years for these spent fuels to degrade to safe levels of radioactivity, and existing storage sites are filling up.

The US Department of Energy began looking for a more permanent, large, long-term geological repository for nuclear waste in the late 1970s, and in 1987 a primary site was identified: Yucca Mountain, near the Nevada nuclear test site about 100 miles northwest of Las Vegas. The site was chosen because it is in a remote area of low seismic activity; was judged to be well above the water table; and is composed of volcanic rocks that would help to block the escape of radioactivity to the surface.

The large backlog of temporarily stored spent fuel from around the country was supposed to have begun being stored inside Yucca Mountain starting in 1998, but construction and tunneling did not begin until 2002 because of public opposition, legal challenges, and political infighting. Despite miles of initial tunnels having now been drilled, public and political opposition to the facility have further delayed the start date, and today the future of the facility is still in legal and political doubt, and its license for nuclear waste storage is still under review.

SEE ALSO Industrial Revolution (c. 1830), Radioactivity (1896), Controlling the Nile (1902), Nuclear Power (1954), Wind Power (1978), Solar Power (1982), Chernobyl (1986), Hydroelectric Power (1994), End of Fossil Fuels? (~2100)

Main image: *Part of an initial exploratory tunnel dug beneath Yucca Mountain as part of the initial assessment studies of the site as a possible long-term repository of nuclear waste.*
Inset: *A 2006 photo of the crest of Yucca Mountain, southern Nevada.*

Light Pollution

To our ancestors, the night sky was a source of reverence, inspiration, and wonder. On a clear, moonless night it was possible even from ancient cities to see thousands of stars with the unaided eye, including the grand, sweeping arch of the Milky Way. But the advent of modern civilization, and especially the growth of major cities and urban centers and their widespread use of electricity for artificial illumination, have significantly changed our relationship with the night sky. Rather than thousands of stars, most people in cities and towns in industrialized countries can now usually see only hundreds of stars on a clear night. Residents of major cities might be lucky to see just ten or twenty stars and a slew of airplanes. The night sky has lost its wonder for most people, becoming instead a dull, faintly glowing, and featureless part of the background.

The culprit in this nocturnal cosmic dulling is light pollution, the alteration of natural outdoor light levels by artificial light sources. Light pollution obscures fainter stars for people living in cities or suburbs, interferes with astronomical observations of faint sources, and can even have an adverse effect on the health of nocturnal ecosystems. It's also economically inefficient—the point of lighting a home or building isn't to spend money and kilowatts lighting up the night sky.

In recognition of the growing global problem of light pollution, in 1988 a group of concerned citizens formed an organization known as the International Dark-Sky Association (IDA), with the mission "to preserve and protect the nighttime environment and our heritage of dark skies through quality outdoor lighting." IDA now has about five thousand members worldwide who work with city and local governments, businesses, and astronomers to raise awareness about the value of dark skies, to help implement lighting solutions that are more energy-efficient and economical, and to create less light pollution.

Despite some notable successes establishing ordinances and building codes that are decreasing light pollution, the effect on astronomy continues to limit the utility of major observatories near large cities (such as the Mount Wilson Observatory, perched above Los Angeles). New telescopes are now usually built in remote deserts or on isolated dark mountaintops to escape the night sky's growing glow.

SEE ALSO Population Growth (1798), Industrial Revolution (c. 1830), Birth of Environmentalism (1845), The Anthropocene (c. 1870), Earth Selfies (1966), Earth Day (1970)

A map of artificial night-sky brightness for part of the Western Hemisphere, from the US Defense Meteorological Satellite Program. The reddest places, mostly in the eastern and western United States, are where light pollution makes the night sky nearly 10 times brighter than the natural sky.

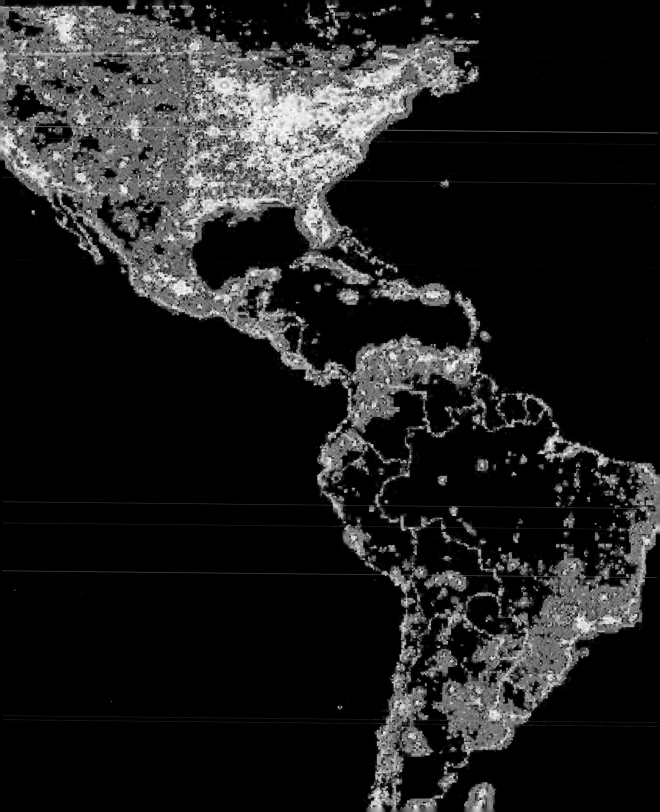

Chimpanzees

Jane Goodall (b. 1934)

Among the primates, and specifically among the hominid subspecies, chimpanzees are the closest living genetic relatives to modern humans (we share about 99 percent of the same DNA), with our species having diverged from each other and a common hominid ancestor only about 5 to 10 million years ago. It is perhaps not surprising, then, that the study of chimpanzees has attracted so much interest among the medical, genetics, sociological, and zoological communities.

Chimpanzees live in the wild in Central and West Africa, and their closely related species bonobos also live in the Congo jungle. Researchers estimate that there could be somewhere between 200,000 and 300,000 chimpanzees and bonobos living in the wild. Little was known about their lives until anthropologists like Mary and Louis Leakey began taking an interest in the origins of humans and other hominid species in the 1940s and 1950s. In 1957, they hired a young British woman named Jane Goodall to provide secretarial help for their research in Tanzania. Goodall soon developed a strong interest in doing field research of her own, and in 1960 spent two years in Gombe National Park learning about the lives of chimpanzees.

Since obtaining her formal education at Cambridge (her 1966 PhD thesis was titled "Behavior of free-living chimpanzees"), Goodall has become the leading chimpanzee researcher in the world, spending decades living among, studying, and writing about them. In her popular 1988 book *My Life with the Chimpanzees*, Goodall discovered that chimps have individual personalities (controversial at the time), that they experience emotions like joy and sorrow, and that they practice familiar habits like kissing, hugging, and even tickling. They live in complex social hierarchies and have complex sex lives. Perhaps most surprisingly, she discovered that they construct and use tools, and that they are not vegetarians. Indeed, her observations of chimpanzee hunting behavior revealed that, like humans, chimps appear to have a darker, more brutal side to their nature.

There is a population of about a thousand chimpanzees and bonobos living in research labs, primarily in the US. While lab chimps have been useful in the past for developing treatments for human medical problems, Goodall and others have now joined the debate about whether they are still needed for research or should be released to sanctuaries.

SEE ALSO Primates (c. 60 Million BCE), First Hominids (c. 10 Million BCE), *Homo sapiens* Emerges (c. 200,000 BCE), Tracing Human Origins (1948), Tropical Rain/Cloud Forests (1973), *Gorillas in the Mist* (1983).

British anthropologist Jane Goodall and a chimpanzee friend.

Biosphere 2

1991

The Earth's *biosphere* is the self-contained, self-regulating zone of all of the planet's living organisms, and its characteristics are the result of the ways that those organisms interact with their individual ecological habitats (biomes) as well as with our planet's other important zones, the *atmosphere, hydrosphere,* and *lithosphere.* Understanding the interrelationships and interdependencies of all of the various components of the biosphere is critical for understanding the origin, evolution, and sustainability of life on our planet. And if we ever hope to expand life beyond Earth, understanding the intricate details of how our biosphere works here will be absolutely critical to establishing new, artificial biospheres on other worlds.

That was the guiding philosophy behind the establishment of the privately funded technological and sociological experiment known as Biosphere 2, which began in 1991 inside a specially designed facility near Tucson, Arizona. After four years of construction and testing of a more than three-acre indoor structure consisting of seven different biome regions, a crew of six "Biospherians" entered to see if they could survive for two years completely cut off from the resources of the outside world, except for sunlight. The biomes in the facility—ocean, rainforest, coral reef, wetlands, savanna, and desert—were selected to try to help the participants achieve the ecosystem balance needed to survive, but the crew would have to monitor and recycle their own food and oxygen carefully.

The Biosphere 2 crew encountered many unanticipated challenges. For example, outgassing and absorption from the concrete in the facility significantly influenced CO_2 levels, and unexpected levels of condensation of water vapor significantly affected soil moisture, especially in the "desert" environment. Some plants grew more aggressively than expected, blocking sunlight and thus upsetting the expected energy balance. Oxygen levels slowly dropped (perhaps due to soil microbes), and had to be boosted from the outside several times, provoking some criticism from the media and skeptics of the experiment.

Do the problems encountered during the 1991–1993 Biosphere 2 experiment mean that it was a failure? Hardly. The crew and the scientific support staff learned an enormous amount about the complexity of large closed ecosystems—including the Earth's ("Biosphere 1") from this unprecedented and visionary experiment.

SEE ALSO Sahara Desert (c. 7 Million BCE), Tropical Rain/Cloud Forests (1973), Temperate Rainforests (1976), Tundra (1992), Boreal Forests (1992), Grasslands and Chaparral (2004), Temperate Deciduous Forests (2011), Savanna (2013)

A 2003 photo from inside Biosphere 2, showing part of the lake and rainforest biomes.

Mount Pinatubo Eruption

Most of our planet's most intense earthquake and volcanic activity occurs along subducting plate boundaries—areas where typically an oceanic plate subducts underneath a continental plate. Most of these zones around the world are sparsely inhabited, partly because of the high risk of frequent earthquakes and volcanoes. Some, however, like the region around the complex collisional boundary between the Pacific plate and the Eurasian plate where the Philippine Islands are located, contain some of the highest population density regions in the world.

Thus, the 1991 cataclysmic eruption of one of the many stratovolcanoes formed above that subduction zone—Mount Pinatubo—had the potential to cause enormous loss of life for the approximately 6 million people who live within that volcano's zone of influence. Pinatubo had lain dormant for more than 500 years, allowing a dense tropical rainforest as well as dense urban settlements to build up along its base and flanks. Fortunately, the Philippine government, with assistance from other agencies like the US Geological Survey, actively monitored seismic activity associated with the volcano and concluded in spring 1991 that a major eruption was imminent. Between April and early June, more than 60,000 people within about 20 miles (32 kilometers) of the summit were ordered to evacuate.

Then, on June 15, the volcano erupted, launching a massive plume of ash, dust, and toxic gases high into the stratosphere. The near-simultaneous arrival of a major tropical typhoon drenched the area, mobilizing landslides and lahar mud flows that destroyed roads, villages, and structures even large distances from the summit. The stratospheric particles circled the globe, creating spectacular sunsets for months after the eruption, heating the upper atmosphere and lowering the average temperature of the surface by 0.4°C (0.7°F). As the first colossal volcanic eruption in the era of modern space-age sensors, Pinatubo proved that volcanoes can alter Earth's climate.

The eruption was the most powerful volcanic explosion of the twentieth century to occur in an urban area, with a Volcanic Explosivity Index of 6. While nearly 1,000 people were killed by the eruption, predication and monitoring of the volcano by local and international geologists certainly saved tens of thousands of lives.

SEE ALSO Plate Tectonics (4–3 Billion BCE?), Cascade Volcanoes (c. 30–10 Million BCE), The Andes (c. 10 Million BCE), Pompeii (79), Huaynaputina (1600), Mount Tambora (1815), US Geological Survey (1879), Krakatoa (1883), Exploring Katmai (1915), Mount St. Helens (1980), Volcanic Explosivity Index (1982), Eyjafjallajökull (2010), Yellowstone Supervolcano (~100,000)

Photograph of the enormous eruptive plume of Mount Pinatubo on the morning of June 12, 1991. The eruption injected more ash and dust into the stratosphere than any volcanic eruption since Krakatoa in 1883.

Tundra

Depending on how different organizations of biologists and ecologists classify them, the world can be divided into between one and two dozen distinct ecological regions, or *biomes*. One type of biome that is common to all classification schemes because it occurs in many places around the world is *tundra*—regions of relatively low temperatures and short growing seasons where tree growth is limited and vegetation consists mostly of shrubs, grasses, mosses, and lichens. Tundra occurs in three main regions of our planet: in the Arctic, in the Antarctic, and in high-elevation alpine terrains. In mountainous regions, tundra is the ecological zone just above the tree line.

Vegetation dominates the biodiversity of the tundra; only a small number of land mammals and fish species dwell permanently in the tundra (for example, Arctic reindeer, rabbits, foxes, and polar bears), though millions of birds migrate there each year. In Arctic and Antarctic tundra regions (and some alpine tundra), the shallow subsurface is characterized by *permafrost*, or permanently frozen soil. These soils can hold enormous amounts of freshwater, and when frozen they trap large quantities of CO_2 and methane (from decaying plants and animals), which are also potent greenhouse gases. Tundra thus plays an important potential role in the world's climate, because melting of its permafrosted soils could significantly enhance global warming.

Partly because of their remote locations and enormous global extent, scientists don't really know how much water, CO_2, or methane is stored in the world's tundra soils, nor is there yet a full accounting of the biologic diversity of tundra regions. Understanding the detailed nature and inventory of this particular biome, as well as the others, was thus a major driver for a global Biodiversity Action Plan (BAP) developed by the United Nations as part of the landmark 1992 Convention on Biologic Diversity. The major goal of this international treaty is to guide national strategies for the conservation and sustainable use of biological diversity. Major fractions of tundra in Canada and Russia, for example, are protected by BAP stipulations. All UN member states have signed the treaty, but only one has not yet ratified it: the United States.

SEE ALSO Sahara Desert (c. 7 Million BCE), Tropical Rain/Cloud Forests (1973), Temperate Rainforests (1976), Boreal Forests (1992), Grasslands and Chaparral (2004), Temperate Deciduous Forests (2011), Savanna (2013)

Beautiful trails through the high-altitude tundra environment in the Rocky Mountains, Colorado.

1992

Boreal Forests (Taiga)

By far the largest of the Earth's ecological regions in terms of land surface coverage is the *boreal* forest, also known as *taiga*. Taiga is characterized by low average annual temperatures (typically –5°C to 5°C, or 23°F to 41°F); moderate precipitation mostly in the form of snow; and thin, relatively nutrient-poor soils. Taiga summers are short, moist, and moderately warm, and winters are long, cold, and dry. Most taiga occurs in the northern hemisphere, between about 50° and 60° north latitude. About two-thirds of the world's boreal forests are found in Siberia, with the rest in Scandinavia, Alaska, and Canada.

Vegetation in the boreal forests is dominated by cold-tolerant evergreen trees such as pine, fir, and spruce. Because their thick canopies absorb so much of the relatively weak incoming sunlight at those latitudes, underbrush is typically limited to moss, lichen, and mushrooms. Only a small number of animal species thrive in the taiga, including bears, moose, wolf, fox, deer, and a number of small mammals. A few bird species live in the boreal forests (such as woodpeckers and hawks), and hundreds of others migrate seasonally into the taiga to nest in the summer months.

Because they span such an enormous fraction of the surface, boreal forests are the source of significant potential natural resources for human activities. Much of the world's lumber, for example, is cut from boreal forests. Numerous mining and oil/gas extraction projects also occur in taiga regions. The potential for mismanagement of taiga resources compelled the creation of advocacy groups like the Taiga Action Network, formed in 1992 and composed of more than 200 non-governmental organizations, indigenous peoples, and individuals working to defend the world's boreal forests. Governmental stewardship efforts relevant to taiga resource extraction were also formalized in 1992 through the United Nations Convention on Biologic Diversity, which stipulates forest stewardship, respect for aboriginal peoples, compliance with environmental laws, forest worker safety, education and training, and other environmental, business, and social requirements.

Stewardship and monitoring of the health of the world's boreal forests is important because they play an important role in the global climate: taiga stores enormous quantities of carbon (more than temperate and tropical forests combined) in wetlands and peat bogs.

SEE ALSO Sahara Desert (c. 7 Million BCE), Tropical Rain/Cloud Forests (1973), Temperate Rainforests (1976), Tundra (1992), Grasslands and Chaparral (2004), Temperate Deciduous Forests (2011), Savanna (2013)

A 2017 photo of a mossy boreal spruce forest in Sweden.

Oceanography from Space

Kathryn D. Sullivan (b. 1951)

The idea of flying an airplane or satellite over continents or islands to study and monitor their geology, weather, plant health, patterns of land use, and so forth seems like a straightforward and reasonable application of modern sensors and Earth science capabilities. But remote sensing of the oceans? What could there be to see?

Plenty, it turns out. The first satellite dedicated to remote sensing of the oceans was a 1978 NASA mission called Seasat, launched as a pilot mission to demonstrate that useful information about the oceans could be obtained from space. Seasat operated for only 106 days but demonstrated the ability to measure sea-surface winds, temperatures, wave heights, internal waves, and sea ice features.

The success of the brief Seasat demonstration mission helped to justify astronaut-tended oceanographic observations from the NASA Space Shuttle, which began flying in the 1980s. An important participant in these efforts has been oceanographer and astronaut Kathryn Sullivan, who flew on three shuttle missions in 1984, 1990, and 1992 and who participated in or led a number of important experiments to study the Earth from space. Sullivan had participated in oceanographic ship expeditions for her doctoral research, studying the Mid-Atlantic Ridge, the Newfoundland Basin, and fault zones off the Southern California coast. After leaving NASA, Sullivan was able to promote significant advances in the study of oceanography from space as the Administrator of the National Oceanic and Atmospheric Administration (NOAA) from 2013 to 2017.

Perhaps the largest advance in oceanography from space came in 1993, however, when the first large-scale oceanographic research satellite, a joint NASA–French Space Agency mission called TOPEX/Poseidon, went into operation. That mission, which operated until 2006, provided the first continuous global coverage of ocean surface topography and revealed the previously unknown patterns of ocean circulation that— because the oceans hold most of Earth's heat from the Sun—are the driving force of the climate on our planet. Follow-on oceanography satellite missions like the Jason series, as well as additional astronaut observations from the Space Shuttle and the International Space Station, continue to dramatically expand our knowledge of the oceans.

SEE ALSO Earth's Oceans (c. 4 Billion BCE), Airborne Remote Sensing (1858), Exploring the Oceans (1943), Humans in Space (1961), Earth Science Satellites (1972), Geology on the Moon (1972), Ocean Conservation (1998).

Main image: *A 2016 composite image, from NASA's Suomi ocean science satellite, of the distribution of phytoplankton (green) in the California Current.*
Inset: *Astronaut Kathy Sullivan views the Earth through the Space Shuttle Challenger's windows in 1984.*

Hydroelectric Power

1994

Hydropower—the conversion of the energy of rivers and streams into mechanical energy—has been used since prehistoric times to, for example, grind flour and pump water uphill. Starting in the late nineteenth century, hydropower began to be used to generate electricity, by using the flow of water to turn the blades of a turbine generator, for example. In order to guarantee a steady flow of water through dry seasons, many rivers and streams were dammed up to create a large reservoir just upstream of the dam, guaranteeing a steady flow year-round. After the Industrial Revolution, numerous industrial cities grew around these hydropower centers of electricity production.

Milestones in the development of hydroelectric power include the 1928 construction of the Hoover Dam on the Colorado River, which produced 1,345 megawatts of electricity, soon eclipsed by the Grand Coulee Dam on the Columbia River, which began producing more than 6,800 megawatts in 1942. The Itaipu Dam on the Paraná River in South America became the largest hydroelectric facility in the world at 14,000 megawatts in 1984, but all these facilities were dwarfed by the 22,500 megawatt capacity of the Three Gorges Dam on the Yangtze River in China—which began construction in 1994 and has become not only the largest hydroelectric facility in the world, but the largest electricity-generation facility of any kind on Earth.

Hydropower now accounts for about 16 percent of the world's total electricity production (and growing), and about half the world's total sustainable (non-fossil-fuel) energy production. There has been a social and environmental cost associated with the adoption of hydropower as a source of renewable energy, however. These include the loss of land and potentially unique ecosystems that get submerged under human-made reservoirs, displacement of people and wildlife in those areas, disruption of aquatic ecosystems downstream of a dam, decreases of water volume and loss of new nutrients and sediment for downstream agriculture, increased methane production from reservoirs, and the risk of major losses of life and property from the potential for dam failure (including abandoned dams).

The risks versus advantages of hydropower must constantly be assessed for specific regions and specific applications. Governments have assumed responsibility for environmental stewardship associated with hydropower facilities, with significant input from citizen advocacy groups.

SEE ALSO The Grand Canyon (c. 6–5 Million BCE), The Pyramids (c. 2500 BCE), Aqueducts (c. 800 BCE), Great Wall of China (c. 1370–1640), Civil Engineering (c. 1500), Controlling the Nile (1902), Nuclear Power (1954), Wind Power (1978), Solar Power (1982), End of Fossil Fuels? (~2100)

A 2013 photograph of locks at the Three Gorges Dam on the Yangtze River, near Sandouping, Hubei Province, China.

Earthlike Exoplanets

Surprisingly, the first planets discovered around another star than the Sun—extrasolar planets—were found orbiting a pulsar (a rapidly spinning neutron star) in 1992. This discovery of what are most likely uninhabitable planets in such a strange and hostile environment compelled astronomers to search for evidence of extrasolar planets around more "normal" Sun-like stars.

For decades, it was known that binary stars (two stars that orbit each other) can show "wobbles" in their slow motion across the sky, because both stars are actually orbiting the system's center of mass. In theory, the same kinds of wobbles—though much smaller—should be seen if a giant Jupiter-like (or larger) planet is in orbit around a single star. A breakthrough came when astronomers realized that they didn't need to measure the precise position of the star over time; instead, they could use the Doppler shift—the change in the star's spectrum as it moves toward and away from us—to deduce its wobbling motion. This is known as the radial velocity method of planet hunting.

In 1995, using the radial velocity method, the first extrasolar planet around a Sun-like star was discovered orbiting the nearby star 51 Pegasus (51 Peg). Planet "51 Peg b" was inferred to be a gas giant many times the size of Jupiter and orbiting very close—only 5 percent of the Earth's distance from the Sun—to its own star. More than 750 other planets around other nearby stars have been found since then using the radial velocity method. Most of these planets are known as "hot Jupiters" because they are large and they also orbit extremely close to their parent stars.

Other ways to find extrasolar planets are to watch for them passing in front of (transiting) their parent stars (the goal of NASA's Kepler mission), to detect them by gravitational lensing, or to just directly image them through the glare of their parent stars. Indeed, more than 3,000 other extrasolar planets have now been discovered using these methods around nearby stars. Many of them are earth-sized worlds, including three found around the star Gliese 581 in 2007 and seven found in 2017 orbiting the star TRAPPIST-1, only about 40 light-years from Earth. It seems that the galaxy, and probably the entire Universe, may be teeming with earthlike planets!

SEE ALSO Laws of Planetary Motion (1619), Extremophiles (1967).

Artist's impression of the planetary system around the red dwarf Gliese 581. Astronomers have found evidence for three "earthlike" planets having masses of 5, 8, and 15 times that of Earth in orbit around that star. Unlike our own terrestrial planets, however, these planets orbit extremely close to their Sun.

Large Animal Migrations

Prior to the invention of agriculture and the establishment of cities, humans were a nomadic species, moving with the seasons to follow food and water resources. Similarly, other members of the animal kingdom have established seasonal migratory patterns (moving from one habitat to another), many of which continue today. Bird migrations are perhaps the best known and involve the largest number of (non-insect) individual animal migrations, but significant populations of fish and aquatic mammals (like salmon, sardines, whales, and dolphins), land mammals (like zebras, wildebeest, springbok, and oryx), and even reptiles and crustaceans also embark on seasonal migrations that can span great distances.

The drive to migrate appears to be triggered by a variety of both environmental and instinctual or inherited cues. Environmentally, seasonal changes in the weather and the ensuing changes in the availability of food or water are of course immediately compelling reasons to seek a different place to be. However, in some migrating species, there appears to be a sort of inherited map of some kind, perhaps keying in on specific orientations relative to the Sun's path across the sky, or to the Earth's magnetic field, or to specific ocean currents, that seemingly must be somehow reinforced through natural selection. Human encroachment and habitat destruction can thus wreak havoc on the instinctual plans of many migrating species. This was a primary motivation for the United Nations' establishment in 1997 of the Convention on the Conservation of Migratory Species of Wild Animals, designed to help protect the cross-national habitats through which migratory animals pass.

Some biologists have attempted to define the typical characteristics of a large animal migration in order to better understand its origin and to better predict how to help conservationists protect this critical animal behavior. These characteristics include prolonged linear movements to a new habitat; special advance preparations or arrival behaviors like overfeeding; special or stored allocations of energy; and the ability to avoid distractions and focus on getting to the new destination. How information on the path and destination is transmitted from generation to generation (sometimes during the migration itself) is not fully understood by biologists, nor is it clear how the group comes to the initial consensus that, indeed, it's time to go.

SEE ALSO Invention of Agriculture (c. 10,000 BCE), Insect Migration (1975), Magnetic Navigation (1975), Tundra (1992), Boreal Forests (1992), Grasslands and Chaparral (2004), Temperate Deciduous Forests (2011), Savanna (2013)

Zebras are an example of a large animal species that makes substantial seasonal migrations.

Ocean Conservation

Sylvia Earle (b. 1935)

The oceans dominate the surface of the Earth (the continents cover only about 30 percent), and energy transfer from the Sun to the oceans, which drives ocean-current circulation, is the main driver of the global climate and weather patterns on our planet. The oceans were intimately involved in the origin and evolution of early life on our planet, and today the ocean hosts or helps to directly support the survival of countless species, including our own.

So when a leading marine biologist and ocean explorer like American conservationist Sylvia Earle says "If the ocean is in trouble, we're in trouble," people listen. And indeed, there is evidence that the oceans are in trouble. That evidence comes in the form of massive die-offs of coral reefs in places like the Great Barrier Reef northeast of Australia; massive reductions or disappearance of marine species from overfishing or habitat encroachment; ocean acidification from the absorption of increasing atmospheric CO_2; habitat destruction and species decimation from pollution (oil spills, effluent discharge, accumulation of plastics); and other clues.

Since the mid-1960s, Earle has been on a mission to promote ocean conservation. An experienced SCUBA and submarine diver who has set numerous dive records, Earle has traveled the world's oceans, documenting their biodiversity and the threats that they, along with the organisms and animals that rely on the ocean, are facing. She and her colleagues have made major innovations in deep-sea exploration engineering and technology, and have helped advocate for the formation of the US National Marine Sanctuary system in 1972, which provides special environmental protections to more than 783,000 square miles (2,030,000 square kilometers) of ocean and shoreline. She is an outspoken expert in the assessment of the environmental impacts of oil spills like the Exxon Valdez Alaskan spill in 1989 and the Deepwater Horizon disaster in the Gulf of Mexico in 2010. During the 1990s, Earle was the first woman to serve as Chief Scientist of the US National Oceanic and Atmospheric Administration.

In 1998, *Time* magazine named Earle its first "Hero of the Planet" for her tireless work in ocean conservation. She and others like her continue to work hard to make ocean conservation an international, human priority.

SEE ALSO Women in Earth Science (1896), Geology of Corals (1934), Exploring the Oceans (1943), Mapping the Seafloor (1957), Mariana Trench (1960), Deep-Sea Hydrothermal Vents (1977), Great Barrier Reef (1981), Oceanography from Space (1993)

Marine biologist and explorer Sylvia Earle (in the SCUBA gear), examining a damaged coral reef sample.

Earth's Spin Slows Over Time

Our home planet spins on its axis once per day. For most of the history of astronomy, it has been sufficient to know that the rotation rate of our planet relative to the distant "fixed" stars is about 23 hours and 56 minutes. The fact that the Earth spins at this rate about 365 and a quarter times during every trip around the Sun has led to a variety of creative ways to add leap years to the calendar, culminating in the modern leap-year method developed during the Gregorian Calendar reform of 1582.

In our modern era of digital computers, global positioning system satellites, and interplanetary space probes, it has become much more important to be able to record time, including the Earth's rotational rate, to a much higher degree of precision. Atomic clocks began to be used in the 1950s and 1960s to precisely reckon the passing of time, using the frequency of stable atomic-energy-level transitions in elements such as cesium. An internationally agreed-upon timekeeping system called Coordinated Universal Time ("UTC time") was developed based on these atomic clocks. Using modern technology, it is now possible to measure the length of the day to almost one part in 10 billion.

The problem for astronomers, navigators, and timekeepers, however, is that the Earth's rotation is not constant. Tidal friction with the Moon and Sun is slowing down our planet's spin ever so slightly each year. In addition, very slight changes in the distribution of mass on the Earth's surface (like glacial melting) and in its interior can also have tiny effects on our planet's spin rate. Thus, since 1972, to keep UTC time precisely aligned with the passage of time as reckoned by the motion of the Sun in the sky, an organization known as the International Earth Rotation and Reference Systems Service has had to occasionally add extra leap seconds to UTC time.

In the 26 years from 1972 to 1998, 22 leap seconds were added to keep UTC time in sync with the Earth's slowing spin. In the 20 years since 1999, however, the rate of slowing of Earth's spin has decreased, and only five leap seconds have had to be added.

SEE ALSO The Pyramids (c. 2500 BCE), Earth Is Round! (c. 500 BCE), Origin of Tides (1686), Proof that the Earth Spins (1851)

The astronomical clock in the town square of Prague, Czech Republic. Analog clocks such as this one, tracking hours and minutes and the motions of the Sun and Moon, have been replaced in scientific research by precise digital clocks and an internationally regulated system of timekeeping.

Torino Impact Hazard Scale

While evidence for all but a few hundred terrestrial impact craters has been erased by our planet's dynamic geology and hydrology, we can tell just by looking at the ancient, heavily scarred surface of our planetary neighbor—the Moon—that significant numbers of asteroids and comets have hit Earth in the past. These high-speed impact events released huge amounts of energy; and geologic evidence and the fossil record suggest that they occasionally significantly altered the planet's climate and biosphere.

The impact rate has decreased exponentially with time over Earth's history; but even in modern times, that rate is not zero—consider, for example, the 1908 explosion of a comet or asteroid in the atmosphere above Siberia (the Tunguska event) and the observation of several large atmospheric fireball explosions every year by military and civilian planetary monitoring satellites, including the huge fireball over Chelyabinsk, Russia, in 2013.

Fueled by public and political interest in understanding the risks associated with cosmic impacts, the rate of discovery of small asteroids and comets has increased over the past few decades, especially within the population known as Near-Earth Objects (NEOs). Dedicated telescopic surveys have identified more than half a million Main Belt asteroids and more than 1,000 NEOs. The few hundred NEOs that could potentially cause a threat to life on our planet get a special acronym: PHAs, for Potentially Hazardous Asteroids.

As the rate of PHA discovery increased, it became clear that there was no systematic or simple way to understand and communicate the risk of PHA impacts. So, in 1999, a group of planetary astronomers developed an index called the Torino Impact Hazard Scale to quantify the risks. Torino values for newly discovered PHAs range from 0 (no chance of impact) to 10 (certain impact with likely catastrophic consequences).

All currently known PHAs have Torino values of 0. About 50 have had nonzero values (most were downgraded to 0 with follow-on observations), with the record so far being an original value of 4 (a 1 percent or greater chance of collision) for the asteroid 99942 Apophis, which will next pass very close to the Earth on April 13, 2029. The Apophis risk has since been downgraded to a 0, but astronomers still monitor it carefully.

SEE ALSO Dinosaur-Killing Impact (c. 65 Million BCE), Arizona Impact (c. 50,000 BCE), The Tunguska Explosion (1908), Apophis Near Miss (2029)

A dark, cratered, near-earth asteroid—perhaps like 99942 Apophis—approaches Earth in this painting by planetary scientist and artist William K. Hartmann.

Vargas Landslide

People tend to think of the geological process of *erosion* as a rather slow thing—mountains slowly being worn down by glaciers, waves slowly breaking down cliffs by the sea. But sometimes erosion can be rapid, violent, and deadly. Prime examples come from the history of landslides, which are the rapid and sudden failure or collapse of surface regions. Gravity plays a critical role in landslide formation, of course, as topography provides the potential energy for materials to move downhill. Other contributing factors include the stability of sloped surfaces (are they anchored by vegetation, for example?), the composition and thickness of soils or bedrock in the region, and local weather and seismic conditions.

Landslides occur in many different ways. They can be shallow, with just a thin soil or dust cover slipping downhill; or deep-seated, involving the motion of enormous quantities of soil and bedrock, often along pre-existing planes of weakness like faults or sedimentary layers. Debris *flows* are landslides that occur in water-saturated soils; they can be fast-moving and very fluid (mudflows), or slow-moving and more viscous (*earthflows*). Finally, more rocky examples of landslides include *debris slides* that form from the downslope motion of rocks, soil, and debris mixed with water and/or ice, and *rock avalanches* that, as the name implies, form from the rapid downhill motion of large rocks and boulders.

Tropical mountainous regions that receive a lot of rainfall and that have rugged topography are among the most susceptible terrains on Earth for large landslides. For example, several days of torrential rain triggered massive mudflows and debris slides along the rugged mountains of the state of Vargas, Venezuela, north of Caracas, on December 15, 1999. Overnight, and without warning, landslides killed between 15,000 and 30,000 people and destroyed or damaged nearly 100,000 homes. Several towns were completely destroyed, and others were buried under as much as 10 feet (3 meters) of mud, rocks, and debris. Landslides with comparable or even greater numbers of casualties and damage have occurred in recorded history in China, Colombia, Peru, and elsewhere and often include events triggered by the seismic shaking of earthquakes or volcanoes. Better geologic mapping and increased education and awareness of landslide risks could help to save many lives in the future.

SEE ALSO Soil Science (1870), Galveston Hurricane (1900), San Francisco Earthquake (1906), Tri-State Tornado (1925), Valdivia Earthquake (1960), Sumatran Earthquake and Tsunami (2004)

View of part of the urban devastation caused by a catastrophic landslide in the city of Caraballeda, Vargas state, Venezuela, on December 15, 1999.

Sumatran Earthquake and Tsunami

Tsunamis are massive, long-period ocean waves generated from a substantial disturbance on the seafloor or along the coast. They are most often triggered by earthquakes, when plates slip along undersea faults and their motion displaces huge amounts of ocean water. Tsunamis can also be triggered by volcanic eruptions, landslides, and (conceivably) large asteroid or comet impact events. Intuitively, the largest earthquakes have the potential to cause the largest tsunamis, and indeed the largest earthquake ever recorded, the 1960 Valdivia earthquake off the western coast of South America, created a large tsunami that traveled across the Pacific, devastating some coastal towns and killing hundreds of people. However, the specific geology, geography, and seafloor topography of a region also play critical roles in the destructive potential of tsunamis.

A tragic case in point is the catastrophic tsunami that was created immediately following the December 26, 2004, magnitude 9.1 earthquake that occurred under the seafloor of the Indian Ocean off the western coast of the Indonesian island of Sumatra. Comparable enormous earthquakes have occurred in this area since prehistoric times, as it is part of the Pacific Ring of Fire where the Indian and Australian plates are rapidly subducting under part of the Eurasian plate, forming the Sunda trench and volcanic island arc. The 2004 Sumatran earthquake occurred about 19 miles (30 kilometers) below the seafloor, and caused the longest-measured crustal rupture to date, along a zone about 250 miles (400 kilometers) long and 60 miles (100 kilometers) wide. The seafloor uplifted by about 6 feet (2 meters), displacing an estimated 7.2 cubic miles (30 cubic kilometers) of water and triggering the tsunami.

People along the coasts of Sumatra and Thailand had little warning because the tsunami began so close to shore. As the wave approached shallower coastal waters, it slowed down and grew in height. In some parts of Indonesia, for example, coastal communities were battered by waves as high as 80 to 100 feet (24 to 30 meters) tall. The massive devastation to these communities, plus damage and casualties in many other coastal communities across the nations bordering the Indian Ocean (many of which did not have adequate tsunami warning systems or evacuation plans), killed an estimated 280,000 people—the deadliest tsunami in history.

SEE ALSO Plate Tectonics (c. 4–3 Billion BCE?), Mount Tambora Eruption (1815), Krakatoa Eruption (1883), Galveston Hurricane (1900), San Francisco Earthquake (1906), Tri-State Tornado (1925), Island Arcs (1949), Valdivia Earthquake (1960)

Flooded homes along the coast of Thailand, immediately after the tsunami of December 2004 washed ashore.

Grasslands and Chaparral

There are more than 12,000 different species of grasses on Earth, and while they occur around the world, many of them dominate specific ecologic regions that have come to be known as *grasslands*. Grasslands typically occur in regions with modest rainfall (24 to 59 inches [600 to 1,500 mm] or less annually) and average mean annual temperatures of –5° to 20°C (23° to 68°F). Perhaps more importantly, grasslands dominate certain regions because of specific environmental conditions that thwart the encroachment of substantial numbers of woody plant species. These include areas prone to wildfires—which thus don't allow trees and other woody shrubs to establish well—and regions where the soil or bedrock composition does not support the nutrient levels needed by most trees.

Because of environmental factors like this, certain special ecological regions have evolved where grasses and some drought-tolerant woody shrubs can coexist, and where occasional wildfires have actually become an important part of the life cycle of their plant communities. Such *shrublands* occur around the world, including a special subclass called *chaparral* (from the Spanish word for the small, shrubby scrub oak) that occurs in so-called Mediterranean climate environments like the southern California coast. Chaparral plant communities are also found in Chile, South Africa, western Australia, and parts of the Mediterranean—places with winter rains and summer droughts, occasional wildfires, and calcium-rich, limestone-dominated soils.

People from prehistoric to modern times have often worked hard to clear shrublands like chaparral and convert them to "useful" environments like grasslands that can serve as hunting grounds or grazing lands for domesticated animals. As well, longer and more frequent droughts in many chaparral regions are naturally converting more of the terrain to grasslands, as most woody shrub species can tolerate occasional, but not frequent, fires. Loss of shrublands like chaparral reduces biodiversity; displaces other plant, insect, and animal species that have evolved to thrive in those environments; and can make regions more prone to wider-spreading wildfires, soil erosion, and landslides. As a result, in 2004, the non-profit California Chaparral Institute began its efforts to educate governments and the public about the role and value of shrublands in the global ecosystem. Similar organizations worldwide monitor and advocate for the health and preservation of grasslands and shrublands.

SEE ALSO First Land Plants (c. 470 Million BCE), Sahara Desert (c. 7 Million BCE), Big Burn Wildfire (1910), Tropical Rain/Cloud Forests (1973), Temperate Rainforests (1976), Tundra (1992), Boreal Forests (1992), Temperate Deciduous Forests (2011), Savanna (2013)

A 2016 photo of grassland-and-chaparral environment in the Santa Catalina Mountains, near Tucson, Arizona.

Carbon Footprint

A footprint is literally a permanent mark left on something, but it is also widely used as a figurative concept to depict the impact of objects, individuals, events, organizations, or societies on a specific topic. We think about the footprint, for example, that a computer or a printer might occupy in our work space. More broadly, since the early 1990s, ecologists, economists, and others have been thinking about the *ecological footprint* that human society has left on our planet. That is, what is the fraction of humanity's demands on nature relative to nature's full ecological capacity? This is obviously a difficult parameter to calculate precisely, partly because it's not clear exactly what to measure.

In an attempt to try to create a specific, measurable quantity that could be used to assess environmental or ecologic impact, in 2007 the ecological footprint concept was narrowed to the idea of just the *carbon footprint* by government officials developing the energy plan for the US city of Lynnwood, Washington. They chose this metric because the quantity of carbon emissions by various activities (fuels used in transportation, creation of electricity, heating/cooling, food production, etc.) could be calculated or at least reasonably estimated, as could the total inventory of carbon (CO_2, methane, and organic molecules) in the atmosphere, hydrosphere, lithosphere, and biosphere. Considerations have to be made in such calculations for both the *direct carbon emissions* created by individual decisions on transportation, heating/cooling, and so forth, as well as the *indirect carbon emissions* created when products that we use or consume (food, clothing, etc.) are created. Indirect sources are easy to ignore because they probably occurred far away from where and when we use the products and services, but they can amount to significant fractions of the total carbon footprint of a product or activity.

While it is still a difficult parameter to calculate accurately, the carbon footprint concept has proved useful in at least educating people, companies, and governments on their likely or potential roles in adding or removing carbon to or from the environment, and thus on potentially influencing bigger-picture issues like global warming and climate change. The concept of the carbon footprint has also proved useful in efforts to monitor and curb the global release of CO_2 into the atmosphere from human activities.

SEE ALSO Domestication of Animals (c. 30,000 BCE), Invention of Agriculture (c. 10,000 BCE), Population Growth (1798), Industrial Revolution (c. 1830), Birth of Environmentalism (1845), Deforestation (c. 1855–1870), The Anthropocene (c. 1870), The Greenhouse Effect (1896), Nuclear Power (1954), Earth Day (1970), Wind Power (1978), Solar Power (1982), Hydroelectric Power (1994), Rising CO_2 (2013), End of Fossil Fuels? (~2100)

A poster, from the "green energy" company RENERGY, that captures the many ways that people stamp their "carbon footprint" on the Earth's atmosphere.

Global Seed Vault

The concept of "seed corn" has been well known to farmers since prehistoric times. That is, some fraction of the seeds from a crop that is grown has to be set aside and saved so that it can be planted for crops the next season, or perhaps even much later in case drought or disease result in multi-year crop failures.

A similar philosophy has driven the creation of a number of *seed banks* around the world, created to preserve the genetic diversity of the world's wild and cultivated plants against the threat of extinction by either natural or human causes. The world's largest seed bank, organized by the Royal Botanical Gardens, is the Millennium Seed Bank Project. Begun in 1996 and housed in a large underground frozen storage facility in West Sussex County, in southern England, it has nearly 2 billion seeds inventoried so far, representing less than 15 percent of the world's wild plant species. Millennium's goal is to reach 25 percent representation of known flora by 2020.

But even seed banks need insurance seed banks, in case of accidents or catastrophes, which was the motivation for the establishment of arguably the most remote, safe, and secure seed repository in the world, the Svalbard Global Seed Vault, on the Norwegian island of Spitsbergen, at 78°N latitude, only about 800 miles (1300 kilometers) from the North Pole. Built inside a mountain and with blast doors designed to withstand a nuclear war, the cold and dry facility opened in 2008 with seeds from the Nordic Gene Bank that had been stored in a less-secure facility on Spitsbergen since 1984. After a decade of operations and international collections/contributions, the so-called "doomsday vault" in Svalbard now houses seeds from around a million plant samples (consisting of many hundreds of millions of individual seeds) from around 6,000 species of plants—still a small fraction of the genetic diversity of plants, but at least preserving seeds from major global agricultural crops spanning 13,000 years of organized farming.

Other major seed banks also exist in Australia, Russia, India, the United States, and elsewhere. The world has not, in fact, put all of its seeds into one basket.

SEE ALSO Invention of Agriculture (c. 10,000 BCE), Natural Selection (1858–1859), The Anthropocene (c. 1870), Soil Science (1870), Terraforming (1961), Genetic Engineering of Crops (1982), Plant Genetics (1983)

A 2015 photo of the entrance to the Global Seed Vault, in Svalbard on the Norwegian island of Spitsbergen.

Eyjafjallajökull Eruption

Volcanic eruptions with a rating of 4 (out of 8) on the Volcanic Explosivity Index (VEI) scale occur about every year or two somewhere on Earth; in terms of their impact, the relatively modest amounts of ash, steam, and/or lava released by such events generally has only local, rather than global, consequence. It is perhaps surprising, then, that the VEI = 4 eruption of Iceland's Eyjafjallajökull volcano in April 2010 will go down in history as having a substantial impact on millions of people around the world.

Eyjafjallajökull is a stratovolcano built up over millions of years as part of the creation of the island of Iceland from lavas erupted from the mid-Atlantic ridge spreading center. Given Iceland's high northerly latitude, the high summits of volcanoes like Eyjafjallajökull are usually covered in snow and/or glaciers. Thus, when eruptions do occur, the interaction of the hot lava and gases with snow and ice cause intense steam explosions that can help eject rapidly cooled, sharp, tiny glassy volcanic shards to great heights.

That was exactly the case with the eruption of Eyjafjallajökull in mid-April 2010. A violent plume of ash and dust was lifted into the stratosphere, to elevations of more than 26,000 feet (8,000 meters). Coincidentally, the eastward-flowing polar jet stream just happened to be located over Iceland at the same time, and so Eyjafjallajökull's ash cloud was quickly swept up in that upper atmospheric flow and volcanic dust was quickly spread to Great Britain, Scandinavia, and across much of Europe. Because volcanic smoke and ash can reduce visibility substantially, and because the glassy and abrasive dust could cause significant damage to jet engines attached to aircraft that would need to fly through it, more than 100,000 commercial airline flights were cancelled within and to/from Europe and North America over the course of about eight days, until the dust literally settled.

Eyjafjallajökull may have been relatively small on the scale of historical volcanic eruptions, but it had a profound effect on the way people move about the planet. An estimated 10 million passengers had to change their plans, and the airline industry lost about $200 million per day because of the eruption cloud. Perhaps the only silver lining was that despite all the hassles, there were no casualties from this particular natural disaster.

SEE ALSO Plate Tectonics (c. 4–3 Billion BCE?), The Atlantic Ocean (c. 140 Million BCE), Pompeii (79), Huaynaputina Eruption (1600), Mount Tambora Eruption (1815), Krakatoa Eruption (1883), Exploring Katmai (1915), Mount St. Helens Eruption (1980), Volcanic Explosivity Index (1982), Yellowstone Supervolcano (~100,000)

Steam and ash cloud from the April 17, 2010 eruption of Eyjafjallajökull volcano, Iceland.

Building Bridges

Historical knowledge and descriptions of bridges of wood, rope, brick, and stone built by Greek, Roman, Indian, Chinese, Incan, and other societies date back many thousands of years; exactly when they were first built by prehistoric societies is unknown. Clearly, however, people have needed to solve the pragmatic problem of crossing an obstacle by building a structure over it for a long time. As technology and science have advanced and stronger materials used in innovative ways have become more common, the overall length, length of each supported span, and carrying capacity of bridges have all grown.

The first bridges were likely simple *beam bridges* with a flat span (usually made from wood) supported at the ends. Practical limits on the strength of wood limited such spans to about 30 feet (9 meters) long, although multiple wooden spans could be combined and supported by piers to enable longer overall bridges called *viaducts*. Modern beam bridges made of steel can span from 50 to 250 feet (15 to 75 meters) and can be strung together into extremely long structures, like the Lake Pontchartrain Causeway in Louisiana, spanning 24 miles (38 kilometers) over water, or the Bang Na Expressway in Thailand, spanning 34 miles (54 kilometers) over land. The longest bridge of any type in the world, the Danyang–Kunshan Grand Bridge for high-speed rail in China, is a viaduct over 100 miles (160 kilometers) long that was opened in 2011.

The use of stone or brick to build *arch bridges* provided a way to increase span length with easily available materials. Arch bridges were common in ancient Greece and Rome (especially for Roman aqueducts); more modern stone arch bridges have spans up to 295 feet (90 meters), which is the length of the Friedensbrücke Bridge, built in 1905, which crosses the valley of the Syrabach River in Germany.

Modern bridges use iron and steel to enable longer spans and new designs, like the use of cables to suspend or support arch or beam spans, perhaps also supported by tall piers at the ends (like San Francisco's famous Golden Gate Bridge). The currently longest main span in a *suspension bridge*, the Akashi Kaikyo Bridge in Japan, is a stunning 6,500 feet (1990 meters) long.

SEE ALSO Stonehenge (c. 3000 BCE), The Spice Trade (c. 3000 BCE), The Pyramids (c. 2500 BCE), Aqueducts (c. 800 BCE), Great Wall of China (c. 1370–1640), Civil Engineering (c. 1500), Hydroelectric Power (1994)

Part of the 33-mile (54-kilometer) Bang Na Expressway bridge in Thailand, the longest car bridge and the sixth-longest bridge overall in the world.

Temperate Deciduous Forests

Roughly 30 percent of Earth's land area is covered by forest. Boreal and tropical/temperate rain forests dominate that coverage, but a significant percent consists of *temperate deciduous forests*, which are broad-leaf forests composed of trees that shed their leaves in the winter. Primarily found in North America, Europe, southern Scandinavia, and east Asia, temperate deciduous forests are commonly dominated by oak, elm, beech, and maple trees, along with shrubs and forest-floor plants specially adapted for long, shady growing seasons while the tree canopy is open. As the name implies, these kinds of forests occur in temperate climatic conditions: moderate rainfall (30–60 inches or 75–150 centimeters), warm moist summers, and cool winters.

The seasonal opening and growth of the leaf canopy in the spring, and then the loss of those leaves in the fall (the origin of "fall" as an alternative to "autumn" partly comes from the season when deciduous leaves fall to the ground), is the defining characteristic of these forests. The fact that many leaves turn from green to remarkably beautiful hues of yellows, reds, and oranges as their photosynthetic chlorophyll pigments break down as winter approaches also makes these forests extremely popular tourist destinations, especially during "peak fall colors."

Temperate deciduous forests support a thriving ecosystem of associated plants, insects, and animals. So-called *spring ephemeral* plants, for example, flower just before the tree canopy opens and the forest floor becomes shady. Squirrels and birds like woodpeckers, hawks, and cardinals thrive high in the canopy, and larger mammals like raccoons, foxes, deer, and bears have adapted to the dramatic change of seasons (some, like bears, by hibernating). Most of the large predators that used to dominate these environments, like wolves and cougars, have been driven out by urbanization.

The 30 percent total forest cover mentioned above is slowly decreasing with time globally, mostly due to the actions of people encroaching on or harvesting forest habitats. In recognition of the critical role that forests play in the health and sustainability of the biosphere, as well as the threats that they constantly face, the United Nations declared 2011 the International Year of Forests, "to raise awareness of sustainable management, conservation, and sustainable development of all types of forests."

SEE ALSO Photosynthesis (c. 3.4 Billion BCE), First Land Plants (c. 470 Million BCE), Tropical Rain/Cloud Forests (1973), Temperate Rainforests (1976), Tundra (1992), Boreal Forests (1992), Grasslands and Chaparral (2004), Savanna (2013)

A 2013 photo of deciduous beech trees displaying spectacular fall colors in William Penn State Forest, Pennsylvania.

Lake Vostok

Antarctica is a continent almost completely covered by ice. Beneath the ice, however, lie mountains, canyons, river valleys, basins, sediments, and fossils, reflecting the rich geologic history of this ancient crustal terrain. Antarctica has also moved significantly since it formed; there is evidence, for example, that during the time of the Pangean supercontinent, Antarctica was in a temperate-to-tropical climate zone.

It is perhaps not surprising, then, that geologists have found evidence for ancient freshwater lakes preserved under the deep glacial ice. The first of these *subglacial lakes* was discovered by Soviet geologists participating in the International Geophysical Year in the late 1950s to early 1960s by sending seismic (shock) waves through the ice and measuring the reflections that came back from the continental surface below. This first one to be discovered was named Lake Vostok, after the USSR's Antarctic research station. More than 400 other subglacial lakes have since been discovered under the Antarctic ice.

Almost as soon as it was discovered, some scientists began calling for a drilling campaign to sample the waters of Lake Vostok. By some estimates, the lake has been isolated from the rest of the world for perhaps 15 to 25 million years, providing a unique opportunity to study extant or extinct organisms that flourished before the continent was buried by glaciers. Because of the uniqueness of that environment, however, other scientists and environmentalists have called for the lake's water to remain pristine, and for sampling to occur only after the international scientific community agrees on strict contamination controls for the drilling/coring process.

Noting the concerns but pressing on despite the high risk of contamination, in the 1990s Russian scientists began drilling the longest ice core in history, temporarily piercing Lake Vostok in February 2012, some 12,400 feet (3,768 meters) below the glacial ice. Biologists were able to identify microbes similar to extant surface species, but the contamination concerns make it ambiguous as to whether they are actually from the lake. Future efforts to re-open the borehole and sample the waters more cleanly will reveal if indigenous organisms live in that cold, dark, high-pressure, low-nutrient environment. If so, the prospects for life in even more extreme places, like Jupiter's ocean moon Europa, will improve.

SEE ALSO Antarctica (c. 35 Million BCE), The Caspian and Black Seas (c. 5.5 Million BCE), The Dead Sea (c. 3 Million BCE), Lake Victoria (c. 400,000 BCE), End of the Last "Ice Age" (c. 10,000 BCE), The Great Lakes (c. 8000 BCE), International Geophysical Year (IGY) (1957–1958), Extremophiles (1967)

NASA RADARSAT image of the surface of Antarctica's Lake Vostok (the smooth area in the center). Satellite radar signals are able to penetrate fairly easily through the more than 12,000 feet (3650 meters) of overlying ice and snow to reveal the buried lake below.

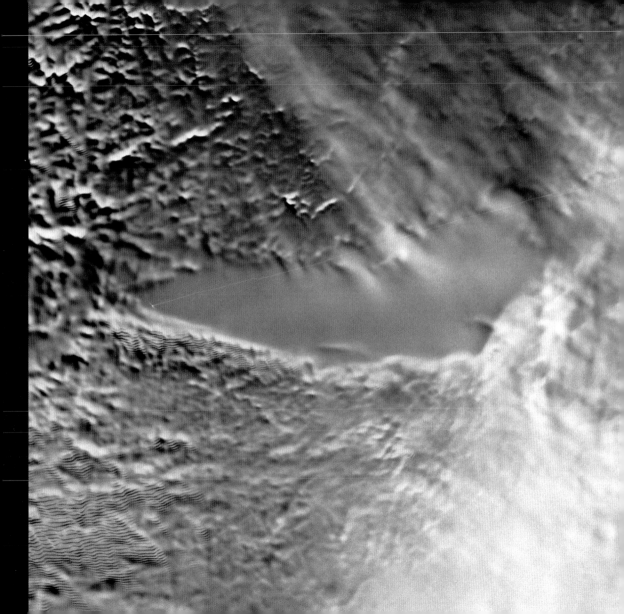

Savanna

Nestled ecologically between open grasslands and densely populated shrublands like chaparral is an important Earth ecosystem known as *savanna*. Savanna is defined as an area where grasslands and woodlands are mixed, but where the woodland shrubs or trees are either not closely spaced enough, or their canopies not dense enough, to block most sunlight from reaching the surface. Ample sunlight thus allows grasses and other non-woody plants to thrive in a continuous expanse along with woody plants.

Savanna zones occur around the world in distinct ecological and geographic areas that include tropical, subtropical, temperate, Mediterranean, and montane savannas. The common factors among all savanna environments are relatively large rainfall variations from year to year (monsoonal climate; including occasional droughts), and the common occurrence of wildfires during dry periods. Indeed, many anthropologists and cultural historians have noted that fire appears to have been a tool used by many indigenous peoples around the world to promote the expansion of savanna terrains, which have historically provided better hunting, gathering, planting, and/or grazing grounds than close-canopy forests. Whether natural or human-induced, savanna fires do help to suppress the growth of new trees, maintaining the overall structure of the ecosystem.

In terms of biodiversity, while large herbivores and carnivores (such as zebras, giraffes, elephants, and lions) of iconic savanna regions like the Serengeti Plains of Tanzania are the most familiar, savanna regions actually support as diverse a mix of plant, mammal, bird, amphibian, and insect species as do other global ecosystems, and thus are deserving of similar active monitoring and ecological health management for their sustainability. In addition, since prehistoric times people have commonly used savannas as grazing pastures and rangelands for domesticated animals such as cattle, sheep, and goats, often making them important parts of local and regional economies.

An example of national-scale joint governmental and non-governmental advocacy on behalf of savanna ecosystems was initiated in 2013 in northern Australia, which contains one of the largest expanses of undeveloped tropical savanna in the world. The "Kimberly to Cape Initiative" is designed to work "in the coordination, policy, and communications space to support local people and identify and promote the benefits of a healthy and intact savanna."

SEE ALSO First Land Plants (c. 470 Million BCE), Domestication of Animals (c. 30,000 BCE), Invention of Agriculture (c. 10,000 BCE), Tropical Rain/Cloud Forests (1973), Temperate Rainforests (1976), Tundra (1992), Boreal Forests (1992), Grasslands and Chaparral (2004), Temperate Deciduous Forests (2011)

A 2008 photo of typical savanna terrain, from Kiang West National Park, The Gambia, West Africa.

Rising CO$_2$

The Earth formed from both rocky and *volatile* (highly active) compounds; the latter escaped from the interior and formed the atmosphere. These volatiles included carbon dioxide (CO$_2$), which started out as a major component of the atmosphere but which began to be significantly depleted by CO$_2$-consuming organisms soon after the invention of photosynthesis. While CO$_2$ is now only a relatively minor trace gas in our atmosphere, it remains important because it is a strong *greenhouse gas* that absorbs heat and helps to warm the atmosphere. In fact, if there were no greenhouse gases like CO$_2$ in our atmosphere, Earth's surface temperature would be so cold that the oceans would freeze solid. A little greenhouse warming is a good thing, it turns out.

Knowing how much CO$_2$ is in the atmosphere is important because data from ancient sedimentary rocks, ice-core studies from Greenland and Antarctica, tree-ring studies, and modern temperature sensors show that the Earth's average surface temperature is directly correlated to the amount of CO$_2$ in the atmosphere. As CO$_2$ increases, Earth's average surface temperature increases, and vice versa. For example, during the height of the last glacial maximum around 2 million years ago, there were about 180 CO$_2$ molecules per million molecules of air (180 parts per million, or ppm) in Earth's atmosphere. That number very slowly went up as the climate warmed and the glaciers receded around 12,000 years ago, averaging about 280 ppm up until the start of the Industrial Revolution, around 1830.

Since then, however, atmospheric CO$_2$ levels have risen dramatically, at a rate of increase higher than ever previously seen or inferred from past data. In 2013, CO$_2$ passed a new milestone, going above 400 ppm in Earth's atmosphere, higher than at any time in at least the last 800,000 years. During this same time, Earth's average surface temperature has increased by 0.5°C to 1.0°C (0.9°F to 1.8°F). The proximate causes for the CO$_2$ increase since the beginning of the Industrial Revolution are the burning of fossil fuels and deforestation. The fact that humans are warming the climate has spurred significant interest in alternate energy sources (wind, hydropower, solar, nuclear) as well as raising significant concerns about rising sea levels, increased storm intensities, and more frequent droughts.

SEE ALSO Earth Is Born (c. 4.54 Billion BCE), Earth's Oceans (c. 4 Billion BCE), Photosynthesis (c. 3.4 Billion BCE), The Great Oxidation (c. 2.5 Billion BCE), Snowball Earth? (c. 720–635 Million BCE), Cambrian Explosion (c. 550 Million BCE), Industrial Revolution (c. 1830), Deforestation (c. 1855–1870), The Anthropocene (c. 1870), The Greenhouse Effect (1896), Nuclear Power (1954), Wind Power (1978), Solar Power (1982), Hydroelectric Power (1994), End of Fossil Fuels? (~2100)

A 1942 photo of smoke billowing from a US World War II production factory.

Long-Duration Space Travel

Scott J. Kelly (b. 1964), **Mark E. Kelly** (b. 1964)

Since the 1960s, space-agency doctors have been tracking the effects on astronauts of weightlessness and other environmental changes. These studies have mostly been to maintain crew health and safety, but their purpose is also to understand the potential for very long-term exposure of people to the space environment, as would be experienced on a trip of many years to Mars, for example. Major milestones of these studies included the 84-day flight of the three-person NASA crew of *Skylab-4* in 1973–74, the year-long flight of the two-person Russian crew of the *Mir* space station in 1987–88, and the 437-day mission of Russian Valery Polyakov on *Mir* in 1994–95, which is still the record for the longest human time spent in space. Doctors observed various degrees of bone and muscle deterioration, changes in body fluid distribution, disruptions of vision and taste sensations, and other physiological effects in these long-duration space travelers.

Perhaps the most robust of these studies, however, was the NASA "Twins Study" conducted on twin-brother astronauts Scott and Mark Kelly. Scott is a veteran of five space missions (riding the Space Shuttle and Soyuz launchers to the ISS, the International Space Station), and Mark of four, also on the Shuttle and ISS. While they never flew together in space, they did work closely together, albeit remotely, when Scott embarked on an 11-month mission to the ISS in 2015–16. While Scott was in space maintaining a specific regimen of diet and exercise and other activities, Mark remained on Earth and tried to follow a similar regimen, acting as a "control experiment" against which to measure changes of his twin brother while in space.

Upon Scott's return to Earth in 2016, both men went through exhaustive medical examinations. Like most people who travel in space, Scott came back (temporarily) a few inches taller than Mark without the effects of gravity having compressed his spine. In addition to some of the symptoms experienced by earlier long-duration astronauts, Scott also exhibited a surprising change that had not previously been studied: about 7 percent of his DNA had changed permanently, compared to his twin brother's. The reasons for these changes are not yet understood, but they will surely be of interest to doctors and future space travelers planning to spend even more time off-planet.

SEE ALSO Exploring the Oceans (1943), Sputnik (1957), Humans in Space (1961), Geology on the Moon (1972), Settlements on Mars? (~2050)

NASA astronauts and twin brothers Mark (left) and Scott (right) Kelly, participants in a landmark 2016 medical study on the effects of long-duration spaceflight on the human body.

North American Solar Eclipse

Eclipses occur when one celestial body passes in front of another, as viewed from a particular location. Solar or lunar eclipses are the kinds of eclipses most familiar to people, because they occur frequently enough, and are often dramatic enough, that they are memorable (or sometimes portentous) events.

A *lunar eclipse* occurs at Full Moon, when the Sun, Earth, and Moon line up (in that order), and the Moon passes exactly behind the Earth from the Sun and thus through the Earth's shadow. The Moon's orbit is tilted relative to the Earth's orbit around the Sun, however, so the Moon only occasionally passes precisely through the Earth's shadow. Most months, at Full Moon the Moon passes a little above or a little below the Earth's shadow and, sadly, there is no eclipse.

A *solar eclipse* occurs at New Moon when the Sun, Moon, and Earth line up (in that order), and the Moon passes exactly between the Earth and the Sun. Again, when the geometry is just right (rarely), the Moon's shadow can fall on the Earth. It is an incredible cosmic coincidence that the apparent angular size of the Moon in the sky is almost the same as the apparent angular size of the Sun (the Sun's diameter is about four hundred times larger than the Moon's, but the Moon is about four hundred times closer to the Earth). The result is that very rarely the Moon's disk can completely cover the Sun's disk in the sky, resulting in a *total solar eclipse*.

Total solar eclipses are rare—any particular spot on Earth experiences one only every 350 to 400 years, on average. Some people, including many astronomers, are "eclipse chasers," traveling to the predicted path of the Moon's shadow to view or take scientific data during these rare celestial events. The element helium, for example, was discovered during a total solar eclipse in 1868 because of the improved view to astronomers of the Sun's extended atmosphere, or corona, which is much more visible when the main part of the Sun is blocked by the Moon.

During the August 21, 2017, total solar eclipse, the Moon's shadow swept across the United States, from Oregon to South Carolina. Astronomers on the ground and in special airborne observatories used the opportunity to study the Sun's corona and magnetic field with newer, more sensitive instruments. However, the real legacy of the "Great American Eclipse" may be the fact that millions of people were educated and profoundly inspired by the awe-inspiring sight of the solar corona during those fleeting few minutes of totality.

SEE ALSO The Earth Is Round! (c. 500 BCE), Last Total Solar Eclipse (~600 Million)

Top: *View of the corona (the Sun's outer atmosphere) from Madras, Oregon, during the August 21, 2017 eclipse.*
Bottom: *Path of the Moon's shadow across America during the total solar eclipse.*

Apophis Near Miss

Dedicated telescopic surveys since the 1990s have identified hundreds of thousands of new asteroids zipping around the solar system. Most are in the Main Asteroid Belt between Mars and Jupiter, but many are in other populations as well, including in three different populations of near-earth asteroids (NEAs)—the *Atens* (orbiting closer to the Sun than Earth), *Amors* (orbiting farther out from Earth), and *Apollos* (with orbits that cross Earth's orbit). All three classes of NEAs pose potential impact hazards for Earth.

One of the most closely watched members of the NEA population is a small asteroid named 99942 Apophis. First discovered in 2004, its orbital parameters were calculated from follow-up telescope observations, including super-accurate measurements using the Arecibo planetary radar facility in Puerto Rico. Then, like hundreds of other NEAs, its parameters were entered into an automated computer program developed by astronomers to predict the future trajectories and earth-impact probabilities of these asteroids. Apophis quickly set off some alarms because the calculations came back showing an approximately 1 in 37 chance that the asteroid would impact Earth on April 13, 2029. Apophis set the record for greatest impact risk yet identified, with a Torino Impact Hazard Scale score of 4 out of 10.

Astronomers quickly organized more observing campaigns to refine the predictions of Apophis's orbit. The new data showed that the asteroid will get very close to Earth, only about two to three Earth diameters away—well within the orbits of Geosynchronous Satellites—but that it will not actually impact our planet. Apophis will pass close to Earth again in 2036, but its odds of hitting Earth then are down to less than 1 in 1 billion, and thus its Torino Scale score has been downgraded to a 0.

Still, prudence is advised: an impact by a rocky asteroid 1,000 feet (300 meters) in diameter would not be devastating globally, but it could still have bad consequences (for example, a giant impact-generated tsunami). Apophis is named after the Egyptian god of destruction—let's hope that this hazardous little asteroid's name doesn't turn out to be accurate.

SEE ALSO Dinosaur-Killing Impact (c. 65 Million BCE), Arizona Impact (c. 50,000 BCE), The US Geological Survey (1879), Hunting for Meteorites (1906), The Tunguska Explosion (1908), Geosynchronous Satellites (1945), Understanding Impact Craters (1960), Torino Impact Hazard Scale (1999)

Main image: *A trajectory plot, to scale, of the positions of the Earth and Moon and the predicted path of the near-Earth asteroid 99942 Apophis during its April 13, 2029, close approach.*
Inset: *The detail shows a zoom of how close Apophis is predicted to come to our planet (the white bar represents the trajectory uncertainty).*

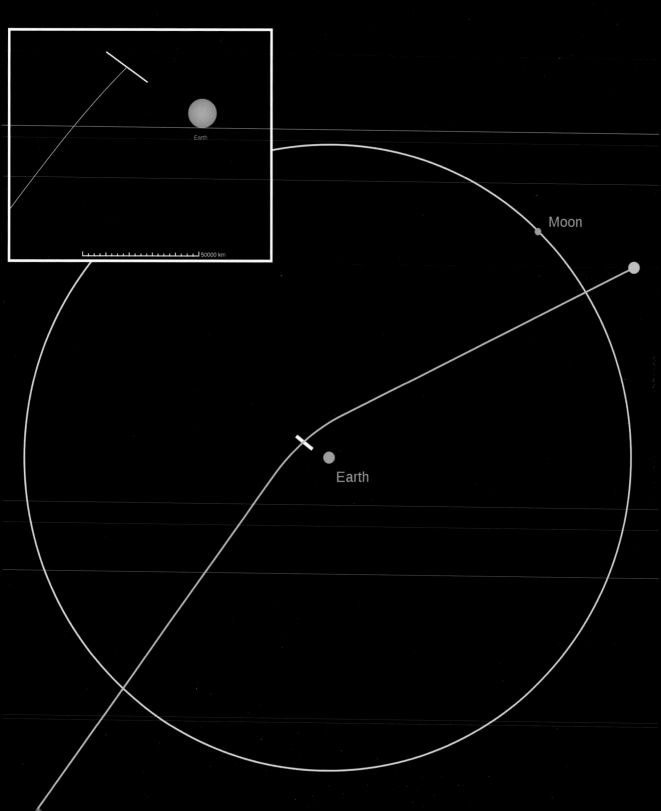

Earth

Moon

50000 km

Earth

Settlements on Mars?

Humans are specifically and uniquely adapted to the environment and constraints of planet Earth, and yet some fraction of us yearns to leave, to explore, to push the capabilities of our species. Perhaps it is yet another manifestation of the way evolution drives us to adapt. Whatever the motivation, people have already begun to travel into space, to view our planet from a unique perspective from outside the biosphere, and (for a very lucky few) to leave Earth's gravity and even walk and work on the Moon.

For many fans and practitioners of space exploration, Mars is the next giant step in the human exploration of space. Dozens of robotic flybys, orbiters, landers, and rovers over the past 50-plus years have revealed that Mars is the most earthlike planet in the solar system besides the Earth itself. It has a thin, cold CO_2 atmosphere, a day about the same length as an Earth day, polar caps made of CO_2 ice as well as water ice that come and go with the seasons, and a fascinating geologic history that rivals that of our own world. Perhaps most compelling is that Mars *used to be* even more like the Earth, with a thicker atmosphere and warmer climate, liquid water flowing on the surface into lakes and seas, and a magnetic field protecting the surface from solar radiation. Mars was a habitable world early in its history. What happened? Was it ever, or is it still, inhabited? Can we learn about the origin of life on our own planet by exploring there?

It will take people going there to really find out. Geologists, astrobiologists, chemists, and meteorologists are all going to have to explore the place in detail, and will need engineers, programmers, pilots, doctors, and others to support the voyages. At first, just small crews on short (3-plus-year) round trips, perhaps in the 2030s; but then settlers will go, establishing the first bases, perhaps as soon as the 2050s. They may be astronauts from NASA or other space agencies, or possibly employees of private "NewSpace" companies who want in on the adventure—or maybe a mix of all the above. Our evolutionary drive to explore, to learn, and to expand our species ever outward will push us to Mars, and beyond.

SEE ALSO (Geo) Science Fiction (1864), Humans in Space (1961), Earth Selfies (1966), Leaving Earth's Gravity (1968), Geology on the Moon (1972), Long-Duration Space Travel (2016)

Artist's concept of the kind of deployable habitat (with airlock) and pressurized rover that could be part of the first human mission to visit the Red Planet in the 2030s, 2040s, or 2050s.

End of Fossil Fuels?

More than 80 percent of the world's energy needs—for food production, heating/cooling, transportation, manufacturing, etc.—come from the burning of coal, oil, and natural gas that was formed millions to hundreds of millions of years ago by the burial, decay, and chemical transformation of ancient plants and other organisms. Those *fossil fuels* are identified by geologic mapping and remote sensing methods, extracted from sedimentary layers in the shallow subsurface by drilling or digging/tunneling, and refined into a variety of petroleum-based products by a massive worldwide manufacturing network. But by their very nature, *fossil fuels* are a finite resource that will eventually become too expensive to extract, or will simply run out.

Predictions by economists and energy experts about when the world would reach "peak oil" production, after which the supply of fossil fuels would slowly decline and be phased out, have varied widely. Predictions of peak oil being reached around the turn of the twenty-first century proved premature, thwarted by new exploration and resource extraction technologies (such as *hydraulic fracturing* to extract oil and gas from shale deposits) that have been able to offset decreases in earlier-identified sources. While technologies will certainly continue to improve, the consensus now appears to be that perhaps by the middle of this century, and almost certainly by the start of the next, the economically viable extraction of fossil fuels will have been tapped out.

Many advocates of more sustainable energy solutions think the end of fossil fuels will come even sooner. Just as technology is currently helping to keep fossil fuels economically viable, so too is it advancing the economic and social appeal of more environmentally friendly renewable energy sources, like solar, wind, hydroelectric, and nuclear power options. Substantial advances in solar and wind energy generation in the last few decades, in particular, have those cleanest of the renewable energy options poised to expand dramatically over the next few decades, with some predicting that solar power alone could account for more than 50 percent of the world's energy needs by mid-century. Maybe the world's reliance on fossil fuels won't end because the supply ends, but instead because their use is simply much more expensive (in cost and environmental impact) compared to the alternatives.

SEE ALSO First Land Plants (c. 470 Million BCE), Population Growth (1798), Industrial Revolution (c. 1830), The Anthropocene (c. 1870), The Greenhouse Effect (1896), Nuclear Power (1954), Wind Power (1978), Solar Power (1982), Hydroelectric Power (1994)

Some day, maybe within 50–100 years or perhaps less, human civilization will run out of easily extractable fossil fuel resources like coal (shown here), oil, and natural gas. Then what?

Next Ice Age?

Evidence from the geologic and fossil records tells us that our planet has gone through at least five large-scale periods of extensively cold climate conditions—*ice ages*—and that during any particular ice-age period, our climate has gone through dozens or even hundreds of shorter-term glacial maximum (*glacial*) and minimum (*interglacial*) periods. For the past 12,000 years or so, for example, we have been in an interglacial period called the Holocene that is just one of many interglacial periods that have occurred within the current ongoing ice age, known as the Quaternary Glaciation, which began about 2.6 million years ago. So, when will the next glacial period begin in our current ice age? And when will this ice age end and the next begin?

Climate modelers have figured out that small, long-term changes in Earth's orbital parameters (like the tilt of the spin axis, or the eccentricity of the orbit) that influence the amount of solar energy reaching our planet have had a large influence on the timing of glacial and interglacial periods. Indeed, such orbital effects are predictable into the near future, at least over the next few million years. It turns out that over the next 100,000 years, those orbital variations will happen to be rather small, and so other effects, like the abundance of CO_2 in the atmosphere, will be the dominant factors controlling global temperature and climate. Thus, with human-induced CO_2 levels continuing to rise, eventually above levels experienced in the last million years, the climate is likely to continue to warm and the current interglacial period is likely to continue. By some estimates, the combination of orbital parameters and assumptions about potentially decreasing future levels of atmospheric CO_2 might mean that the next glacial period won't occur for 50,000 years or more.

If that turns out to be true, then it may not be just an interglacial period that we're in, but perhaps a transition to the end of the Quaternary ice age itself, with the eventual complete disappearance of the Greenland and Antarctic ice caps. How long will that take, and then how long until the next true ice age begins? These are the subject of much controversy and scientific research.

SEE ALSO Snowball Earth? (c. 720–635 Million BCE), End of the Last "Ice Age" (c. 10,000 BCE), The Little Ice Age (1500), Discovering Ice Ages (1837), The Greenhouse Effect (1896), Rising CO_2 (2013)

An artistic depiction of Earth during the last glacial maximum of the current ice age, around 2 million years ago. Is this what Earth will look like once again, perhaps 50,000 years from now?

Yellowstone Supervolcano

The most powerful volcanic explosion in modern history was the 1815 eruption of Mount Tambora in Indonesia. Even though it led to the deaths of more than 100,000 people and significantly changed the Earth's climate for several years, Tambora is still rated only a 7 out of 8 on the Volcanic Explosivity Index (VEI). To be rated an 8, the volcano has to be a monster that geologists call a *supervolcano*.

Supervolcano eruptions are defined as those that eject more than ~250 cubic miles (~1,000 cubic kilometers, or about one-quarter the volume of the Grand Canyon) of ash, dust, and gas onto the surface of the Earth and into its atmosphere. Plumes from supervolcanoes push these massive amounts of dust and other Sun-blocking aerosols into the stratosphere and can thus have a profound and long-lasting influence on the Earth's climate. Because they are so intense, they are also extremely rare, with only one or two such eruptions occurring every 100,000 years. No supervolcanoes have erupted in recorded history; evidence from the geologic record shows that the last few erupted about 25,000 years ago (Taupo, New Zealand) and 74,000 years ago (Toba, Indonesia).

There are dormant supervolcanoes among us today, however. Among the most famous is the 34-by-45-mile-wide (55 by 72 kilometers) caldera that encompasses Yellowstone National Park in the US state of Wyoming. Even though it is far from an active tectonic plate margin, Yellowstone has been the site of significant geothermal activity and prehistoric volcanism for millions of years because the region overlies a mantle *hotspot*. Unlike the hotspot in the middle of the Pacific plate under the Hawaiian Islands, however, the Yellowstone hotspot is under a deep layer of continental crust, and its heat has to percolate through much more rock before it can be released in volcanic eruptions.

The Yellowstone supervolcano last erupted about 630,000 years ago, covering most of North America with thick ash and likely cooling the global climate for years. Luckily, there is no evidence for an imminent eruption in the near future. Still, because the consequences of such an eruption would be so extreme, the region is very carefully monitored for seismic or other evidence indicating potential future activity.

SEE ALSO Plate Tectonics (c. 4–3 Billion BCE?), The Atlantic Ocean (c. 140 Million BCE), Hawaiian Islands (c. 28 Million BCE), Pompeii (79), Huaynaputina Eruption (1600), Mount Tambora Eruption(1815), Krakatoa Eruption (1883), Exploring Katmai (1915), Mount St. Helens Eruption (1980), Volcanic Explosivity Index (1982)

Artist's concept of cross-section of magma rising through the crust and erupting at the site of the "Yellowstone supervolcano" in central North America.

Loihi

The long northwest-to-southeast-trending chain of islands and seamounts in the Central Pacific Ocean that ends at the Big Island of Hawaii is a primary piece of evidence that the Pacific plate is moving under a relatively stable mantle hotspot beneath the oceanic crust. These hotspots are thought to be the manifestation of enormous plumes of heat and molten rock that churn inside Earth's mantle, helping the deep interior to release its heat. Geologists believe that there are 50 or 60 such mantle hotspots of varying sizes and intensities that underlie oceanic and continental crust around the world. Other famous ones include the Yellowstone hotspot, and the hotspots that formed the Deccan Traps in India and the Galápagos Islands off South America.

The Hawaiian hotspot has been active for at least 28 million years (the age of the oldest seamounts in the chain), and recent eruptions of the Mauna Loa and Kilauea volcanoes provide evidence that it is still active today. At the same time, the Pacific plate continues to creep along to the northwest at an average rate of about 2 to 4 inches (5 to 10 centimeters) a year. The expectation, then, is that a new island should appear southeast of the island of Hawaii sometime soon (geologically speaking, that is).

Indeed, seafloor mapping has revealed the presence of a tall seamount about 22 miles (35 kilometers) off the southeastern coast of Hawaii, called Loihi ("low-ee-hee," meaning "long" in Hawaiian, for its elongated shape). It is currently an impressive undersea mountain that rises more than 10,000 feet (3,000 meters) above the seafloor, taller than many stratovolcanoes like Mount St. Helens. The flanks of the submarine volcano support a number of deep-sea hydrothermal vents and an impressive diversity of microorganisms and other marine life.

Loihi is the focus of significant seismic activity, and undersea volcanic eruptions have been active since modern records were first kept in the late 1950s. Sampling of lavas from the base of this growing mountain reveals that it started forming about 400,000 years ago on the seafloor. Depending on the rate and intensity of future eruptions, it is likely that Loihi will have breached the surface to become a new Hawaiian Island by about 100,000 to 200,000 years from now.

SEE ALSO Earth's Mantle and Magma Ocean (c. 4.5 Billion BCE), Plate Tectonics (c. 4–3 Billion BCE?), Deccan Traps (c. 66 Million BCE), Hawaiian Islands (c. 28 Million BCE), Galápagos Islands (c. 5 Million BCE), Mapping the Seafloor (1957), Deep-Sea Hydrothermal Vents (1977), Mount St. Helens Eruption (1980)

A 2013 photo of an erupting undersea volcano forming a brand-new island off the coast of Nishinoshima, Japan. The new Hawaiian island of Loihi will look like this in the not-too-distant future, geologically speaking.

Next Big Asteroid Impact?

It took centuries for geologists to figure it out, but impact cratering is now understood to be a major force of geologic change, not just in the history of the Moon and other heavily cratered planets, moons, and asteroids out there, but also of the Earth. A giant impact between the early Earth and a Mars-sized protoplanet likely formed our Moon. An impact certainly played a major role in the extinction of the non-avian dinosaurs and many other species at the end of the Cretaceous period 65 million years ago, and impacts may have played roles in other mass extinction events in Earth's history as well. Well-preserved young craters like Meteor Crater in Arizona and small impact events like those in Tunguska in 1908 or Chelyabinsk in 2013 remind us that there are still potentially threatening impacts waiting to happen one day.

So when will an impactor large enough to potentially cause a major climatic and/or biologic catastrophe next hit the Earth, and could we do anything to prevent it? Astronomers, planetary scientists, engineers, and planetary defense experts are working to try to figure that out. An important first step is completing a census of the Potentially Hazardous Asteroids that occasionally pass near the Earth, including their size, composition, and internal strength (are they solid boulders, or rubble piles of pebbles and sand?), so that their threat potential can be assessed. By 2011, surveys had found 90 percent of the asteroids large enough to cause a global catastrophe; continuing observations over the next decade will attempt to expand that survey to include smaller asteroids that could still cause local or regional catastrophes.

Statistically, a potential "dinosaur-killer"-sized asteroid (about 6 miles [10 kilometers] in diameter) is predicted to hit Earth about once every 100 million years, so perhaps we're safe from that fate for another 35 million years or more. However, an asteroid around 0.6 miles (1 kilometer) diameter strikes the Earth every 500,000 years or so, releasing more than 3,000 times the energy of a World War II atomic bomb and thus causing substantial local devastation and climate disruption. Will we have to wait 500,000 years for that next big impact, or will it come sooner? Will we be ready?

SEE ALSO Birth of the Moon (c. 4.5 Billion BCE), Dinosaur-Killing Impact (c. 65 Million BCE), Arizona Impact (c. 50,000 BCE), The US Geological Survey (1879), Hunting for Meteorites (1906), The Tunguska Explosion (1908), Geosynchronous Satellites (1945), Understanding Impact Craters (1960), Torino Impact Hazard Scale (1999), Apophis Near Miss (2029)

The orbits of known members of the Near-Earth Asteroid population (blue) plotted along with the relatively circular orbits of Mercury, Venus, Earth (white), and Mars, in a view looking down onto the north pole of the solar system.

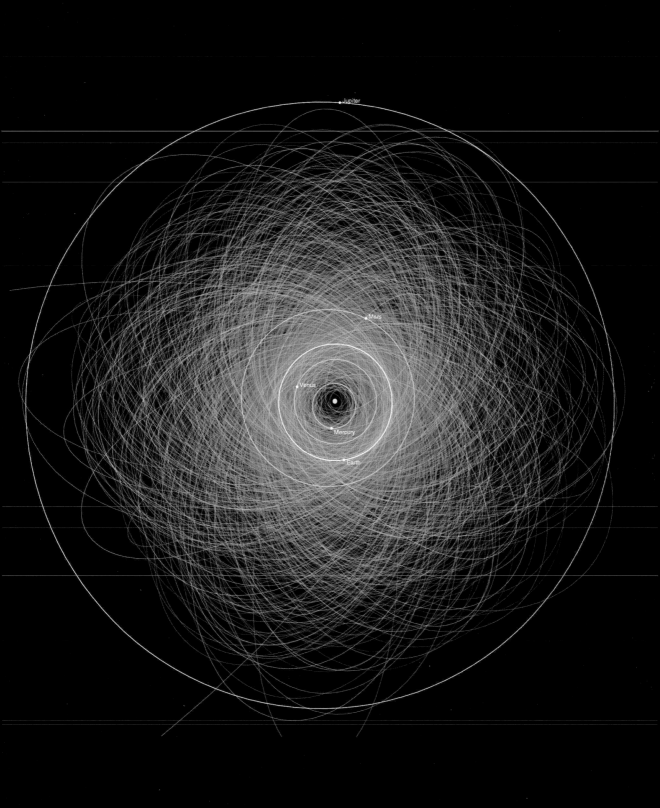

Pangea Proxima

The breakup of the supercontinent Pangea starting around 200 million years ago led to the geographic distribution of continental land masses that we see on the Earth today. However, the Earth's several dozen large tectonic plates are still in motion relative to each other, and so the map of the world today is merely a snapshot in time of a dynamic, changing planet.

What will the Earth look like in the future? Geologists can use the currently measured relative velocities of Earth's tectonic plates, enter them into a computer model, and predict the positions of the plates in the far future. For example, the African and Eurasian plates are moving toward each other today in a slow-motion collision that is uplifting the Alps and shrinking the Mediterranean basin. Eventually, the Mediterranean will be pinched off entirely, and a new Himalaya-like mountain range will begin to form in the former cradle of civilization.

Some models of the future predict that the collision of Africa and Eurasia; more vigorous seafloor spreading in the Pacific and the eventual subduction of the Atlantic seafloor underneath the eastern coasts of North and South America; and the (re)merging of Antarctica, Australia, and Southeast Asia will ultimately bring all the major continents back together again, in a grand reunion of Pangea that has been dubbed *Pangea Proxima* ("next Pangea"), around 250 million years from now. The prediction is speculative, however, and other potential future paths for the continents are possible because plate motions can change based on unpredictable changes in the intensity and distribution of huge convection cells of heat in the mantle.

Regardless, there appear to have been and will continue to be *supercontinent cycles* on the Earth's surface—the breakup of one supercontinent leading to the eventual formation of another, which then breaks up, forms another, etc. About a dozen super-continents older than Pangea are hypothesized to have formed over the ~4 billion years or so since continental crust first formed, based on preserved magnetic orientation data and similarities among ancient fossils and metamorphosed rocks that are now, literally, a world apart. Watching a time-lapse movie of a map of the Earth over billions of years might be like seeing the continuous assembly and disassembly of a jigsaw puzzle!

SEE ALSO Continental Crust (c. 4 Billion BCE), Plate Tectonics (c. 4–3 Billion BCE?), Pangea (c. 300 Million BCE), The Atlantic Ocean (c. 140 Million BCE), The Alps (c. 65 Million BCE), Continental Drift (1912), Mapping the Seafloor (1957), Seafloor Spreading (1973)

If the plates continue to move in the directions that they're moving today, within about 250 million years the continents are predicted to come together once again, creating what some have dubbed "Pangea Proxima" ("next Pangea") or "Pangea Ultima" ("last Pangea").

250 Million Years in Future

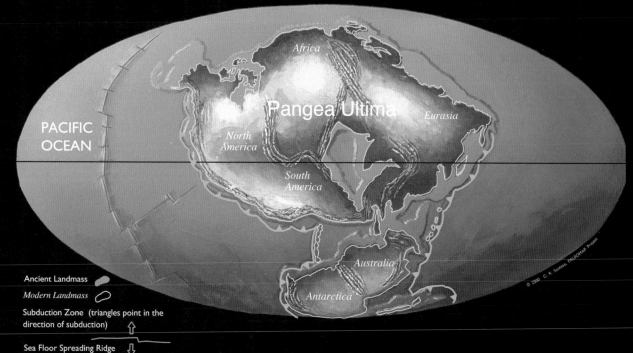

Africa

Pangea Ultima

Eurasia

PACIFIC
OCEAN

North
America

South
America

Australia

Antarctica

© 2000 C. R. Scotese, PALEOMAP Project

Ancient Landmass

Modern Landmass

Subduction Zone (triangles point in the
direction of subduction)

Sea Floor Spreading Ridge

Last Total Solar Eclipse

One of the simplest experiments on the surface of the Moon by the Apollo astronauts was to deploy a set of mirrors. The mirrors were pointed back at the Earth, so that astronomers could shine high-powered lasers at the Moon and measure precisely how long it took the laser light to reflect back to the Earth. Knowing the travel time and the speed of light, the distance to the Moon can be measured to better than 0.04 inches (1 millimeter). Over time, these kinds of measurements have revealed that the Moon is slowly spiraling away from the Earth, at a rate of about 1.5 inches (3.8 centimeters) per year.

Why is the Moon receding from the Earth? The answer is the conservation of *angular momentum*. The Moon's gravity raises tidal "bulges" in the Earth's oceans that produce a drag force on the Earth's crust that is (gradually) slowing down the spin of our planet. Earth's loss of angular momentum is the Moon's gain. As the Moon speeds up, it moves farther away, according to Kepler's Laws of Planetary Motion. The fact that the Moon is moving away today means it was much closer very long ago when it formed. Indeed, the Moon may once have appeared as much as fifteen times bigger than it does today in our night sky!

One implication of the Moon's spiraling-out motion is that it is slowly (*very* slowly) getting smaller in the sky as seen from the Earth. Today, the apparent angular size of the Moon in the sky varies (depending on where the Moon is in its slightly elliptical orbit) from just a little less than the apparent angular size of the Sun to just a little more—when *total solar eclipses* are possible. As the Moon gets smaller in the sky over time, however, eventually it won't be able to completely cover the Sun's disk. Some day, about 600 million years from now, our far-distant descendants (whoever or whatever they are), will gather to watch the last total solar eclipse visible from the Earth's surface. From then on, all similar events will be *annular solar eclipses*, displaying only a glorious annulus or ring of sunlight around the dark face of the Moon.

SEE ALSO The Earth Is Round! (c. 500 BCE), Laws of Planetary Motion (1619), Tides (1686), Gravity (1687), Geology on the Moon (1972), Global Positioning System (1973), North American Solar Eclipse (2017)

In the far future, the Moon will be too far away from the Earth to completely block the disk of the Sun, and all solar eclipses will be "annular" eclipses like this one, from May 20, 2012.

Earth's Oceans Evaporate

The life cycle of a so-called "main sequence" star like the Sun is quite predictable. Astronomers in the early twentieth century figured out the basic evolutionary track of stars like the Sun by observing countless similar stars at different stages of development. By the middle of the century, theories had also been worked out about the insides of the stars and the nuclear fusion processes that make them shine. And thanks to meteorite research and radioactive dating methods, we know the precise age of the Sun (4.567 billion years) and thus can predict the next milestones of our star's life.

Hydrogen is converted into helium at the enormous temperatures and pressures in the Sun's core. Over time, then, the Sun's hydrogen supply is slowly decreasing. To keep its balance of gravitational (inward) versus radiational (outward) pressure— and thus to stay on the main sequence—the Sun's core is slowly getting hotter. This increases the rate of nuclear fusion in the core, offsetting the effect of the decreasing hydrogen supply and increasing the Sun's brightness over time. Astronomers estimate that the Sun's energy output is increasing by about 10 percent per billion years because of the decreasing supply of hydrogen.

Such a dramatic change in the Sun's energy output will have a correspondingly dramatic change on the Earth's climate. In tens to hundreds of millions of years, it will become warm enough that the oceans will start to permanently evaporate, turning our planet into a steamy world. Scientists further predict that within about a billion years, the slow breakdown of all that atmospheric water by sunlight and the subsequent escape of the liberated hydrogen will turn our planet into a bone-dry, inhospitable desert world. Unfortunately, the future is looking too bright.

Some long-term climate modelers think that our planet will become uninhabitable long before the oceans are completely dried up. As the climate gets hotter, more carbon dioxide will get trapped in carbonate rocks, leaving less for plants to use in photosynthesis. Within maybe a half billion years, then, much of the base of the food chain could collapse, making the biosphere overall unsustainable. It's not a cheery long-term prognosis, but maybe by then our species (or whatever it has become) will have found one or more new beautiful blue water worlds to call home.

SEE ALSO Earth Is Born (c. 4.54 Billion BCE), Earth's Oceans (c. 4 Billion BCE), Radioactivity (1896), The End of the Earth (~5 Billion)

An artist's conception of a "hot Jupiter," among the most common kind of extrasolar planets initially discovered in the solar neighborhood. A billion years from now, as the Sun continues to mature and grow hotter, Earth's oceans will evaporate, and our planet will become a "hot earth."

Earth's Core Solidifies

The interior of the Earth is hot because of Earth's original heat of formation, as well as the heat released from the decay of radioactive elements over time. The deep Earth is hot enough, in fact, for the outer core of iron-nickel metal to melt (temperatures above 4,000°C to 6,000°C, or 7,200°F to 10,800°F), although the inner core remains solid. Earth's molten, spinning, electrically conducting outer iron core creates a strong magnetic dynamo with a field that extends deep into space around our planet, helping to protect the surface and atmosphere from harmful solar and cosmic radiation. Without the molten outer core and the protective magnetic field it creates, life on Earth might never have been able to form. And even if it did, it would be difficult or impossible for it to survive on the radiation-bombarded surface.

The Earth's interior is slowly cooling, however, partly because the abundance of radioactive elements is decreasing with time, and partly because heat is transferred from the core to the mantle to the crust and then to space, via mantle convection, volcanic eruptions, and thermal radiation. As the interior cools, the liquid outer core is slowly freezing (solidifying), enlarging the diameter of the solid inner core by about 0.04 inches (1 millimeter) per year. Inexorably, over time as the planet continues to cool, Earth's liquid outer core will completely solidify. By most geophysical estimates, this will happen between about 2 and 3 billion years from now.

Once the core solidifies, the magnetic dynamo engine deep inside the Earth will stop working, and the Earth's magnetic field and magnetosphere will dissipate relatively quickly. Decreasing heat flow will slow mantle convection and could potentially cease plate tectonics on the surface. Without the protection from the magnetic field, the solar wind (the stream of high-energy radiation coming from the Sun) will begin to directly impact and erode the Earth's upper atmosphere, breaking apart CO_2, O_2, H_2O, and other molecules and driving light elements like hydrogen and other volatiles off into space. Loss of the magnetic field is the primary explanation for how early Mars lost its once-thicker and warmer atmosphere; a similar fate likely awaits our own world in the distant future.

SEE ALSO Earth's Core Forms (c. 4.54 Billion BCE), Earth's Mantle and Magma Ocean (c. 4.5 Billion BCE), Solar Flares and Space Weather (1859), Radioactivity (1896), The Inner Core (1936), Earth's Radiation Belts (1958), Reversing Magnetic Polarity (1963), Extremophiles (1967), Earth Science Satellites (1972), The Oscillating Magnetosphere (1984)

Artist's rendering of a solar storm hitting Mars and stripping ions from the planet's upper atmosphere. Does the same fate await the Earth billions of years in the future, when our planet's magnetic field shuts down?

The End of the Earth

The Sun's fate is sealed, and it is humbling and perhaps a little sad to realize that our glorious star will not shine forever. The Milky Way has billions of so-called "main sequence" stars with the same basic properties as our Sun, and we can study them all around us in different stages of their predictable life cycles. The destiny of a star is dictated by its initial mass; in the case of a star with the mass of our Sun, its destiny is a short, violent, energetic youth followed by a relatively long, stable, 10-billion-year middle age, and then a relatively gentle, quiet death.

Radioactive dating of primitive meteorites as well as analysis of solar wind particles collected by missions like NASA's *Genesis* spacecraft tell us that the Sun is about 4.567 billion years old, or about halfway through its life cycle as a typical low-mass star. As it goes through middle age and uses up more of its hydrogen nuclear-fusion fuel supply, our star is slowly getting hotter—in about a billion years it will be hot enough to evaporate Earth's oceans. In 5 billion years or so, all the Sun's hydrogen will be used up and the core will contract and heat up further, expanding the Sun's outer atmospheric layers until it becomes a red giant star.

The red-giant Sun will eventually swell to about 250 times its present size, engulfing and destroying the inner planets, most likely including the Earth. As the Sun's helium and other heavier elements are also depleted, enormous pulsating death throes will jettison the Sun's outer layers (including all the atoms of the former inhabitants of the now-vaporized Earth) into deep space as a planetary nebula, to be recycled into new stars. The remaining ember of the Sun's core will become a white dwarf star that will slowly cool, eventually fading into the background oblivion of cold space.

Earth will be gone; but will life survive? If we can survive our current challenges and first become a multi-planet and then a multi–solar system species, then perhaps our distant descendants—whatever species they become—will find new habitable worlds around other, younger, Sun-like stars to call home.

SEE ALSO Earth Is Born (c. 4.54 Billion BCE), Earth's Oceans (c. 4 Billion BCE), Radioactivity (1896), Earth's Oceans Evaporate (~1 Billion)

Main image: *Space artist Don Dixon's imagined view of the swollen, red-giant Sun about to engulf the Earth and Moon about 5 billion years in the future.*
Inset: *Spitzer Space Telescope infrared image of the Helix Nebula, a shell of debris formed during the death throes of a star that was once similar to our Sun.*

GSA Geologic Timescale

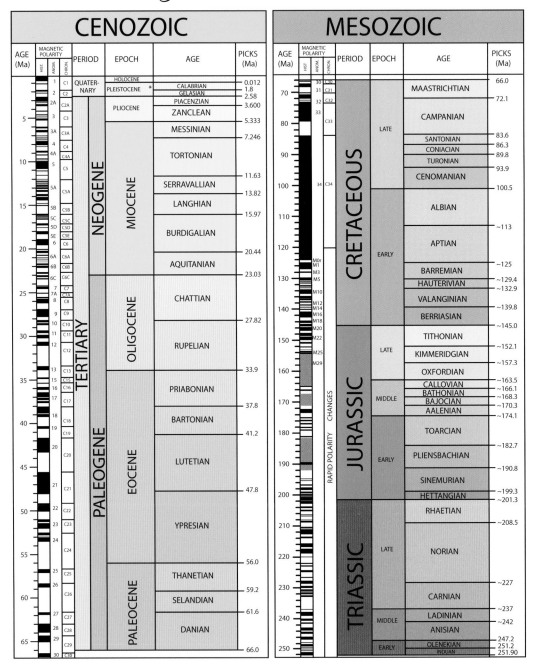

CENOZOIC

AGE (Ma)	MAGNETIC POLARITY (HIST / ANOM / CHRON)	PERIOD	EPOCH	AGE	PICKS (Ma)
	C1	QUATER-NARY	HOLOCENE		0.012
	C2		PLEISTOCENE	CALABRIAN *	1.8
				GELASIAN	2.58
	C2A	NEOGENE	PLIOCENE	PIACENZIAN	3.600
5	C3			ZANCLEAN	5.333
	C3A		MIOCENE	MESSINIAN	7.246
10	C4 / C4A / C5			TORTONIAN	11.63
	C5A			SERRAVALLIAN	13.82
15	C5B / C5C / C5D / C5E			LANGHIAN	15.97
	C6			BURDIGALIAN	20.44
20	C6A / C6B / C6C			AQUITANIAN	23.03
25	C7 / C7A / C8 / C9	PALEOGENE	OLIGOCENE	CHATTIAN	27.82
30	C10 / C11 / C12			RUPELIAN	33.9
35	C13 / C15 / C16 / C17		EOCENE	PRIABONIAN	37.8
40	C18 / C19			BARTONIAN	41.2
45	C20 / C21			LUTETIAN	47.8
50	C22 / C23 / C24			YPRESIAN	56.0
55	C25 / C26		PALEOCENE	THANETIAN	59.2
60	C27			SELANDIAN	61.6
65	C28 / C29 / C30			DANIAN	66.0

TERTIARY

MESOZOIC

AGE (Ma)	MAGNETIC POLARITY (HIST / ANOM / CHRON)	PERIOD	EPOCH	AGE	PICKS (Ma)
70	C30 / C31 / C32	CRETACEOUS	LATE	MAASTRICHTIAN	66.0
80	C33			CAMPANIAN	72.1
				SANTONIAN	83.6
90				CONIACIAN	86.3
				TURONIAN	89.8
100	C34			CENOMANIAN	93.9
					100.5
110			EARLY	ALBIAN	~113
120				APTIAN	~125
	M0r / M1 / M3 / M5			BARREMIAN	~129.4
130	M10			HAUTERIVIAN	~132.9
140	M12 / M14 / M16 / M18 / M20			VALANGINIAN	~139.8
	M22			BERRIASIAN	~145.0
150	M25 / M29	JURASSIC	LATE	TITHONIAN	~152.1
				KIMMERIDGIAN	~157.3
160				OXFORDIAN	~163.5
			MIDDLE	CALLOVIAN	~166.1
170				BATHONIAN	~168.3
				BAJOCIAN	~170.3
				AALENIAN	~174.1
180			EARLY	TOARCIAN	~182.7
190				PLIENSBACHIAN	~190.8
				SINEMURIAN	~199.3
200				HETTANGIAN	~201.3
210		TRIASSIC	LATE	RHAETIAN	~208.5
220				NORIAN	~227
230				CARNIAN	~237
240			MIDDLE	LADINIAN	~242
				ANISIAN	247.2
250			EARLY	OLENEKIAN	251.2
				INDUAN	251.90

RAPID POLARITY CHANGES

PALEOZOIC

AGE (Ma)	PERIOD	EPOCH	AGE	PICKS (Ma)
		Lopin-gian	CHANGHSINGIAN	251.90
			WUCHIAPINGIAN	254.14
260		Guada-lupian	CAPITANIAN	259.1
	PERMIAN		WORDIAN	265.1
			ROADIAN	268.8
280		Cisura-lian	KUNGURIAN	272.95
			ARTINSKIAN	~283.5
			SAKMARIAN	290.1
300			ASSELIAN	295.0
				298.9
		LATE	GZHELIAN	303.7
			KASIMOVIAN	307.0
	PENNSYL-VANIAN	MIDDLE	MOSCOVIAN	
320		EARLY	BASHKIRIAN	315.2
	CARBONIFEROUS	LATE	SERPUKHOVIAN	323.2
				330.9
340	MISSIS-SIPPIAN	MIDDLE	VISEAN	
		EARLY	TOURNAISIAN	346.7
360				358.9
		LATE	FAMENNIAN	
				~372.2
380	DEVONIAN		FRASNIAN	
		MIDDLE	GIVETIAN	~382.7
			EIFELIAN	~387.7
400			EMSIAN	~393.3
		EARLY	PRAGIAN	~407.6
			LOCHKOVIAN	~410.8
420		PRIDOLI		~419.2
		LUDLOW	LUDFORDIAN	~423.0
			GORSTIAN	~425.6
	SILURIAN	WENLOCK	HOMERIAN	~427.4
			SHEINWOODIAN	~430.5
440		LLANDO-VERY	TELYCHIAN	~433.4
			AERONIAN	~438.5
			RHUDDANIAN	~440.8
			HIRNANTIAN	~443.8
		LATE	KATIAN	~445.2
			SANDBIAN	~453.0
460	ORDOVICIAN	MIDDLE	DARRIWILIAN	~458.4
			DAPINGIAN	~467.3
		EARLY	FLOIAN	~470.0
480			TREMADOCIAN	~477.7
		FURON-GIAN	AGE 10	~485.4
			JIANGSHANIAN	~489.5
			PAIBIAN	~494
500		Epoch 3	GUZHANGIAN	~497
			DRUMIAN	~500.5
	CAMBRIAN		AGE 5	~504.5
		Epoch 2	AGE 4	~509
			AGE 3	~514
520			AGE 2	~521
		TERRE-NEUVIAN		~529
540			FORTUNIAN	541.0

PRECAMBRIAN

AGE (Ma)	EON	ERA	PERIOD	BDY. AGES (Ma)
				541
			EDIACARAN	635
750		NEOPRO-TEROZOIC	CRYOGENIAN	720
			TONIAN	
1000				1000
			STENIAN	
1250	PROTEROZOIC	MESOPRO-TEROZOIC	ECTASIAN	1200
				1400
1500			CALYMMIAN	
				1600
1750			STATHERIAN	
				1800
2000		PALEOPRO-TEROZOIC	OROSIRIAN	2050
			RHYACIAN	
2250				2300
			SIDERIAN	
2500				2500
		NEOARCHEAN		
2750				2800
3000		MESO-ARCHEAN		
	ARCHEAN			3200
3250				
		PALEO-ARCHEAN		
3500				3600
3750		EOARCHEAN		
4000				4000
	HADEAN			

Notes and Further Reading

During the research for this book I consulted many different resources, including a variety of general historical and encyclopedic sources, to verify the factual information (or the range of accepted consensus), as well as a variety of web sites for additional details and follow-up.

In the list below, I have attempted to indicate at least one major additional resource—a book, newspaper or magazine article, or a web site—where you can go to find more in-depth background or knowledge on the topic at hand. I use the http://tinyurl.com/ short-cut to keep the web links pithy; hopefully these links will stay valid for a long time to come.

As I described in the Introduction, selecting just 250 milestones in the entire history of earth science is a daunting task, and my selections naturally reflect my own biases, knowledge, and experience. Someone else writing this book would surely have picked a different set of milestones, although I like to believe that there would be significant overlap with the ones I have chosen. Still, I would be delighted to consider suggestions for other topics to swap in for future editions of this book, and of course would also welcome any corrections or suggestions in general about the content. Please feel free to write to me by email at Jim.Bell@asu.edu.

General Reading: Earth Sciences

Allaby, M., *A Dictionary of Geology & Earth Sciences*. Oxford, U.K: Oxford University Press, 2013.

Diagram Group, *Earth Science: An Illustrated Guide to Science*. Langhorne, PA: Chelsea House, 2008.

Kolbert, E. *The Sixth Extinction*. New York: Henry Holt & Co., 2014.

McPhee, J: *Annals of the Former World*. New York: Farrar, Straus and Giroux, 2000.

Reed, C., *Earth Science: Decade by Decade*, Facts on File, 2008.

Reynolds, S. and J. Johnson, *Exploring Earth Science*. New York: McGraw-Hill, 2015.

Sagan, C., *Pale Blue Dot*. New York: Ballantine, 1997.

General-Interest Earth Science Web sites

Earth Science Picture of the Day: *epod.usra.edu*

Earth science policy issues, from the American Geophysical Union: *science-policy.agu.org*

National Earth Science Teacher's Association: *serc.carleton.edu/nesta*

Want a degree in earth sciences? Check out *www.geosciences.com*

Wikipedia earth sciences portal: *wikipedia.org/wiki/Portal:Earth_sciences*

c. 4.54 Billion BCE, Earth Is Born
Dalrymple, G. B., *The Age of the Earth*. Stanford, CA: Stanford Univ. Press, 1994.

c. 4.54 Billion BCE, Earth's Core Forms
See a recent article on new core formation research from the French Institut Nationale des Sciences de l'Univers at *tinyurl.com/y6vkbwns*.

c. 4.5 Billion BCE, Birth of the Moon
Canup, R. M., and K. Righter, eds., *The Origin of the Earth and Moon*. Tucson: Univ. of Arizona Press, 2000.

c. 4.5 Billion BCE, Earth's Mantle and Magma Ocean
See "Giant Magma Ocean Once Swirled Inside Early Earth," LiveScience's C.Q. Choi at *tinyurl.com/y7ffze6c*.

c. 4.5–4 Billion BCE, The Hadean
K. Chang, "A New Picture of the Early Earth," *New York Times*, Dec. 1, 2008, *tinyurl.com/6k6s38*.

c. 4.1 Billion BCE, Late Heavy Bombardment
The giant planets likely played a significant role in the late heavy bombardment: *tinyurl.com/csg6zh4*.

c. 4 Billion BCE, Continental Crust
Wikipedia's page on "Bowen's Reaction Series" at *tinyurl.com/y9fl79dd* is a fine place to start exploring the ways felsic continental rocks form from mafic volcanic rocks.

c. 4 Billion BCE, Earth's Oceans
Did the formation of the global ocean really lead to a reducing atmosphere for the early Earth? *See tinyurl.com/ybnf622s* to dive into the debate.

c. 4–2.5 Billion BCE, The Archean
C. J. Allégre and S. H. Schneider, "Evolution of Earth," *Scientific American*, July 2005, *tinyurl.com/y9tkkssh*.

c. 4–3 Billion BCE?, Plate Tectonics
While incomplete, the Wikipedia page on "Plate Tectonics" at *tinyurl.com/hmby9d4* is a great jumping-off point to begin to learn more about this major Earth process.

c. 3.8 Billion BCE?, Life on Earth
A. Ricardo and J. W. Szostak, "The Origin of Life on Earth," *Scientific American*, Sep. 2009.

c. 3.7 Billion BCE, Stromatolites
See The *Guardian's* article "Oldest fossils on Earth discovered in 3.7bn-year-old Greenland rocks" at *tinyurl.com/zpt393k*.

c. 3.5 Billion BCE, Greenstone Belts
C. Zimmer, "The Oldest Rocks on Earth," *Scientific American*, March 2014.

c. 3.4 Billion BCE, Photosynthesis
Check out some of the latest advances in "artificial photosynthesis" in "Bionic Leaf Makes Fuel from Sunlight, Water and Air," by D. Biello, *Scientific American*, June 2016, tinyurl.com/yb3ezbcz.

c. 3–1.8 Billion BCE, Banded Iron Formations
View a wonderful 3-minute video on BIFs from a segment of the PBS show *NOVA* at tinyurl.com/yalbg4oy.

c. 2.5 Billion BCE, The Great Oxidation
D. E. Canfield, *Oxygen: A Four Billion Year History*. Princeton, NJ: Princeton Univ. Press, 2014.

c. 2 Billion BCE, Eukaryotes
C. Zimmer, "Scientists Unveil New 'Tree of Life'," *New York Times*, April 11, 2016, tinyurl.com/y9kmcczq.

c. 1.2 Billion BCE, The Origin of Sex
L. Margulis and D. Sagan, *Origins of Sex: Three Billion Years of Genetic Recombination*. New Haven: Yale Univ. Press, 1990.

c. 1 Billion BCE, Complex Multicellular Organisms
A. Knoll, *Life on a Young Planet*. Princeton, NJ: Princeton Univ. Press, 2003.

c. 720–635 Million BCE, Snowball Earth?
P.H. Hoffman and D.P. Schrag, "Snowball Earth," *Scientific American*, Jan. 2000.

c. 550 Million BCE, Cambrian Explosion
D. Quammen, "When Life Got Complicated," *National Geographic*, March 2018.

c. 500 Million BCE, Roots of the Pyrénées
Wikipedia's page on the Pyrénées at tinyurl.com/ofm6zqn provides lots of facts, figures, and links about these mountains.

c. 480 Million BCE, The Appalachians
The US Geological Survey's "Appalachian Highlands Province" page provides more details and links: tinyurl.com/y74lr7vm.

c. 470 Million BCE, First Land Plants
W. N. Stewart and G. W. Rothwell, *Paleobotany and the Evolution of Plants*. Cambridge, UK: Cambridge Univ. Press, 1993.

c. 450 Million BCE, Mass Extinctions
For a perspective on the potential role of life itself to cause the End Ordovician mass extinction, *see* tinyurl.com/ya2cb7dx.

c. 375 Million BCE, First Animals on Land
K. Westenberg, "The Rise of Life on Earth: From Fins to Feet," *National Geographic*, May 1999.

c. 320 Million BCE, The Ural Mountains
A lovely illustrated summary of the many exotic gems and minerals found in the Ural Mountains can be found at tinyurl.com/y84zqrco.

c. 320 Million BCE, Reptiles
See the National Geographic Society's "Reptiles" page for more history and details on this fascinating group of animals: tinyurl.com/yc2zktcq.

c. 300 Million BCE, The Atlas Mountains
Is a hot mantle plume helping to uplift the Atlas mountains? Check out some recent research at tinyurl.com/ybcdhxoa.

c. 300 Million BCE, Pangea
More details about the formation, history, and breakup of Pangea can be found at tinyurl.com/ybydbe5w.

c. 252 Million BCE, The Great Dying
N. St. Fleur, "After Earth's Worst Mass Extinction, Life Rebounded Rapidly, Fossils Suggest," *New York Times*, Feb. 16, 2017, tinyurl.com/y86zo2r2.

c. 220 Million BCE, Mammals
See photos of more than 4,000 kinds of modern mammals from the American Society of Mammologists at tinyurl.com/y8xyuwuz.

c. 200 Million BCE, Triassic Extinction
R. Smith, "Dark Days of the Triassic: Lost World," *Nature*, Nov. 16, 2011, tinyurl.com/y7t97o92.

c. 200–65 Million BCE, Age of the Dinosaurs
G. Paul, ed., *Scientific American Book of Dinosaurs*. New York: St. Martin's Press, 2003.

c. 160 Million BCE, The First Birds
2018 is the National Geographic Society's "Year of the Bird." Check it out at tinyurl.com/yc962k3j.

c. 155 Million BCE, The Sierra Nevada
The US Geological Survey's "Geology and National Parks" page provides more details and links, tinyurl.com/m75ne35.

c. 140 Million BCE, The Atlantic Ocean
Wikipedia's page on "The Opening of the North Atlantic Ocean" provides a good place to start more detailed research on this milestone event in Earth history, tinyurl.com/y9hzmtss.

c. 130 Million BCE, Flowers
B. Kanapaux, "Evolution of the first Flowers," *Natural History*, Aug. 8, 2009, tinyurl.com/p2jry3d.

c. 80 Million BCE, The Rockies
A. Minard, "Rockies Mystery Solved by New Mountain-Creation Story?" *National Geographic News*, March 3, 2011, tinyurl.com/yaboem42.

c. 70 Million BCE, The Himalayas
For some fun animations of the India–Eurasia continental collision, *see* Univ. California geologist Tanya Atwater's page at tinyurl.com/y84souk5.

c. 66 Million BCE, Deccan Traps
L. Karlstrom and J. Byrnes, "The Meteorite That Killed the Dinosaurs May Have Also Triggered Underwater Volcanoes," *Smithsonian*, Feb. 8, 2018, tinyurl.com/y7ocvfuw.

c. 65 Million BCE, The Alps
O. A. Pfiffner, *Geology of the Alps*. West Sussex, UK: Wiley-Blackwell, 2014.

c. 65 Million BCE, Dinosaur-Killing Impact
For more details and references about the controversy surrounding the impact hypothesis, a great place to start is Wikipedia's K-Pg extinction event page, tinyurl.com/mm2dz.

c. 60 Million BCE, Primates
C. Palmer, "Fossils Indicate Common Ancestor for Old World Monkeys and Apes," *Nature*, May 16, 2013, tinyurl.com/y7qls56q.

c. 35 Million BCE, Antarctica
A. B. Ford and D. L. Schmidt, *The Antarctic and its Geology*, USGS Report 1978-261-226/50, tinyurl.com/yamntv8s.

c. 30 Million BCE, East African Rift Zone
E. Biba, "A Superplume Is the Reason Africa Is Splitting Apart," July 15, 2014, tinyurl.com/yagfqnxt.

c. 20–30 Million BCE, Advanced C4 Photosynthesis
K. Bullis, "Supercharged Photosynthesis," *MIT Technology Review*, tinyurl.com/yc8w2hd3.

c. 30–10 Million BCE, Cascade Volcanoes
For background and details on the latest activity, visit the USGS Cascades Volcanic Observatory site at *tinyurl.com/y8qzjoyn*.

c. 28 Million BCE, Hawaiian Islands
The Wikipedia page on "Hawaii hotspot" is a great place to start to learn more about the work of Wilson and others to decipher the geologic history of the Hawaiian Islands, *tinyurl.com/y8tdl655*.

c. 10 Million BCE, The Andes
The Smithsonian Institution's Global Volcanism Program keeps track of active volcanoes in the Andes and the rest of the world, *volcano.si.edu*.

c. 10 Million BCE, First Hominids
For a set of informative articles about hominids in *National Geographic*, see *tinyurl.com/y8364st9*.

c. 7 Million BCE, Sahara Desert
"Sahara Desert Formed 7 Million Years Ago, New Study Suggests," *Sci News*, Sept. 20, 2014, *tinyurl.com/lrmalp8*.

c. 6–5 Million BCE, The Grand Canyon
A. Witze, "Debate Over Grand Canyon's Age May Finally Be Over," *Huffington Post*, Jan. 27, 2014, *tinyurl.com/y9m3t peh*.

c. 6–5 Million BCE, The Mediterranean Sea
K. J. Hsü, "When the Mediterranean Dried Up," *Scientific American*, Dec. 1972.

c. 5.5 Million BCE, The Caspian and Black Seas
N. Romeo, "Centuries of Preserved Shipwrecks Found in the Black Sea," *National Geographic*, Oct. 26, 2016, *tinyurl.com/ydaax4hy*.

c. 5 Million BCE, Galápagos Islands
See spectacular images and videos from *National Geographic* that demonstrate why the Galápagos are a United Nations World Heritage Site, *tinyurl.com/ y9jpqmnr*.

c. 3.4 Million BCE TO 3300 BCE, The Stone Age
Wikipedia's "Stone Age" entry is a great place to start to explore the history, limitations, and controversies over the division of human prehistory into the Stone, Bronze, and Iron Ages: *tinurl.com/y9qvt2rr*.

c. 3 Million BCE, The Dead Sea
J. Hammer, "The Dying of the Dead Sea," *Smithsonian Magazine*, Oct. 2005, *tinyurl.com/ybq4z8aa*.

c. 2 Million BCE, Death Valley
View a very nice short video by the National Geographic Society on the geology and ecology of Death Valley at *tinyurl.com/y9xrj4ws*.

c. 400,000 BCE, Lake Victoria
C.K. Yoon, "Lake Victoria's Lightning-Fast Origin of Species," *New York Times*, Aug. 27, 1996, *tinyurl.com/ycwqbgz4*.

c. 200,000 BCE, *Homo sapiens* Emerges
Seed magazine reporter Holly Capelo wrote an interesting summary of recent evidence that Paleolithic cave art may capture some aspects of ancient astronomical and celestial lore, *tinyurl.com/cvgtd6q*.

c. 70,000 BCE, The San People
E. Yong, "Africa's genetic diversity revealed by full genomes of a Bushman and a Tutu," *National Geographic Blog*, Feb. 17, 2010, *tinyurl.com/mu843al*.

c. 50,000 BCE, Arizona Impact
The University of New Brunswick in Canada maintains an online database of the nearly two hundred known and suspected impact-crater sites on the Earth, *tinyurl.com/y7dub47f*.

c. 40,000 BCE, The First Mines
A history of the Ngwenya Mine by the Sawiland National Trust Commission can be found at *tinyurl.com/ybqqkz7b*.

c. 38,000 BCE, La Brea Tar Pits
Check out more details on the history of the tar pits at the L.A. Tar Pits & Museum web site at *tarpits.org*.

c. 30,000 BCE, Domestication of Animals
B. Hare and V. Woods, "We Didn't Domesticate Dogs. They Domesticated Us," *National Geographic*, March 3, 2013, *tinyurl.com/yafxbckj*.

c. 10,000 BCE, Invention of Agriculture
P. E. L. Smith, "Stone Age Man on the Nile," *Scientific American*, August 1976.

c. 10,000 BCE, End of the Last "Ice Age"
A. C. Revkin, "When Will the Next Ice Age Begin?," *New York Times*, Nov. 11, 2003, *tinyurl.com/cjcaek*.

c. 9000 BCE, Beringia Land Bridge
The US National Park Service helps celebrate the cultural and geologic history of Beringia at *tinyurl.com/y7z3d365*.

c. 8000 BCE, The Great Lakes
K. A. Zimmermann, "Great Facts About the Five Great Lakes," *LiveScience*, June 29, 2017, *tinyurl.com/yapx8n3z*.

c. 7000 BCE, Fermentation of Beer and Wine
J. Klein, "How Pasteur's Artisti Insight Changed Chemistry," *New York Times*, June 14, 2017, *tinyurl.com/yaez54nl*.

c. 6000 BCE, Fertilizer
"Manure was used by Europe's first farmers 8,000 years ago," Univ. of Oxford web site, School of Archaeology, July 15, 2013, *tinyurl.com/ybqvquah*.

c. 3300–1200 BCE, The Bronze Age
B. Keim, "Bronze Age Woman Had Surprisingly Modern Life," *National Geographic*, May 21, 2015, *tinyurl.com/yc26gr6z*.

c. 3200 BCE, Synthetic Pigments
I. Shih, "Ancient Egyptian pigment provides modern forensics with new coat of paint," *The Conversation*, May 30, 2016, *tinyurl.com/y9gwgzjl*.

c. 3000 BCE, Oldest Living Trees
K. Goldbaum, "What Is the Oldest Tree in the World?," *LiveScience*, Aug. 23, 2016, *tinyurl.com/ycglunzf*.

c. 3000 BCE, Stonehenge
C. A. Newham, *The Astronomical Significance of Stonehenge*. Warminster, UK: Coates & Parker, 1993.

c. 3000 BCE, The Spice Trade
G. P. Nabhan, *Cumin, Camels, and Caravans: A Spice Odyssey*. Berkeley, CA: Univ. California Press, 2014.

c. 2500 BCE, The Pyramids
I treasure my early edition of E. C. Krupp's *Echoes of the Ancient Skies: The Astronomy of Lost Civilizations* (Mineola, NY: Dover, 2003), which provides a fascinating account of how much the objects and motions of the sky meant to our distant ancestors.

c. 2000 BCE, Magnetite
Photos and other information about magnetite can be found on the magnetite page of minerals. net at *tinyurl.com/z422vxs*.

c. 1200–500 BCE, The Iron Age
T. A. Wertime and J. D. Muhly, *The Coming of the Age of Iron*. New Haven: Yale Univ. Press, 1980.

c. 800 BCE, Aqueducts
I. Rodà, "Aqueducts: Quenching Rome's Thirst," *National Geographic History*, Nov./Dec. 2016, *tinyurl.com/y7ew4kac*.

c. 600 BCE, First World Maps
G. Miller, "Bizarre, Enormous 16th-Century Map Assembled for First Time," *National Geographic*, Dec. 7, 2017, *tinyurl.com/y7lxle7y*.

c. 500 BCE, The Earth Is Round!
In case Pythagoras, Eratosthenes, and the modern space program have not convinced you that you're living on a rotating sphere, you can always stick your head in the sand and join other nonbelievers from the Flat Earth Society, *tinyurl.com/346e6c8*.

c. 500 BCE, Madagascar
Learn more about the global quest to save Madagascar's threatened lemur species from the International Union for the Conservation of Nature's "Lemur Conservation Network," *tinyurl.com/ybhzecp2*.

c. 300 BCE, Quartz
National Geographic's "Minerals and Gems" site showcases quartz and other important minerals in earth science, *tinyurl.com/ya93c8tw*.

c. 300 BCE, Great Library of Alexandria
A. Lawler, "Raising Alexandria," *Smithsonian Magazine*, April 2007, *tinyurl.com/y8gvoa28*.

c. 280 BCE, A Sun-Centered Cosmos
Kragh, H. S., *Conceptions of Cosmos—From Myths to the Accelerating Universe: A History of Cosmology*. New York: Oxford Univ. Press, 2007.

c. 250 BCE, Size of the Earth
Since 2000, the teaching module "Follow the Path of Eratosthenes" has enabled students to reproduce his more than 2,200-year-old experiment on their own, *tinyurl.com/y8rrtym5*.

79, Pompeii
The 1987 National Geographic documentary "In the Shadow of Vesuvius" can be watched online at *tinyurl.com/y979y99n*.

c. 700–1200, Polynesian Diaspora
The Wikipedia page on the 1947 (and subsequent) Kon-Tiki Expedition provides excellent links and details about modern-day attempts to recreate many of the epic voyages of the Polynesian diaspora, *tinyurl.com/yaa4mzbk*.

c. 1000, Mayan Astronomy
A high-resolution version of the complete Dresden Codex can be downloaded from *tinyurl.com/d5f38vq*. See also Prof. Anthony

Aveni's *Conversing with the Planets: How Science and Myth Invented the Cosmos*, Kodansha International, 1994.

c. 1370–1640, Great Wall of China
The Great Wall is a World Heritage Site within the United Nations Educational, Scientific, and Cultural Organization (UNESCO). Read more at *tinyurl.com/ltktlfy*.

c. 1400, Native American Creation Stories
A collection of Native American creation stories can be found online at *tinyurl.com/yaanrfyd*. See also A. Shumov, "Creation Myths from Around the World," *National Geographic*, *tinyurl.com/yax5xlwv*.

c. 1500, The Little Ice Age
B. Handwerk, "Little Ice Age Shrank Europeans, Sparked Wars," *National Geographic*, Oct. 5, 2011, *tinyurl.com/y8ssxklq*.

c. 1500, Civil Engineering
What is civil engineering? Find out from the Institution of Civil Engineers at *tinyurl.com/y7g5ystp*.

1519, Circumnavigating the Globe
Learn more about Magellan and his voyage on the History Channel's "Ferdinand Magellan" page: *tinyurl.com/ydgf9dr6*.

1541, Amazon River
D. Stone, "Amazon Tribes Stand Up for Their Survival," *National Geographic*, June 23, 2017, *tinyurl.com/y8zxypc3*.

1600, Many Earths
Wikipedia's Giordano Bruno page (*tinyurl.com/ayqfd*) provides a starting point for more detailed study of the controversial friar, philosopher, and astronomer.

1600, Huaynaputina Eruption
A. Witze, "The Volcano That Changed the World," *Nature*, Apr. 11, 2008, *tinyurl.com/ya6cnndc*.

1619, Laws of Planetary Motion
More about Johannes Kepler can be found in C. Wilson's "How Did Kepler Discover His First Two Laws?" (*Scientific American*, March 1972) and O. Gingerich's *The Great Copernicus Chase and Other Adventures in Astronomical History* (Cambridge, MA: Sky Publishing, 1992).

1669, Foundations of Geology
C. Gaylord, "How Nicolas Steno changed the way we *see* the world, literally," *Christian Science Monitor*, Jan. 11, 2012: *tinyurl.com/ya9uwk6o*.

1686, Tides
Excellent introductory discussions of tides can be found at "How Tides Work" on E. Siegel's blog *Starts with a Bang!* (*tinyurl.com/yapavo4f*) and "Tidal Misconceptions" by D. Simanek (*tinyurl.com/yaqwr4qq*), as well as pages 265–274 in V. D. Barger and M. G. Olsson's *Classical Mechanics: A Modern Perspective*. New York: McGraw-Hill, 1973.

1687, Gravity
Hawking, S., *On the Shoulders of Giants: The Great Works of Physics and Astronomy*. Philadelphia: Running Press, 2002.

1747, Feldspar
What is feldspar and how is it used economically? Find out about it (and many other minerals) from the Industrial Minerals Association: *tinyurl.com/ybmfk6ru*

1769, Transit of Venus
A great popular-level account of the history of Venus transit observations is in W. Sheehan and J. Westfall's *The Transits of Venus*. Amherst, NY: Prometheus, 2004.

1788, Unconformities
Sketches and photos showing examples of the kinds of unconformities observed in Earth's geologic record can be found on the Wikipedia "Unconformity" page: *tinyurl.com/yd4752kg*.

1789, Olivine
L. Geggel, "Earth's Mantle is Hotter than Scientists Thought," *Scientific American*, March 4, 2017: *tinyurl.com/zwr77yr*.

1791, Desalination
The website for Jefferson's home, Monticello, contains more details on his efforts to research and widely disseminate desalination methods: *tinyurl.com/ybgoddh5*.

1794, Rocks from Space
Smith, C., S. Russell, and G. Benedix, *Meteorites*. Buffalo: Firefly, 2011.

1798, Population Growth
D. Dimick, "As World's Population Booms, Will Its Resources Be Enough for Us?" *National Geographic*, Sept. 21, 2014: *tinyurl.com/y92vfb5q*.

1802–1805, Platinum Group Metals
B. Griffith, "Two Men, Two Centuries, Four Metals," *Chemistry World*, March 1, 2005: *tinyurl.com/yccquxe5*.

1804, Charting North America
S. E. Ambrose, *Undaunted Courage: Meriwether Lewis, Thomas Jefferson, and the Opening of the American West*. New York: Simon & Schuster, 1996.

1811, Reading the Fossil Record
For more on her life and work, *see* "Mary Anning, the Fossil Finder" by Charles Dickens (1865), at *tinyurl.com/ybm9da8q*.

1814, Sunlight Deciphered
An online biography of Joseph von Fraunhofer's life and achievements is at *tinyurl.com/y7pqdc4w*.

1815, Mount Tambora Eruption
W. J. Broad, "A Volcanic Eruption That Reverberates 200 Years Later," *New York Times*, Aug. 24, 2015: *tinyurl.com/yakpwgza*.

1815, Modern Geologic Maps
S. Winchester, *The Map That Changed the World: William Smith and the Birth of Modern Geology*. New York: HarperCollins, 2001.

1830, Uniformitarianism
For a fascinating geologic and philosophical history of Lyell's promotion of uniformitarianism, check out Stephen J. Gould's *Time's Arrow, Time's Cycle: Myth and Metaphor in the Discovery of Geological Time*. Cambridge, MA: Harvard Univ. Press, 1987.

c. 1830, Industrial Revolution
"The Third Industrial Revolution." *The Economist*, Apr. 21, 2012: *tinyurl.com/yd4cwscu*.

1837, Discovering Ice Ages
To learn more about glaciology and the glaciers of the world today, visit Glaciers Online at *swisseduc.ch/glaciers*.

1845, Birth of Environmentalism
A. Wulf, *The Invention of Nature: Alexander von Humboldt's New World*. New York: Vintage, 2016.

1851, Proof that the Earth Spins
California Academy of Science: *tinyurl.com/y9mjyr9w*.

c. 1855–1870, Deforestation
See Jared Diamond's *Collapse* (New York: Penguin, 2005) for more details on the deforestation of Easter Island and other societies.

1858–1859, Natural Selection
Wikipedia's page on "Natural Selection" at *tinyurl.com/opvzf5g* is a great place to start exploring the details of the history and modern practice of evolutionary biology.

1858, Airborne Remote Sensing
"History of Aerial Photography," Professional Aerial Photographers Association: *tinyurl.com/l4debno*.

1859, Solar Flares and Space Weather
"NASA Science News: A Super Solar Flare" (May 6, 2008): *tinyurl.com/32v6amx*.

1862, The Age of the Earth
Burchfield, J. D., *Lord Kelvin and the Age of the Earth*. Chicago: Univ. of Chicago Press, 1990.

1864, (Geo) Science Fiction
Learn more about the history and current state of science fiction at *museumofsciencefiction.org*.

1869, Exploring the Grand Canyon
If you're excited by the Grand Canyon, John Wesley Powell's original 1875 book *The Exploration of the Colorado River and Its Canyons* (published by the Smithsonian Institution) is a must-read!

c. 1870, The Anthropocene
L. E. Edwards, "What is the Anthropocene?" *Eos*, Nov. 30, 2015: *tinyurl.com/yb5qdxpj*.

1870, Soil Science
National Geographic's collection of articles about soil at *tinyurl.com/y7gcbcgp* is a great educational resource for this critical component of the Earth's surface.

1872, National Parks
Start your exploration of hundreds of the most wild and scenic places in the US at the National Park Service's web site: *www.nps.gov*.

1879, The US Geological Survey
For more about the history and current scientific research being conducted by the USGS, visit online at *usgs.gov*.

1883, Krakatoa Eruption
S. Winchester, *Krakatoa: The Day the World Exploded, August 27, 1883*. New York: Penguin, 2003.

1892, The Sierra Club
Learn more about the history and mission of the Sierra Club online at *sierraclub.org*.

1896, The Greenhouse Effect
United Nations' Intergovernmental Panel on Climate Change Fifth Assessment Report (2014): *tinyurl.com/hyfm99k*.

1896, Radioactivity
Hedman, M., *The Age of Everything: How Science Explores the Past*. Chicago: Univ. of Chicago Press, 2007.

1896, The Structure of the Atmosphere
Wikipedia's "Atmosphere of Earth" page (*tinyurl.com/yahwpxsj*) is a great starting point for more detailed information and history on the field of aerology.

1896, Women in Earth Science
M. A. Holmes and S. O'Connell, "Where are the Women Geoscience Professors?" (*tinyurl.com/yacv2fwt*). See also *tinyurl.com/y9jmzst6*.

1900, Galveston Hurricane
For a sad and morbid list of natural disasters around the world sorted by death toll, visit *tinyurl.com/pydp2x7*.

1902, Controlling the Nile
T. Lippmann, "Excess Water Is a Problem as Aswan Dam Tames Nile," *Washington Post*, Nov. 12, 1978: *tinyurl.com/y9d7xf9e*.

1906, San Francisco Earthquake
For more details on the earthquake from the US Geological Survey, *see tinyurl.com/y9weemsc*.

1906, Hunting for Meteorites
W. G. Hoyt, *Coon Mountain Controversies: Meteor Crater and the Development of Impact Theory*. Tucson: Univ. of Arizona Press, 1987.

1908, The Tunguska Explosion
Artist and planetary scientist W. K. Hartmann has put together a fascinating account of eyewitness stories and artistic impressions about the Tunguska event at *tinyurl.com/95pjc2t*.

1909, Reaching the North Pole
R. M. Bryce, *Cook & Peary: The Polar Controversy Resolved*. Mechanicsburg, PA: Stackpole, 1997.

1910, Big Burn Wildfire
S. J. Pyne, *Fire: A Brief History*. Seattle: Univ. of Washington Press, 2001.

1911, Reaching the South Pole
"Tragedy and Triumph: The Heroic Age of Polar Exploration," *Scientific American* (digital edition), July 2012.

1911, Machu Picchu
K. Hearn and J. Golomb, "Machu Picchu," *National Geographic* online: *tinyurl.com/y8j5ycsu*.

1912, Continental Drift
H. E. Le Grand, *Drifting Continents and Shifting Theories*. Cambridge, UK: Cambridge Univ. Press, 1989.

1913, The Ozone Layer
A. Witze, "Antarctic Ozone Hole Is on the Mend," *Nature*, July 1, 2016: *tinyurl.com/yd7x4juq*.

1914, The Panama Canal
"Make the Dirt Fly," Smithsonian Library Digital Exhibition on the Panama Canal: *tinyurl.com/dx74h*.

1915, Exploring Katmai
J. Fierstein, "Katmai National Park Volcanoes," US National Park Service: *tinyurl.com/yc3xpznz*.

1921, Russian Famine
"The Great Famine," PBS film from *The American Experience*, Season 23, Episode 8, 2011.

1925, Tri-State Tornado
P. S. Felknor, *The Tri-State Tornado: The Story of America's Greatest Tornado Disaster*. Ames, IA: Iowa State Univ. Press, 1992.

1926, Liquid-Fueled Rockets
Goddard's original 1919 book on rocketry, *A Method to Reach Extreme Altitudes* (Smithsonian Institution Press), can be downloaded from *tinyurl.com/9tha5jc*.

1926, Exploration by Aviation
R. E. Goerler, *To the Pole: The Diary and Notebook of Richard E. Byrd, 1925–1927*. Columbus, OH: Ohio State Univ. Press, 1998.

1933, Angel Falls
R. Robertson, "Jungle Journey to the World's Highest Waterfall," in *Worlds to Explore: Classic Tales of Travel and Adventure* from National Geographic, 2007.

1934, Geology of Corals
Dorothy Hill (1907–1997), *Australian Academy of Science Biographical Memoirs*: *tinyurl.com/yb7otso8*.

1935, Dust Bowl
Ken Burns's 2012 film *The Dust Bowl* provides a fascinating history of this important time in US history.

1936, The Inner Core
S. Kruglinski, "Journey to the Center of the Earth," *Discover*, June 8, 2007: *tinyurl.com/37twlm*.

1937, Landfills
The US National Park Service provides more details about the Fresno Sanitary Landfill at *tinyurl.com/ya88mfhj*.

1943, Exploring the Oceans
J. Cousteau and F. Dumas, *The Silent World: A Story of Undersea Discovery and Adventure*. New York: Harper, 1953.

1943, Sky Islands
W. Heald, *Sky Island*. New York: Van Nostrand, 1967.

1945, Geosynchronous Satellites
Arthur C. Clarke's 1945 prophetic *Wireless World* magazine article appears in a volume edited by space historian J. Logsdon: *Exploring the Unknown: Selected Documents in the History of the US Civil Space Program*, published by the NASA History Office, Washington, DC (*tinyurl.com/bruoxsd*).

1946, Cloud Seeding
J. Sanburn, "Scientists Create 52 Artificial Rain Storms in Abu Dhabi Desert," *Time*, Jan. 3, 2011: *tinyurl.com/3yaggmb*.

1947, Weather Radar
R. C. Whiton et al., "History of Operational Use of Weather Radar by US Weather Services. Part I: The Pre-NEXRAD Era," *Weather & Forecasting*, vol. 13, 1998 : *tinyurl.com/y8yzr3j7*.

1948, Tracing Human Origins
Much more about the lives and careers of Mary and Louis Leakey can be found on the Leakey Foundation web site, at *tinyurl.com/y9cvhtb6*.

1949, Island Arcs
Learn more about island arcs at the online course site Study.com: *tinyurl.com/ycd9r2ss*.

1953, Ascending Everest
D. Roberts, "50 Years on Everest," *National Geographic Adventure*, April 2003: *tinyurl.com/y7apnu5f*.

1954, Nuclear Power
US Energy Information Administration, "International Energy Outlook, 2017": *tinyurl.com/yc2c6j5r*.

1957, Mapping the Seafloor
B. Embley, "Seafloor Mapping," *NOAA Ocean Explorer Program*, 2003: *tinyurl.com/yao4ojcm*.

1957, Sputnik
For an entertaining glimpse of the America shocked by Sputnik and then spurred on to reach the Moon, check out H. Hickam's *Rocket Boys*. New York: (Delacorte Press, 1998) and the related 1999 film *October Sky* (Universal Pictures).

1957–1958, International Geophysical Year (IGY)
Wikipedia's "International Geophysical Year" page (*tinyurl.com/y9wa4xlb*) contains lots of details and pointers to much more information about IGY.

1958, Earth's Radiation Belts
More details about the phenomenally successful Explorer small satellite program (with nearly 100 launches between 1958 and 2018) can be found at *tinyurl.com/qp34s*.

1960, Weather Satellites
The Intellicast weather service provides a free site where you can monitor the latest images from all of the weather satellites covering North America: *tinyurl.com/y8h37wv*.

1960, Understanding Impact Craters
D. Levy, *Shoemaker by Levy: The Man Who Made an Impact*. Princeton, NJ: Princeton Univ. Press, 2002.

1960, Mariana Trench
K. Than, "James Cameron Completes Record-Breaking Mariana Trench Dive," *National Geographic News*, March 25, 2012: *tinyurl.com/yav8zvxl*.

1960, Valdivia Earthquake
"The Largest Earthquake in the World," US Geological Survey web site: *tinyurl.com/ybgatuyz*.

1961, Humans in Space
For details on the early Cold War space race, *see* T. Wolfe, *The Right Stuff*. New York: Farrar, Straus, and Giroux, 1979 (and also the 1983 film of the same name).

1961, Terraforming
M. J. Fogg, *Terraforming: Engineering Planetary Environments*, SAE International, 1995.

1963, Reversing Magnetic Polarity
"Magnetic Stripes and Isotopic Clocks," US Geologic Survey, *This Dynamic Earth* web site: *tinyurl.com/y8hv3g53*.

1966, Endosymbiosis
L. Margulis, *Origin of Eukaryotic Cells*. New Haven: Yale University Press, 1970.

1966, Earth Selfies
You can read more about the history of space selfies in my book *The Interstellar Age*, Penguin/Dutton, 2015.

1967, Extremophiles
T. Brock's call for expanding the search for habitable environments on Earth was published in "Life at High Temperatures" (*Science*, vol. 158, Nov. 1967, pp. 1012–1019).

1968, Leaving Earth's Gravity
A. Chaikin, *A Man on the Moon: The Voyages of the Apollo Astronauts*. New York: Viking, 1994.

1970, Meteorites and Life
Rosenthal, A. M., "Murchison's Amino Acids: Tainted Evidence?" (*Astrobiology*, Feb. 12, 2003): *tinyurl.com/y8f8fan8*.

1970, Earth Day
R. Carson, *Silent Spring*. New York: Houghton-Mifflin, 1962.

1972, Earth Science Satellites
For more details, *see* the USGS Landsat web site at landsat. usgs. gov.

1972, Geology on the Moon
The NASA documentary film, "On the Shoulders of Giants," provides lots more details on the science of the *Apollo* missions: *tinyurl.com/y7sdxqct*.

1973, Seafloor Spreading
Tanya Atwater also made significant contributions to the visualization of plate tectonic interactions for the general public. *See*, for example, *tinyurl.com/y7gt8pzc*.

1973, Tropical Rain/Cloud Forests
For a list of some of the world's most amazing tropical rainforests and cloud forests, *see tinyurl.com/yctmkfuj*.

1973, Global Positioning System
B. Parkinson and J. Spilker, *The Global Positioning System*. American Institute of Aeronautics and Astronautics, 1996: *tinyurl. com/yavrszej*.

1975, Insect Migration
F. Urquhart, "Found at Last: The Monarch's Winter Home," *National Geographic*, Aug. 1976.

1975, Magnetic Navigation
S. Johnsen and K. Lohmann, "Magnetoreception in Animals," *Physics Today*, March 1, 2008: *tinyurl.com/ybrw5cq2*.

1976, Temperate Rainforests
Wikipedia's "Temperate Rainforest" page (*tinyurl.com/ya32cj6v*) is a great place to start exploring these beautifully wet and wild regions of the world.

1977, Voyager Golden Record
C. Sagan et al., *Murmurs of Earth*. New York: Random House, 1978. *See* also the 2017 film, *The Farthest—Voyager in Space* (preview at *tinyurl.com/y98cwcse*.)

1977, Deep-Sea Hydrothermal Vents
"Deep Sea Hydrothermal Vents: Redefining the Requirements for Life," National Geographic online: *tinyurl.com/y942a7c3*.

1978, Wind Power
For more on its rich history, *see* "Wind Power," United Nations Food and Agriculture Organization report (1986): *tinyurl.com/csdr7sa*.

1979, A World Wide Web
A great starting point to explore the detailed history of the Internet and the World Wide Web is Wikipedia's "History of the Internet" page (*tinyurl.com/mgavzra*).

1980, Mount St. Helens Eruption
"Eruption of Mt. St. Helens," *National Geographic*, Jan. 1981.

1980, Extinction Impact Hypothesis
W. Alvarez, *T. Rex and the Crater of Doom*. Princeton, NJ: Princeton Univ. Press, 1997.

1981, Great Barrier Reef
J. Bowen and M. Bowen, *The Great Barrier Reef: History, Science, Heritage*. Cambridge, UK: Cambridge Univ. Press, 2002.

1982, Genetic Engineering of Crops
S. Blancke, "Why People Oppose GMOs Even Though Science Says They Are Safe," *Scientific American*, Aug. 18, 2015: *tinyurl. com/gmg3gzs*.

1982, Basin and Range
J. McPhee, *Basin and Range*. New York: Farrar, Straus and Giroux, 1982.

1982, Solar Power
R. Naam, "Smaller, Cheaper, Faster: Does Moore's Law Apply to Solar Cells?" *Scientific American*, March 16, 2011: *tinyurl.com/y9ax8g9v*.

1982, Volcanic Explosivity Index
The history and details/examples of the Volcanic Explosivity Index can be found on Wikipedia's page at *tinyurl.com/y9bdled2*.

1983, *Gorillas in the Mist*
D. Fossey, *Gorillas in the Mist*. New York: Mariner Books, 1983. *See* also A. McPherson, "Zoologist Dian Fossey: A Storied Life With Gorillas," *National Geographic*, Jan. 18, 2014: *tinyurl.com/y7pw2kja*.

1983, Plant Genetics
E. F. Keller, *A Feeling for the Organism: The Life and Work of Barbara McClintock*. New York: W. H. Freeman, 1983.

1984, The Oscillating Magnetosphere
"NASA's THEMIS Sees Auroras Move to the Rhythm of Earth's Magnetic Field," *NASA Press Release*, Sept. 12, 2016:. *tinyurl.com/ydb29khb*

1985, Underwater Archaeology
Travel with Robert Ballard to "The Astonishing Hidden World of the Deep Ocean" via his TED talk, at *tinyurl.com/p4zcobj*.

1986, Chernobyl Disaster
Wikipedia's "International Nuclear Event Scale" page (*tinyurl.com/mayaxyu*) provides details and links to more information about all documented reactor accidents worldwide.

1987, California Condors
Learn more about the California Condor Recovery Program from the US Fish and Wildlife Service: *tinyurl.com/y8dp6dxz*.

1987, Yucca Mountain
"Disposal of High-Level Nuclear Waste," US Government Accountability Office Report, 2017: *tinyurl.com/y7lzxtpf*.

1988, Light Pollution
Learn more about the International Dark-Sky Association (and join!) at *www.darksky.org*.

1988, Chimpanzees
J. Goodall, *My Life with the Chimpanzees*. New York: Minstrel, 1988.

1991, Biosphere 2
To follow the continuing scientific and educational mission of Biosphere 2, check out their web site at *biosphere2.org*.

1991, Mt. Pinatubo Eruption
"In the Path of a Killer Volcano," PBS/NOVA Episode, Feb. 9, 1993: *tinyurl.com/y7h9y7dt*.

1992, Tundra
"Tundra," *National Geographic* online: *tinyurl.com/ydhabh8r*.

1992, Boreal Forests
"Taiga," *National Geographic* online: *tinyurl.com/y8uuxg3o*.

1993, Oceanography from Space
Learn more online about space-based ocean exploration by NASA (*tinyurl.com/ydx25a9c*) and NOAA (*tinyurl.com/yarku9wj*).

1994, Hydroelectric Power
"Hydroelectric Energy," *National Geographic* online: tinyurl.com/y9tqktq8.

1995, Earthlike Exoplanets
Extrasolar Planets Encyclopaedia's "Interactive Extra-solar Planets Catalog," *tinyurl.com/32bozw*.

1997, Large Animal Migrations
"Animal Migrations," *National Geographic* online: tinyurl.com/yckrtpu4.

1998, Ocean Conservation
"Meet the Ocean Explorers," Sea and Sky online: tinyurl.com/j8yumv8.

1999, Earth's Spin Slows Over Time
Wikipedia's "Leap second" page at *tinyurl. com/b4oar* provides a fascinating and detailed account of the history and controversy surrounding this curious feature of modern timekeeping.

1999, Torino Impact Hazard Scale
More details about the Torino Impact Hazard Scale and the more recent Palermo Technical Impact Hazard Scale can be found at *tinyurl.com/kwt3tg* and *tinyurl. com/94lg6dx*, respectively.

1999, Vargas Landslide
R. L. Schuster and L. M. Highland, "Socioeconomic and Environmental Impacts of Landslides in the Western Hemisphere," USGS Open File Report 01-0276, 2001: *tinyurl.com/ycs3yysa*.

2004, Sumatran Earthquake and Tsunami
An Indian Ocean Tsunami Warning System has since been set up to attempt to save future lives. *See tinyurl.com/ybgs2ucg* for links and details.

2004, Grasslands and Chaparral
Learn about the mission of the California Chaparral Institute at *tinyurl.com/yak8orz8*.

2007, Carbon Footprint
Wikipedia's "Carbon Footprint" page (*tinyurl. com/y7d57fv5*) provides more background, details, and links to the history and applications of this concept.

2008, Global Seed Vault
J. Duggan, "Inside the 'Doomsday' Vault," *Time*, April 8, 2017: *tinyurl.com/kr4afx3*.

2010, Eyjafjallajökull Eruption
A detailed description of the effects of the volcano's eruption on commercial and military aviation can be found online at *tinyurl.com/yc249zy9*.

2011, Building Bridges
D. J. Brown, Bridges: *Three Thousand Years of Defying Nature*. Buffalo: Firefly Books, 2005.

2011, Temperate Deciduous Forests
To learn more about forests, and the United Nations' environmental protection programs in general, visit *tinyurl.com/yb9jccpr*.

2012, Lake Vostok
Wikipedia's "Lake Vostok" page (*tinyurl.com/gqeu5yn*) is a great starting point for more details on the history, science, and controversy of this unique environment.

2013, Savanna
Read more about the Kimberly to Cape Initiative and the savanna that those advocates *seek* to protect, at *tinyurl.com/ybcsfo5k*.

2013, Rising CO_2
See R. Kunzig, "Climate Milestone: Earth's CO_2 Level Passes 400 ppm," *National Geographic*, May 12, 2013: *tinyurl.com/ybfrr7o4*.

2016, Long-Duration Space Travel
"NASA Twins Study Confirms Preliminary Findings," *NASA* online, March 15, 2018: *tinyurl.com/yb3hxddu*.

2017, North American Solar Eclipse
Eclipse scientist F. Espenak keeps a detailed, updated set of Web pages on upcoming solar and lunar eclipses and planetary transits at *tinyurl.com/6cqw2c*.

2029, Apophis Near Miss
Apophis's 2036 miss distance from Earth depends on precisely where it passes the Earth and Moon in 2029 and how its trajectory responds to subtle variations in the Earth's and Moon's gravity fields. *See tinyurl.com/cf8xjcr*.

~2050, Settlements on Mars?
When, how, and why will people go to Mars? Follow along with The Planetary Society to keep up and find out! *planetary.org/blogs*.

~2100, End of Fossil Fuels?
T. Appenzeller, "The End of Cheap Oil," *National Geographic*, June 1, 2004: *tinyurl. com/ydeuafkd*.

~50,000, Next Ice Age?
A. C. Revkin, "The Next Ice Age and the Anthropocene," *New York Times*, Jan. 8, 2012: *tinyurl.com/yb54xqu7*.

~100,000, Yellowstone Supervolcano
S. Hall, "A surprise from the supervolcano under Yellowstone," *New York Times*, Oct. 10, 2017: *tinyurl.com/y8327teg*.

~100,000–200,000, Loihi
"Loihi," US Geological Survey Volcano Hazards Program, 2017: *tinyurl.com/ycy55ffu*.

~500,000, Next Big Asteroid Impact?
H. Weitering, "NASA Offers New Plan to Detect and Destroy Dangerous Asteroids," *Scientific American*, June 21, 2018: *tinyurl. com/y9654qlc*.

~250 Million, Pangea Proxima
"Continents in Collision: Pangea Ultima," *NASA Science* online, Oct. 6, 2000: *tinyurl. com/y763zgf6*.

~600 Million, Last Total Solar Eclipse
S. Mathewson, "Earth Will Have Its Last Total Solar Eclipse in About 600 Million Years," Space.com, July 31, 2017: *tinyurl. com/yblg6rr3*.

~1 Billion, Earth's Oceans Evaporate
J. Kasting and colleagues, "Earth's Oceans Destined to Leave in Billion Years": *tinyurl. com/8t28g6x*.

~2–3 Billion, Earth's Core Solidifies
An educational (though perhaps depressing) place to start researching the likely near-term to distant future of our planet is Wikipedia's "Future of Earth" page (*tinyurl. com/gv47pdr*).

~5 Billion, The End of the Earth
J. B. Kaler, *Stars*, Scientific American Library. New York: W. H. Freeman, 1992.

Index

Photo Credits